THEORETICAL INFORMATION STUDIES

Information in the World

World Scientific Series in Information Studies

(ISSN: 1793-7876)

Published:

Vol. 11 *Theoretical Information Studies: Information in the World*
edited by Mark Burgin & Gordana Dodig-Crnkovic

Vol. 10 *Philosophy and Methodology of Information: The Study of Information in the Transdisciplinary Perspective*
edited by Gordana Dodig-Crnkovic & Mark Burgin

Vol. 9 *Information Studies and the Quest for Transdisciplinarity Unity through Diversity*
edited by Mark Burgin & Wolfgang Hofkirchner

Vol. 8 *The Future Information Society: Social and Technological Problems*
edited by Wolfgang Hofkirchner & Mark Burgin

Vol. 7 *Information Theory Models of Instabilities in Critical Systems*
by Rodrick Wallace

Vol. 6 *Information and Complexity*
edited by Mark Burgin & Cristian S Calude

Vol. 5 *Theory of Knowledge: Structures and Processes*
by Mark Burgin

Vol. 4 *An Information Approach to Mitochondrial Dysfunction: Extending Swerdlow's Hypothesis*
by Rodrick Wallace

More information on this series can also be found at https://www.worldscientific.com/series/wssis

World Scientific Series in Information Studies — **Vol. 11**

THEORETICAL INFORMATION STUDIES

Information in the World

Editors

Mark Burgin
University of California, Los Angeles, USA

Gordana Dodig-Crnkovic
Chalmers University of Technology, Sweden & Mälardalen University, Sweden

World Scientific

NEW JERSEY • LONDON • SINGAPORE • BEIJING • SHANGHAI • HONG KONG • TAIPEI • CHENNAI • TOKYO

Published by

World Scientific Publishing Co. Pte. Ltd.
5 Toh Tuck Link, Singapore 596224
USA office: 27 Warren Street, Suite 401-402, Hackensack, NJ 07601
UK office: 57 Shelton Street, Covent Garden, London WC2H 9HE

Library of Congress Cataloging-in-Publication Data
Names: Burgin, M. S. (Mark Semenovich), editor. | Dodig Crnkovic, Gordana, 1955– editor.
Title: Theoretical information studies : information in the world / editors,
Mark Burgin (University of California, Los Angeles, USA) and Gordana Dodig-Crnkovic
(Chalmers University of Technology, Sweden & Mälardalen University, Sweden).
Description: New Jersey : World Scientific, [2020] | Series: World scientific series in
information studies, 1793-7876 ; vol. 11 | Includes bibliographical references and index.
Identifiers: LCCN 2019053550 | ISBN 9789813277489 (hardcover) | ISBN 9789813277496 (ebook) |
ISBN 9789813277502 (ebook other)
Subjects: LCSH: Information theory. | Interdisciplinary research.
Classification: LCC Q360 .T57 2020 | DDC 003/.54--dc23
LC record available at https://lccn.loc.gov/2019053550

British Library Cataloguing-in-Publication Data
A catalogue record for this book is available from the British Library.

For any available supplementary material, please visit
https://www.worldscientific.com/worldscibooks/10.1142/11191#t=suppl

Desk Editors: V. Vishnu Mohan/Tan Rok Ting

Typeset by Stallion Press
Email: enquiries@stallionpress.com

Contents

Introduction

Mark Burgin[*,‡] and Gordana Dodig-Crnković[†,§]
[*] *University of California, Los Angles, CA, USA*
[†] *Chalmers University of Technology, Sweden*
[‡] *mburgin@math.ulca.edu*
[§] *dodig@chalmars.se*

Theoretical Information Studies is the second volume which together with *Philosophy and Methodology of Information* comprises a two-volume edition with the aim of laying out the foundation of the emerging research field of *the Study of Information.* It is based on the Summit "Digitalisation for a Sustainable Society" organized by the International Society for the Study of Information held in Gothenburg in 2017 (http://is4si-2017.org). This volume contains a selection of the best theoretical contributions from the Gothenburg summit, together with a number of invited contributions of leading contemporary researchers in the field of *the Study of Information.*

As people observe the world, they start to understand that information is everywhere, playing a crucial role in nature, society, among human beings and in technological systems. Consequently, information studies involve and penetrate all sciences, playing an indispensable role in every scientific discipline. At the same time, the high-speed development of information technology has caused an explosion of research in the field of information and its applications, opening future paths for the advancement of information technology, scientific exploration of the universe and social organization. However, phenomenal achievements of information studies exploring theoretical, methodological and philosophical issues have caused many new problems in these areas. For instance, in spite of great achievements of information theory and technology, notwithstanding the numerous definitions of information suggested by different authors, there is no general agreement and mutual understanding in this area. The complexity of information phenomena demands development of an advanced theory of information, as well as utilization of various scientific and social disciplines for exploration of the nature of information, discovering new laws of information processes and study of information at all levels, from natural information processes to the information society and information technology, such that they open new vistas that allow for improved applications. At the same time, the language of information is increasingly used in many different disciplines, and a comparison of the usage in different fields contributes to a better understanding of information in its own right allowing more powerful applications of information theory.

In this context, it is necessary to stress that although many think that information theory is the same as Shannon's information theory, which was further developed by different researchers, this is an essentially distorted opinion and misconception. In contrast to this, because Shannon's information theory was aimed at a specific aspect of communication, and to overcome this limitation, many other directions and domains in information theory were created. The most notable of them are as follows:

- statistical information theory, which comprises Claude Shannon's information theory, Ronald Fisher's information theory and some others;
- semantic information theory, originated in the works of Yehoshua Bar-Hillel and Rudolf Carnap;
- algorithmic information theory, originated and developed in the works of Ray Solomonoff, Andrey Kolmogorov and Gregory Chaitin;
- pragmatic information theory, which comprises economic information theory originated in the works of Jacob Marschak and Arrow;
- decision-oriented information theory, originated in the works of Harkevitch, Bongard and some others;
- quantum information theory, originated in physics;
- qualitative information theory, originated by Marian Mazur;
- dynamic information theory, originated and developed in the works of Fred Dretske, Jon Barwise, Jerry Seligman, Alexander Chechkin and Mark Burgin;
- algebraic information theory, originated and developed in the works of Juerg Kohlas and Mark Burgin;

- logical information theory, originated in the works of Keith van Rijsbergen;
- evolutionary information theory, originated by Mark Burgin;
- structural information theory, originated by Emanuel "Maan" Leeuwenberg;
- integrated information theory, originated by Giulio Tononi;
- general theory of information, which comprises all other information theories and was created by Mark Burgin.

Information theory is a component of theoretical information studies, which in turn form a part of *the scientific study of information*, while two other parts are *experimental information studies* and *practical information studies*. Note that in addition to the scientific study of information, the study of information also includes information philosophy and methodology.

Theoretical information studies use means from different mathematical fields, such as probability theory, statistics, mathematical logic, decision-making theory, category theory and functional analysis, comprising three areas:

- *Information theory*,
- *Applications of information theory*,
- *Information-oriented and information-based theories in different areas.*

These areas include the following domains:

- Information theory, which, in turn, includes statistical information theories, semantic information theories, pragmatic information theories and dynamic information theories;
- Theoretical computer science, as it studies information processing by technical devices;
- Theoretical linguistics, as it studies information processing by natural and artificial languages;
- Semiotics, as it studies symbolic information processing;
- Psychology, as it studies information processing by people;
- Pedagogy, as it studies information transmission and knowledge acquisition;
- Application of information theory to theoretical physics, exploring the role of information in the Universe;
- Application of information theory to theoretical computer science, exploring relations between information, computation and networking;

- Application of information theory to the theory of complexity, exploring relations between information and complexity;
- Application of information theory to mathematics, exploring relations between information and structures;
- Application of information theory to theoretical linguistics, exploring relations between information and language;
- Applications of information theory in economics and finance, exploring the role of information in the economy and finance;
- Application of information theory to decision-making, exploring the role of information in decision-making;
- Application of information theory to pedagogy, exploring the role of information in education;
- Applications of information theory in sociology, exploring the role of information in society;
- Applications of information theory in psychology, exploring relations between information and creativity;
- Applications of information theory in biology and medicine, exploring the role of information in the organism;
- Applications of information theory in anthropology, exploring the role of information for human beings and their progress;
- Applications of information theory in ecology, exploring the role of information in the environment;
- Application of information theory to semiotics, exploring relations between information and symbols;
- Application of information theory to epistemology, exploring the role of information in cognition;
- Application of information theory to esthetics, exploring relations between information and beauty;
- Application of information theory to ethics, exploring relations between information and morals;
- Application of information theory to law, exploring the role of information in law;
- Application of information theory in humanities;
- Information-oriented and information-based physical theories;
- Information-oriented and information-based psychological theories;
- Information-oriented and information-based biological theories;
- Information-oriented and information-based theories of decision-making;
- Information-oriented and information-based economical theories;
- Information-oriented and information-based sociological theories;
- Information-oriented and information-based ecological theories;

- Information-oriented and information-based epistemology;
- Information-oriented and information-based esthetics;
- Information-oriented and information-based ethics;
- Information-oriented and information-based anthropological theories.

This volume presents work aimed at establishing theoretical aspects of a new interdisciplinary/transdisciplinary field of study, with new ways of knowing and models of explanation, based on the advanced theory of information (Burgin, 2010), data–information–knowledge ecology (Capurro, 1990; Burgin, 2016; Zhong, 2017; Burgin, 2017; Burgin and Zhong, 2018), reality stratification (Burgin, 2012) and information taxonomy (Burgin and Dodig-Crnković, 2017). It provides theoretical perspectives on information as a phenomenon that connects together, and enables new non-reductive unifications of, phenomena otherwise studied in isolation within domains of particular sciences, humanities, technologies and arts, as well as other fields dedicated to the study of different aspects, uses and roles of information. Currently, specialized academic fields divide human knowledge and experience into silos isolated from each other. Based on the research results of the domain-knowledge, *the study of information* seeks a new, networked approach to enable an inclusive view, from which future research projects transcending disciplinary boundaries will derive, based on the understanding of information structures and processes underlying particular domains. The study of information involves a recursive process of analysis and synthesis through continuous learning about the natural as well as synthetic/created/cultural worlds, and about their actors — living beings and artificial agents. Theoretical information studies form an intrinsic component of the study of information.

To reflect the most important domains of theoretical information studies, the contents of this book are divided into four parts: *Foundations of Information, Information as a Natural Phenomenon, Cognition and Intelligence in Natural and Artificial Systems*, and *Social, Cultural, Ethical and Legal Aspects of Information.*

The *first part* of the book is dedicated to foundations of information, with chapters encompassing a broad spectrum of principal issues.

In their chapter, Mark Burgin and Gordana Dodig-Crnković develop a multiscale, multiaspect taxonomy of information as a natural, social and technological phenomenon. This taxonomy has the form of a unified system of aspect taxonomies of information, the majority of which are developed by the authors, while others are suggested by other researchers.

Each of the aspect taxonomies reflects some aspect of information representing modes of this aspect. In such a way, information is represented in a multidimensional parametric space. This representation has several goals:

- to support the conceptual analysis of information;
- to allow better utilization of information in cognition;
- to improve efficiency of information processing systems, such as search engines;
- to provide better methodological understanding of information and its functioning.

In his chapter, Meir Buzaglo studies regeneration of information as concept expansion in a model of a language. He introduces such concepts as partially and universally regenerative models, treating regeneration as the phenomenon where a whole loses some of its parts and is regenerated (reconstructed or restored) from its remaining parts. To develop a formal model of regeneration, Buzaglo uses the concept of function expansion.

In their chapter, Wu Kun and Jian Wang analyze ethical principles reconciling, and at the same time maintaining, a reasonable tension between opposing factors, such as information and entropy, order and disorder, integrity and reducibility, certainty and uncertainty, determinism and nondeterminism, purpose and randomness, inevitability and contingency. They demonstrate that the syntactic informational approach to ethical problems suggested by Luciano Floridi is too narrow and wide of the mark because the problems of information directly related to ethics and values are essentially in semantics and pragmatics, rather than simply in its syntactic structures, such as information entropy.

In their chapter, Gianmarco Tuccini, Roberta Lanfredini and Luca Baronti perform a philosophical analysis of the qualitative as the basis for carrying out qualitative applied ontology for a more accurate representation of human actions and natural language as a tool for performing complex tasks. The goal is a reconsideration of the relationship between the computational theory of the mind and its qualitative dimension by analyzing the notion of effective procedure and the general notion of information. At the same time, the authors replace computational mind by an adaptive and common sense computational model, able to incorporate the performative aspect of linguistic utilization and in a consistent way to take into account social context, emotional intelligence, openness and adaptability to the environment.

In their chapter, Yixin Zhong and Gordana Dodig-Crnković develop a theory of semantic information based on the ecological approach to information studies developed in Zhong (2014, 2017, 2017a) and Burgin and Zhong (2018). They assume that the ecological system of information is the full process of information–knowledge–intelligence conversion while production of semantic information is based on both syntactic and pragmatic information.

The *second part* of the book contains studies of information as a natural phenomenon.

In his chapter, Paul Benioff further develops his theory of local mathematics. According to this theory, there are different types of mathematics, i.e., unified systems of mathematical structures, at different regions of the physical space–time manifold. Examples of such mathematical structures include number structures of different types, (natural, integer, rational, real and complex), vector spaces, etc. Relations between number structures of each type at different locations are based on two aspects: the distinction between two so far connoted concepts, *number* and *number value*, and the "No information at a distance" principle.

The idea of locality in physics has very deep roots and an interesting history. It started with the works of Riemann, who introduced locality to geometry as a description of the physical space (Riemann, 1892). Later Lorentz introduced the concept of local time in physics (Lorentz, 1895). Similar ideas can be found in works of Poincaré and Einstein (Poincaré, 1898, 1900; Einstein, 1905). There were also local theories in mathematics. For instance, Bell studied local set theories (Bell, 1988). Development of information dynamics brought forth the concept of a Barwise and Seligman introduced and studied local logics (Barwise and Seligman, 1997), while a more general concept of a local axiomatic theory was introduced by Burgin (Burgin, 1995; Burgin and de Vey Mestdagh, 2015). At the same time, the advance of gauge theory gave birth to the idea of a local vector space associated with points of the physical space–time manifold. The principal goal of Paul Benioff is building a coherent theory of physics and mathematics.

The chapter by Rafael Capurro and Mark Burgin consists of two parts. In the *first part*, Rafael Capurro analyzes Weizäcker's ideas about the unity of nature based on his longstanding dialog with Martin Heidegger about Heidegger's interpretation of being as three-dimensional time. In the *second part*, Mark Burgin demonstrates how new scientific discoveries support ideas of Weizäcker about relations between information and the Platonic and Aristotelian concepts of *eidos*, Ideas or Forms. Explication of these relations is based on the discovered structure of the world (Burgin, 2012) and the system theory of time (Burgin, 2002).

The next chapter presents the interview of one of the preeminent thinkers of the 20th century, Carl Friedrich von Weizsäcker, by Rafael Capurro. It covers such issues as relations between physics and philosophy, problems of ecology, meditation as a way to knowledge, unification of modern physics and the problem of time in physics.

In his chapter, Sungchul Ji elaborates the molecular linguistics of DNA based on the hierarchy of letters, words, sentences, texts and their meanings in the context of structural similarities, hylomorphism as they were called by Aristotle, between the cell language and human language. His main objective is to further explore the structure and function of cell language using the recent developments in information theory and the mathematical structures of DNA sequences.

In his chapter, Robert Logan narrows the definition of information, looking at it from the perspective of the original meaning of the word, as the formation of the mind. This brings forth instructional biotic information, which guides all living organisms from the simplest single cell organisms to human beings and other animals in possession of a brain to make choices that are in their own self-interest. Exemplifying information as formation or transformation of the mind shows that the theory of instructional biotic information is a subtheory of the general theory of information, which contends that information for a system is a capacity to cause transformations in this system (Burgin, 2010, 2017).

In their chapter, Jorge Navarro and Pedro Marijuán explore biologically inspired information thinking. According to their approach, information is never detached from a system, but can be only detected through the transformations it generates, symbolically flowing amid the different subjects or agency realms and contributing to adaptively modify their material structures and self-production processes. This strongly supports the definition of information given in the second Ontological Principle of the general theory of information and its more specified versions because this principle asserts that information for a system is a capacity to cause transformations in this system (Burgin, 2010, 2017). This gives additional evidence demonstrating that the general theory of information adequately describes the phenomenon of information, giving a comprehensive, accurate definition of information.

In his chapter, Walter Riofrio studies design of molecular signaling networks, which are synthetic biology devices with components that imitate transmission and processing of biological information in living beings. This work reflects the current situation when the community of biologists is increasingly accepting the concept of information as vital for living systems. Riofrio emphasizes that the results of research into the deepest intracellular level and on upward through the rest of the intracellular, physiological, systemic, interspecies and even ecological levels often utilize such concepts as information, codes, signals, language, meaning and other related terms.

In his chapter, Guenther Witzany analyzes processes of the new genetic information generation. For more than half a century, it has been accepted that new genetic information is mostly derived from random error-based events. However, this explanation has problems in clarification of the sudden emergence of new species, new phenotypic traits and genome innovations as an abrupt single event. As a result, it was recognized that errors could not explain the evolution of genetic information, genetic novelty and complexity. Now, empirical evidence establishes the crucial role of non-random genetic content editors, such as viruses, diversity generating retroelements and other RNA networks, to produce new genetic information, complex regulatory control, inheritance vectors, genetic identity, immunity, new sequence space, evolution of complex organisms and evolutionary transitions.

The *third part* of the book explores information problems of cognition and intelligence in natural and artificial systems.

In his chapter, Mark Burgin explores information processing by structural machines. Structures contain information as matter contains energy and structural machines work with structures of arbitrary order transforming not only elements of these structures or the content of these elements, as conventional models of computation do, but also relations of different orders in the processed structures. Being structurally universal abstract automata, structural machines can directly work with molecular structures, with structures studied and utilized in the topological quantum field theory (TQFT) and with structures of quantum information such as qubits. This allows achieving higher flexibility and efficiency in comparison with regular models of computation, including both conventional and unconventional computing systems. Special attention is paid to information processes such as observation, information transmission and information reception. Superior flexibility of structural machines is demonstrated by proving that they can simulate such advanced computational models as Kolmogorov algorithms (Kolmogorov, 1953) and simple-limit Turing machines (Burgin, 2001). In essence, many important and popular models of computation are special cases of structural machines.

In their chapter, Ron Cottam and Roger Vounckx explore intrinsic relations between consciousness and information. They describe decision-making and action-planning as an unconscious precursor of consciousness, placing consciousness as the explicit component of neural processing through the medium of information.

In his chapter, Rao Mikkilineni demonstrates convergence of natural intelligence and artificial intelligence. He examines the role of cognition in symbolic computing and neural networks, exploring the potential for symbiotic relationships between natural and artificial intelligence, allowing extension of the collective consciousness and development of a unified culture that binds the participants to define common goals and execute processes that transcend human and machine capabilities. It is demonstrated that the cognitive overlay is a crucial control component in defining the workflow process intent, monitoring its execution and correcting deviations as and when they occur due to non-deterministic fluctuations in the environment.

The *fourth part* of the book gives an exposition of various social, cultural, ethical and legal aspects of information.

In their chapter, Gengxian Cao and Donghe Li use information system theory to analyze transformation of Western music in the 20th century, in which "rigid" characteristics were transformed into "flexible" features of music. In such a way, the authors explicate relations of this process to traditional Chinese music, which traditionally possessed a flexible countenance.

In his chapter, Cornelis N.J. de Vey Mestdagh develops an information-based model of complexity for the legal domain. As a qualitative notion, complexity is used to legally and politically analyze and evaluate legal proceedings, while as a quantitative concept, it is used to compare, rank, plan and optimize these proceedings. To achieve these goals, de Vey Mestdagh elaborates a model of problem complexity in the legal domain based on a formal model of legal knowledge describing parameters of the problem complexity of various legal cases.

In his chapter, Ouyang Kang explores the influence of the phenomenon of Big Data on social science research. Assuming that people are now living in the information age and Big Data time, Kang analyzes specific opportunities opened up by the Big Data paradigm and the challenges of Big Data.

The aim of the chapter by Loet Leydesdorff is an analysis and specification of what is communicated in interpersonal communications. Leydesdorff argues that in contrast to transmission of energy and moments among particles or the exchange of molecules in biological processes, codes constructed in and by interpersonal communication can provide feedback as controls of the communication or operate as feedforward loops enhancing the communication. Building his model, Leydesdorff considers the possibility to represent the codes of communication as the eigenvectors of a communication matrix since interpersonal communication is a multidimensional process resulting in a "fractal manifold" of communications.

In his chapter, Jun Liu analyzes the development of Confucian epistemology and ontology by Li Erqu, who was one of the three greatest Confucian philosophers at the beginning of the Qing Dynasty.

In his chapter, Ugo Pagallo treating law as information explores complex and even unpredictable ways in which technological developments of information technology have affected both the requirements and functions of law, namely, what law is supposed to be (requirements), and what it is doing (functions). This is a very important problem because the new realities of the information society have not only increased the difficulty of law enforcement on global technological structures, such as the Internet, but they often lead to the illegitimate condition where states claim to regulate unilaterally extraterritorial conduct by imposing norms on individuals who have no say in the decisions affecting them.

In his chapter, Tomás Sigmund investigates relations between information and openness in society. He tries to answer the questions to what extent information is unchangeable and to what extent it can be changed. To achieve this goal, Sigmund analyzes three aspects of information, namely, its syntactic, semantic and pragmatic dimensions. He shows that on all three levels, information is changeable and serves more as a trigger for transformation than as a hard fact. This implies that information in society is not an objective and independent phenomenon because it is influenced by society, by its culture and historical development. Such an understanding validates the general theory of information as a unified comprehensive information theory (Burgin, 2017) and endorses the ecological approach to studies of information (Zhong, 2017; Burgin and Zhong, 2018).

The book gives up-to-date multiaspect exposition of contemporary theoretical studies in the field of information science and related areas as well as their practical implications. It presents the most recent achievements, ideas and opinions of leading researchers in the domain of information studies as well as physicists, biologists and social scientists.

Collaboration of researchers from different areas and fields opens new perspectives for the innovative sustainable development of science, society and technology.

This work is aimed at readers who conduct research into any aspect of information, information society and information technology. It opens new perspectives for those who develop or implement technological or social applications of scientific theories. It is also for those who are interested in any aspect of information, information society and information technology, and especially for those who are participating in setting the goals for science in general and sciences of information in particular.

References

Bell, J. L. (1988). *Toposes and Local Set Theories: An Introduction*, Oxford University Press, Oxford.

Barwise, J. and Seligman, J. (1997). *Information Flow: The Logic of Distributed Systems.*

Burgin, M. (1995). Logical tools for inconsistent knowledge systems, *Information: Theories & Applications*, 3(10), 13–19.

Burgin, M. (2001). Topological algorithms, in *Proceedings of the ISCA 16th International Conference "Computers and their Applications"*, ISCA, Seattle, Washington, pp. 61–64.

Burgin, M. (2010). *Theory of Information: Fundamentality, Diversity and Unification*, World Scientific Publishing Co. Pty, Ltd., Singapore.

Burgin, M. (2012). *Structural Reality*, Nova Science Publishers, New York.

Burgin, M. (2016). *Theory of Knowledge: Structures and Processes*, World Scientific, New York.

Burgin, M. (2017). The general theory of information as a unifying factor for information studies: The noble eight-fold path, in *Proceedings*, 1(3), 164, p. 6, doi:10.3390/IS4SI-2017-04044.

Burgin, M. and Dodig-Crnković, G. (2017). Prolegomena to information taxonomy, in *Proceedings*, 1(3), 210, p. 4, doi:10.3390/IS4SI-2017-04035.

Burgin, M. and de Vey Mestdagh, C. N. J. (2015). Consistent structuring of inconsistent knowledge, *Journal of Intelligent Information Systems*, 45(1), 5–28.

Burgin, M. and Zhong, Y. (2018). Information ecology in the context of general ecology, *Information*, 9(3), 57.

Capurro, R. (1990). Towards an information ecology, in *Information and Quality*, Wormell, I. (ed.), Taylor Graham, London, pp. 122–139.

Einstein, A. (1905). Zur Elektrodynamik bewegter Körper, *Annalen der Physik*, 17, 891–921.

Kolmogorov, A. N. (1953). On the concept of algorithm, *Uspekhi Mat. Nauk.*, 8(4), 175–176 (in Russian).

Lorentz, H. A. (1895). *Versuch einer Theorie der elektrischen und optischen Erscheinungen in bewegten Körpern*, E.J. Brill, Leiden.

Poincaré, H. (1898). La measure du temps, *Revue de Metaphysique et de Morale*, 1, 1–13.

Poincaré, H. (1900). La théorie de Lorentz et le principe de réaction, *Arch. Néerland. Sci. Ex. Nat.*, 5(2), 252–278.

Riemann, B. (1892). *Gesammelte mathematische Werke und wissenschaftlicher Nachlass*, Leipzig, Teubner.

Zhong Y. X. (2014). *Principles of Advanced Artificial Intelligence*, Science Press (in Chinese).

Zhong, Y. X. (2017). The law of "Information Conversion and Intelligence Creation", in *Information Studies and the Quest for Transdisciplinarity*, World Scientific, New York, USA; London, UK; Singapore.

Zhong, Y. (2017a). Information ecology and information studies, in *Proceedings*, 1(3), 200.

PART I
Foundations of Information

Chapter 1

A Multiscale Taxonomy of Information in the World

Mark Burgin*,‡ and Gordana Dodig-Crnković†,§

** University of California, Los Angeles, CA, USA*
† Chalmers University of Technology, Sweden
‡ mburgin@math.ucla.edu
§ dodig@chalmers.se

We construct a multiscale, multiaspect taxonomy of information. This taxonomy is a unified system of aspect taxonomies of information, the majority of which are developed by the authors while others are suggested by other researchers. Each aspect taxonomy reflects some aspect of information representing modes of this aspect. In such a way, information is represented in a multidimensional parametric space. This representation has several goals: to support conceptual analysis of information, allow better utilization of information in the domain of cognition, enable improvement of efficiency of information processing systems, such as search engines, and provide better methodological approaches to information.

1. Introduction

In the contemporary information era, there are a host of different sorts, kinds and types of information. To organize this huge diversity into a system, it is necessary to classify information with respect to various criteria, thus developing a *multiscale information taxonomy*. In such a taxonomy, each dimension corresponds to an aspect of information. We construct such a multiscale information taxonomy based on the general theory of information (Burgin, 2003, 2004, 2010a) and making use of its principles and technical tools. This work is an extension of the ideas presented in Burgin and Dodig-Crnković (2017).

1.1. *Taxonomies as laws of science*

To start with our study, it is important to recognize that taxonomies are not just auxiliary constructions and representation tools in science but, being scientifically grounded and validated, they also express specific kinds of laws of science.

According to the mathematical form, there are three types of scientific laws: classificational, equational and implicational (Burgin and Kuznetsov, 1993, 1994).

An equational law has the form of an equation, for example, the differential equations that depict the many laws in physics, e.g., $E = mc^2$, chemical equations, such as $-\mathrm{d}c_A = k \cdot c_A \mathrm{d}t$ representing kinetics of a chemical reaction, or the gravity equation of trade in economics — $\mathrm{T}_{A,B} = G \cdot (\mathrm{GDP}_A)^{\alpha}(\mathrm{GDP}_B)^{\beta}/(\mathrm{Dist}_{AB})^{\varsigma}$.

An implicational law has the form of an implication "If A, then B". For instance, in logic, $\neg(p \to q) \Longleftrightarrow \neg(\neg p \vee q)$. In mathematics, if ΔABC is a right triangle, then its sides satisfy the equation $c^2 = a^2 + b^2$. This is a mathematical law called the Pythagorean theorem. An example of implication in chemistry is $3\mathrm{NH}_2 + \mathrm{N2} \to 2\mathrm{NH}_3$, describing a chemical reaction.

A classification law has the form of a classification, typology or taxonomy. The biological taxonomy of Linnaeus (1735) and triadic typologies of Peirce (1839–1914) are examples of classification laws.

In addition to the systematization according to mathematical form, scientific laws can be differentiated with respect to each aspect of the research object in focus. We distinguish between two categories — qualitative and quantitative laws — which either describe the existential features of a research object that are not measurable or countable (qualitative laws) or focus on the extent or characteristics that are measurable or countable (quantitative laws). Qualitative and quantitative are relative terms, depending on how the object is perceived or studied. The color of an object can be characterized as a qualitative property, while it can also be expressed quantitatively by light wavelength. Sweetness can be described as a quality, but can also be expressed as the amount of sugar per unit of volume.

A *quantitative law* describes relations between quantitative characteristics of a given phenomenon. For instance, Newton's Second Law of Motion, $F = ma$, is a quantitative law of physics.

A *qualitative law* describes relations between qualitative characteristics of definite phenomena. Newton's First Law of Motion, presented in the *Philosophiae Naturalis Principia Mathematica*, was actually discovered by Galileo and refined by Descartes (who added "in a straight line"):

> *Every object will remain at rest or in uniform motion in a straight line unless compelled to change its state by the action of an external force.*

It is an example of a qualitative law of physics. Classificational laws are usually qualitative laws that utilize naming procedures for data and information structuring, such as in systems for classifying biological organisms. The Standard Model taxonomy of particle physics, describing all fundamental forces in the universe (the electromagnetic, weak and strong interactions), except gravitation, provides quantitative classification of all known elementary particles based on their quantum numbers (Vayenas and Souentie, 2012). Another example of a quantitative taxonomy is the Mendeleev periodic table of elements (Scerri, 2011). The periodic table organizes elements in rows and columns, classifying them by their atomic number and electron configuration, and explaining the characteristics of each element by its position in the table.

The methodological insight about the fundamental role of classifications as governing laws gives a higher scientific status and importance to the Linnaeus classification, as well as to the taxonomies in general, including the one constructed in this chapter. One of the important features of classifications is their capacity to predict, based on the structure of the taxonomy. This new understanding of scientific laws describes *taxonomies as laws of science.*

While equational and implicational laws have been acknowledged in science from its very origin, classification laws acquired their nomological status only recently in the structure-nominative direction of the methodology of science (Burgin and Kuznetsov, 1994), which was further developed in Burgin (2016a).

There are also descriptive, operational and representational laws (Burgin and Kuznetsov, 1993; Burgin, 2016a), but they are considered elsewhere; here our main emphasis is on *classification laws in the form of taxonomies.*

Thus, the biological taxonomy of Linnaeus (1735) is a law of biology, the Mendeleev periodic table is a law of chemistry and the Standard Model's taxonomy of elementary particles represents physical laws, in the same way as Newton's laws or Maxwell's equations present the laws of physics.

In this chapter, we follow the taxonomic traditions of Linnaeus and Peirce in the study of information and information science. On the one hand, the results of our research connect new information science and technology with classical science demonstrating intrinsic links between information theory and the profound results of Linnaeus and Mendeleev. On the other hand, these results show a unity in the achievements of scientists working in different research fields such as biology (classification of Linnaeus), chemistry (classification of Mendeleev), semiotics (classifications of Peirce), physics (classifications of subatomic particle hadrons by Gell-Mann and Nishijima and Ne'eman, as well as the Standard Model) and information science (such as classifications developed in this chapter).

Here, we explicate classification laws of information using the word *taxonomy* as the generic term that stands also for classifications, typologies and categorizations. We begin with a brief exposition of methodological principles of taxonomy construction (Section 2) and then, in the next sections, apply these principles to the development of basic information taxonomies.

2. Principles of Taxonomy Construction

When we have a multiplicity of objects, it is useful to introduce organization in it in order to study, understand and utilize this multiplicity. Organization is achieved by structuring of the multiplicity. An efficient technique of structuring is construction of taxonomies, classifications, typologies and categorizations. Let us consider the process and basic principles of taxonomy construction.

Classical taxonomies use one scale to induce categorization of the studied collection of objects. Usually, such a taxonomy is hierarchical. However, when objects have a complex nature and a variety of properties, it is more efficient to use different scales, building a *multiscale taxonomy.*

Construction of a multiscale taxonomy is organized in the following way. Taking a multiplicity of objects M, we explicate the objects' properties, molding aspects or amalgamated features. Then we elucidate a criterion for each aspect. This allows us to form a scale for measuring/evaluating each aspect. Such a scale together with the corresponding criterion allows us to build an aspect taxonomy. Combining together all aspect taxonomies, we obtain a multiscale taxonomy of the multiplicity M. This process is described in Figure 1.

$$
\text{Multiplicity } M \rightarrow \text{properties} \rightarrow \begin{cases} \text{aspect } 1 \rightarrow \text{criterion } 1 \rightarrow \text{scale } 1 \rightarrow \text{aspect taxonomy } 1 \\ \text{aspect } 2 \rightarrow \text{criterion } 2 \rightarrow \text{scale } 2 \rightarrow \text{aspect taxonomy } 2 \\ \text{aspect } 3 \rightarrow \text{criterion } 3 \rightarrow \text{scale } 3 \rightarrow \text{aspect taxonomy } 3 \end{cases}
$$

Figure 1. Multiscale taxonomy construction from the multiplicity M.

Note that there is a difference between the multiscale system and the multidimensional system in general and multiscale taxonomy and multidimensional taxonomy in particular.

When mathematicians or physicists speak about *multidimensional* spaces, they have in mind vector spaces such as Euclidean spaces, Hilbert spaces or Banach spaces. In these spaces, all scales are the same, i.e., isomorphic to the space of all real or of all complex numbers. Other (discrete) multidimensional taxonomies have the form of a table of two or more dimensions. The standard form of the periodic table of chemical elements is an example of multidimensional taxonomies. The majority of periodic tables have two dimensions, while there are also periodic tables with three and four dimensions. In contrast to this, in a *multiscale space*, scales can be essentially different as in our information taxonomy. From a mathematical point of view, a multiscale space is not a vector space but a topological manifold.

Besides, *all dimensions in a multidimensional space are independent.* In contrast to this, in a multiscale space, one scale can be the scale of one position in another scale, as in our information taxonomy. Examples are the form/action taxonomy (Section 5) and the fitness taxonomy of emotional (affective) information (Section 14). In what follows, we describe a system of information aspect taxonomies, which together form a hierarchical multiscale, multiaspect information taxonomy.

3. The Existential Taxonomy

We begin with the uppermost level of the taxonomy, which includes a huge diversity of types, kinds, sorts, categories and classes of an omnipresent phenomenon of information (Burgin and Dodig-Crnković, 2011). The structure of the world affects the existence of specific forms of information, which correspond to this structure. The large-scale structure of the world is represented by the *Existential Triad of the World* (Burgin, 2012), which is given in Figure 2.

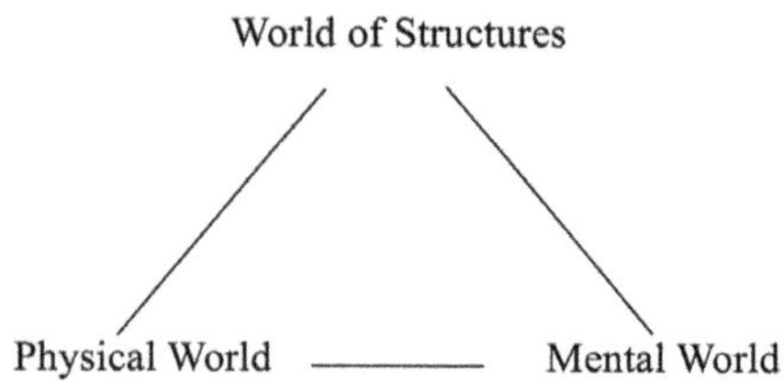

Figure 2. The Existential Triad of the world.

In the Existential Triad, the physical (material) world is conceived as the physical reality studied by natural sciences, the mental world encompasses different levels of mentality studied by cognitive sciences and the world of structures consists of various forms and types of structures studied by formal sciences.

The existential stratification of the world continues the tradition of Plato with his World of Ideas (Plato, 1961) and the tradition of Peirce with his extensive triadic classifications (Peirce, 1931–1935).

This stratification brings us to the phenomenon studied by the general theory of information, called "information in a broad sense" (Burgin, 2011b). According to this approach, information in a broad sense is represented in each of the three worlds. In the physical (material) world, it is called "energy", supporting in such a way the conjecture of von Weizsäcker that "mass is information" and "energy is information" (Weizsäcker, 2014, p. 167, 1974). Situated at the first level of the Mental (Cognitive) World, individual mental energy has been investigated by psychologists such as von Brücke, Freud and Jung. In contemporary neuroscience, it is being addressed as the change of metabolic energy of the brain above its ground level when performing cognitive tasks as it is demonstrated in Gailliot *et al.* (2007) and Camandola and Mattson (2017), who studied the differences in brain metabolism in health, aging and neurodegeneration.

Information in a broad sense, which is situated in the World of Structures, is called *information in the strict sense* according to Burgin (2010a).

In the global existential setting, we have the following taxonomy:

- *Physical world information* or physical energy;
- *Mental world information* or mental/cognitive energy;
- *Structural world information* or information *per se* (which presents *potential* structural energy).

We will not analyze here the first two kinds of information in a broad sense as the first of them belongs to the scope of physics, while the second belongs to the domain of cognitive science/psychology. Our concern is information in the strict sense that we simply term *information.* In the following, we will address taxonomy of information, which according to Figure 1 is the multiplicity M that we study from various aspects.

4. The Developmental/Temporal Taxonomy

The developmental/temporal aspect distinguishes three types of information:

- *Potential information*;
- *Emerging information*;
- *Actualized information.*

Let us consider some examples, from Burgin (2011a).

Example 1. Information in a book before somebody reads it is *potential information.* It is possible to measure potential information by its potential to make changes in the corresponding infological system, where an infological system plays the role of a parameter that discerns different kinds of information and combines them all in one general concept. For instance, measuring potential epistemic information, we estimate (measure) potential changes in the knowledge system.

Example 2. Information in a computer, which processes information, or in the head of a person who thinks about it, is emerging. It is possible to estimate emerging information by its potential to make changes, by transformations it made in the corresponding infological system and by the rate of ongoing transformations. For instance, measuring emerging epistemic information, we estimate what changes in the knowledge system have already been made and reckon the rate of ongoing changes.

Example 3. Information in a book after somebody reads it is *actualized information* for that reader. It is possible to measure actual information by changes it has made in the corresponding infological system. For instance, measuring actualized epistemic information, we determine (measure) changes in the knowledge system made by reception of this information.

5. The Form/Action Taxonomy

Information taxonomy based on the *extended triune model* of the brain has been elaborated in Burgin (2010a) as the further development of the *triune brain model* (MacLean, 1973). It refers to the three basic systems, distributed through different anatomical parts of the brain, representing the following:

- *Rational intelligence/reasoning*;
- *Emotions/affective states*;
- *Will and instinct.*

As a result, the bifocal *form/action taxonomy* can be defined with the following types of information:

- *Epistemic* (form) or *cognitive* (action);
- *Emotional* (form) or *affective* (action);
- *Instructional* (form) or *effective* (action).

In this taxonomy, the first facet (term/name of each class) represents the form of the corresponding infological system and the form of information existence, while the second facet (term/name of each class) mirrors the action category of information existence, as well as the operational quality of the corresponding infological system. It means that the first nominal attribute of the taxonomic classes characterizes form aspects of information, while the second operational attribute of the taxonomic classes characterizes procedural aspects of information.

Note that in this taxonomy, emotional information in the context of the extended triune model of the brain is different from emotional information in the context of emotional intelligence (Goleman, 1995) from the reference taxonomy.

6. The Correctness-based Taxonomy of Information

With respect to correctness of information, the following classification of information can be made about the type of knowledge it generates:

- *Correct information* that gives correct (true) knowledge;
- *Tentative information* that gives unconfirmed knowledge;
- *Incorrect information* that gives incorrect/false knowledge.

It is essential to have correct information. Unfortunately, new technologies, especially the Internet, make it much easier for generation and dissemination of incorrect and misleading information. This urgent situation illustrates the significance of the correctness taxonomy.

Note that correctness is a relative property because what is correct in one situation may be incorrect (wrong) in other circumstances.

7. The Justification-based Taxonomy

With respect to justification, there are three classes of information:

- *Validated or grounded information*, which has been validated by empirical testing or theoretical confirmation;
- *Unconfirmed information*, which does not have validation;
- *Unfounded/unjustified/misleading information* for which there is evidence of it being false.

Now there is a host of unfounded and false information on the Internet. Validation of information is an important and, in some cases, vital task.

These types are applicable to all classes of information delineated in the existential and formation/action taxonomies. It means that this taxonomy is complementary or orthogonal to the existential taxonomy and to the formation/action taxonomy.

8. The Relevance Taxonomy

The property of relevance of information differentiates three classes:

- *Relevant information*;
- *Quasi/partially relevant information*;
- *Irrelevant information*.

This taxonomy describes/classifies information reflecting its estimate with respect to what it is related to.

Example. Let us consider information about the size of the Sun. It is relevant with respect to astronomy, cosmology and astrophysics, partially relevant with respect to quantum physics and irrelevant with respect to sociology.

Relevance is an important property for evaluation of information obtained on the Internet, used in decision-making and found by search engines. For instance, contemporary search engines often offer a lot of irrelevant information to their users.

9. The Precision/Accuracy Taxonomy

Exactness of information gives us the following classes:

- *Exact information*;
- *Approximate information*;
- *Inexact information.*

Example. Information that the Earth has the form of a geoid is exact. Information that the Earth has the form of a ball is approximate. Information that the Earth has the form of a square is inexact.

It is a question of whether it is possible to have absolutely exact information about the physical world. The uncertainty principle in quantum physics shows that physical measurements can provide only approximate information and absolute precision is unachievable. Another example is information about number π, for which it is possible only to calculate increasingly better approximations.

10. The Effectiveness Taxonomy of Instructional or Effective Information

Efficiency as a property of instructional or effective information gives us three types of information:

- *Efficient information*;
- *Quasi-efficient information*;
- *Inefficient information.*

Classification of instructional or effective information depends on the mission, task or goal used as the criterion of information evaluation. That is why the same information can be efficient for achieving one goal, quasi-efficient for another one and inefficient for a third goal.

11. The Domain-based Taxonomy

With regard to domain, on the top level, three domains of information are discerned: nature, society and technology. This gives us the following taxonomy:

- *Natural information* is information available in nature, i.e., in objects in nature and their processes.
- *Social information* is information pertaining to society, i.e., social systems and social processes.
- *Technological information* is information about technological systems and technological processes.

For instance, with respect to the domain, information about the size of the Earth is natural information. Information about the size of a group of people is social information. Information about the size of a building is technological information.

12. The Pragmatic Taxonomy

Property of applicability of information gives us three classes:

- *Applicable information*;
- *Superfluous information*;
- *Prohibited information.*

Pragmatic classification of information depends on the problem, task and domain. In other words, the same information can be applicable to one domain, superfluous for another one and prohibited in the third. Besides, applicability of information depends on the user of the information — what can be applicable for one user can be superfluous for another.

13. The Formative Taxonomy of Cognitive Information

The property of being instrumental/formative for a cognitive task gives us three classes:

- *Constructive information* providing constructive knowledge;
- *Explanatory information* providing explanatory knowledge;
- *Descriptive information* providing descriptive knowledge.

For instance, the basic difference between conventional and intuitionistic mathematics is that the latter accepts only constructive information while the former recognizes all three types of cognitive information.

14. The Fitness Taxonomy of Emotional or Affective Information

The appropriateness of information with respect to emotional/affective behavior gives rise to the following taxonomy:

- *Appropriate affective information*;
- *Quasi- or relatively appropriate affective information*;
- *Inappropriate affective information.*

This classification is relative, depending on norms and values of an individual, group, community or society.

15. The Evaluative Taxonomy of Emotional or Affective Information

Attitudes of people establish three classes of information:

- *Positive affective information* generates positive emotions;
- *Neutral affective information* generates no emotions;
- *Negative affective information* generates negative emotions.

This classification essentially depends on the receiver of the information (Bosmans and Baumgartner, 2005). Indeed, what can cause very positive emotions and feelings in one person can be detrimental for another person. For instance, information that a football team A won in a match with a team B will make fans of A happy and fans of B unhappy.

16. Information Substantiation

Different concepts of information can be treated as an *inventory of forms* of information carriers. Namely, we have the following carrier forms of information appearing to us:

- *Material object*, e.g., newspaper, book, letter;
- *Structure in a hierarchy between data and knowledge*;
- *Symbolic system*, e.g., text;

- *Abstract entity* such as a "difference";
- *Cognitive/Psychological process*, e.g., perception, elimination of uncertainty, fear;
- *Potential cognitive energy in the structural world.*

This shows that there are different views on the nature of information and people often call information carriers by the name information, e.g., a letter is called information. The general approach to this important and extremely complex problem is developed in the general theory of information (Burgin, 2010a).

17. The Evaluative Pragmatic Taxonomy

Utility as a property of information discerns the following classes:

- *Positive or useful information* is valuable to the receiver.
- *Neutral or useless information* does not change the quality of the situation of the receiver.
- *Negative or harmful* information is detrimental for the receiver.

Negative information relates to negative probability, which is extensively utilized in physics (Dirac, 1974; Feynman, 1987; Burgin, 2009; 2010b) and also used in social and behavioral sciences, quantum computing, economics and finance (Hofmann, 2009; Burgin, 2015; 2016b; Burgin and Meissner, 2012).

18. Accessibility Taxonomy

Based on accessibility, knowledge is traditionally divided into two categories (Burgin, 2016a):

- *Esoteric knowledge*, which is available to a small group of specially initiated;
- *Exoteric knowledge*, which is, to the contrary, open to everybody.

This distinction in knowledge accessibility is reflected in the following taxonomy of information as the source of knowledge:

- *Exoteric information*, which gives exoteric knowledge;
- *Impartial information*, which gives knowledge neither inclusive nor exclusive *a priori*;
- *Esoteric information*, which gives esoteric knowledge.

Note that the accessibility type of the information or knowledge depends on an individual because what is esoteric for one person can be exoteric to another.

19. The Linguistic Taxonomy

Linguistics delineates three basic aspects of a text in a language: syntax, semantics and pragmatics. As any text is a carrier of information, it results in the linguistic taxonomy of information:

- *Syntactic information*;
- *Semantic information*;
- *Pragmatic information*.

Examples. *Syntactic or statistical information* theories include Shannon's theory of communication (Shannon, 1948) and Fisher's information theory (Fisher, 1922). *Semantic information theories* include the theory of Bar-Hillel and Carnap (1952) and *pragmatic information* theories include the economic information theory (Arrow *et al.*, 1951; Marschak, 1954).

20. The Carrier Representation-based Taxonomy

According to Landauer (1996) as well as the general theory of information (Burgin, 2010a), information does not exist by itself, but has a representation and a carrier. One of the taxonomic criteria is the *nature of the carrier* of information (Dodig-Crnković, 2016). According to this criterion, we have the following taxonomy:

- *Quantum information* is carried by quantum systems.
- *Molecular information* is carried by molecular systems.
- *Biological information* is carried by biological systems.
- *Cognitive information* is carried by cognitive systems.
- *Social information* is carried by social networks.

With respect to representation, we have the following classes:

- *Symbolic information* is represented by symbols.
- *Digital information* is represented by digits.
- *Genetic information* is represented in DNA and so on.

Any object can be a carrier of information and represent information (Dodig-Crnković and Giovagnoli, 2017). That is why a taxonomy of existing objects also gives the carrier representation-based taxonomy of information.

21. The Receptor-based Taxonomy

Another classification criterion is based on the type of the system that receives information. In humans, we distinguish between *external senses*, which include sight, hearing, touch, smell, taste, etc., and *exteroceptive senses*, perceiving the body's own state. Depending on the receptor of information, we have the following taxonomy:

- *Visual information* is received by eyes;
- *Auditory information* is received by ears;
- *Olfactory information* is received by the nose;
- *Haptic information* is received by skin.

For instance, according to neurophysiological data given, for example, in Burgin (2016a), 80% of all information that people get through their senses is visual information, 10% is auditory information and only 10% of the information people get is through the other senses.

22. The Informational Level-based Functional Taxonomy

One more criterion is the system where the information functions. It gives us the following classes that can be seen as a nested hierarchy of organizational levels (Dodig-Crnković, 2006, 2016, 2017):

- *Physical information* is information *in or about physical systems*;
- *Chemical information* is information *in or about chemical systems*;
- *Biological information* is information *in or about biological systems*;
- *Cognitive information* is information *in or about cognitive systems*;
- *Social information* is information *in or about social systems.*

Functional taxonomy resembles information carrier-based taxonomy, but there is a difference. Physical information refers to the information that is contained in a physical system or the information about physical systems. It is important to make the distinction between these two kinds of information because information in a physical system can be about mentality or structure. For instance, information in a computer can describe somebody's psyche. It is necessary to see the difference between cognitive emotional information, which is cognitive information about emotions, and direct affective, or direct emotive, information, which influences the emotional state of the receiver.

23. The Utilization Domain Taxonomy

According to the domain where information is used, the following classes are distinguished:

- *Physical/chemical/biological/medical information*;
- *Physiological/psychological/sociological/technological information*;
- *Philosophical/Political/Aesthetic information* and so on.

Information can be utilized in any area. That is why a taxonomy of existing areas of reality also gives a utilization taxonomy of information.

24. The Referential Taxonomy

One more classification criterion is what the information is about. It gives us the following taxonomy:

- *Weather information*;
- *Geopolitical information*;
- *Health care information* and so on.

As information can be about any object or process, a taxonomy of existing concepts also gives a referential taxonomy of information.

25. The Operational Taxonomy

Types of information with respect to the receiver determine accessible, available and acceptable information for a system.

- *Accessible information for a system* is the information the system has access to.
- *Available information for a system* is the one that the system can acquire.
- *Acceptable information for a system* is the one that the system can accept.

To see the difference between accessible information and available information, let us consider the situation when you come to a library that has one million books. You know that information you need is in some of these books, but you do not know in which one. If you do not have contemporary search tools to get this information and can only read books to find it, then it will not be available to you. Thus, we can see that even if a person has access to some information, it does not mean that this person can get this information. For example, information about Riemann integration is available only to those who know mathematics although anybody can find this information on the Internet or in a textbook. Moreover, in quantum physics, Heisenberg's uncertainty principle states that there is information about physical reality unavailable to an observer. Gödel's incompleteness theorems (Gödel, 1990) assert that there is information about mathematical structures unavailable to people. In computer science, it has been demonstrated (Burgin, 2005) that for a universal Turing machine, information concerning whether this machine halts, given arbitrary input, is unavailable when Turing machines or other recursive algorithms are used for this purpose.

Finally, even if a person gets information, it does not necessarily mean that the person accepts this information, as it may sound unbelievable or unreliable. For instance, when the great British physicist Paul Dirac theoretically discovered a positive "electron", which was later called the positron, other physicists did not accept Dirac's outstanding result and even mocked at him. Several years later, American physicist Carl Anderson experimentally discovered the positron. Some time passed and both physicists got a Nobel Prize in Physics for this discovery.

26. The Conditionally Operational Taxonomy

Parallel to the operational taxonomy, we can construct a conditional operational taxonomy for a system:

- *Conditionally accessible information for a system* is the information where the system has access to an information carrier.
- *Conditionally available information for a system* is the one that the system can acquire through an information *carrier*.
- *Conditionally acceptable information for a system* is the one that the system can accept as a representation of the information.

To see the difference between accessible information and conditionally accessible information, imagine a book in English. For a person who knows English and has the book, information in this book is accessible. At the same time, for a person who does not know English and has the book, information in this book is only conditionally accessible. There are different conditions for accessibility. One condition is that the person learns English. Another condition is that the person finds an interpreter. One more condition would be that the book is translated from English to the language that the person knows.

27. The Alethic/Truthfulness-based Taxonomy

Based on a criterion of truthfulness (alethic modality), three basic types of information can be identified:

- *True or correct information*;
- *Pseudo-information*;
- *False information or misinformation.*

The problem of false information is an important part of information studies and we need advanced scientific methods to treat this problem in an adequate manner. This is especially necessary because some researchers do not accept existence of false information arguing that it cannot be considered as information. However, a multiplicity of examples show that false information exists, and making a distinction between true and false information can be vital, as false information can cause huge damage.

There is also such a type as *biased information*. This type naturally exists only for information in society. For instance, Mowshowitz and Kumar (2009) wrote the following: *Search engine consolidation raises the specter of biased information and free speech abridgement. . . .* One of the main problems in information management is discerning genuine information, biased information, false information and misinformation.

To have true information relevant to usual understanding, we assess correctness of knowledge or validity of knowledge (Burgin, 2016a). Thus, we can call cognitive information *false* when it decreases the validity of the knowledge it gives. We have false information when its acceptance makes our knowledge less correct. For instance, let us consider people who lived in ancient Greece and accepted ideas of Leucippus and Democritus that all material bodies consist of atoms. Then they read Aristotle's physics that eliminated the idea of atoms. Because we now know that the idea of atoms is true, Aristotle's physics decreased true knowledge about such an aspect of the world as the existence of atoms and thus gave false information about atoms. Remarkably, false information can be communicated by omission of true information. We see that false information is also information because it has a definite impact on the infological system, only this impact is negative.

It is interesting that there is no direct correlation between false information and meaningless information. For instance, Bloch gives examples when false information was meaningful for people, while true information was meaningless for them (Bloch, 1949).

28. The Relative Evaluative Taxonomy

Taking some information measure m, e.g., Shannon's entropy or Kolmogorov complexity, we can use it as the criterion for building the following information taxonomy (Burgin, 2010a). A portion of information I is called the following:

- *Positive information*, relative to the measure m if $m(I) > 0$;
- *Neutral information*, relative to the measure m if $m(I) = 0$;
- *Negative information*, relative to the measure m if $m(I) < 0$.

It is interesting that all information is positive, relative to the initial Shannon entropy, which employs classical probability. However, allowing negative probability (Burgin, 2009, 2010a) and building a generalization of Shannon's entropy based on negative probability, we show that information can be negative relative to the generalized Shannon entropy.

29. The Receiver-centered Value-based Evaluative Taxonomy

This receiver-centered value-based evaluative taxonomy identifies three basic types of information with respect to the receiver's value system:

- *Positive information* that gives positive results;
- *Neutral information* that gives neutral results;
- *Negative information* that gives negative results.

For instance, the same information, advertizing hamburgers as excellent food, is neutral for someone who already has a value system in which hamburgers are not taken as a good food; it could be negative for people who have better food choices, but get persuaded by the commercial; or it can be positive for those who have poorer choices.

30. Information Taxonomies Suggested by Other Authors

Several researchers considered other types of information classifications. Csanyi (1989), Kampis (1991) and Banathy (1995) study three important types of information. With respect to a system R, we can distinguish the following:

- *Referential information* that has meaning in the system R;
- *Non-referential information* that has meaning outside the system R, e.g., information that reflects mere observation of R;
- *State-referential information* that reflects an external model of the system R, e.g., information that represents R as a state transition system and according to Banathy (1995) is the catalog of states contained in the system's model.

Shannon's information theory and its extensions are state referential. The model component of scientific theories is mostly state referential (Burgin, 2016a).

Braman (1989b) classifies information with respect to its roles, into "information-as" the following:

- *Resource*, coming in pieces unrelated to knowledge or information flows into which it might be organized;
- *Commodity*, obtained using information production chains, which create and add economic value to the information;
- *Perception of patterns*, having past and future, affected by motive and other environmental factors, and itself has its effects;
- *Constitutive force in society*, essentially affecting its environment.

However, it is necessary to add the fifth role of information, not mentioned above:

- Information as a *natural phenomenon*.

Buckland (1991) too discusses types of "information-as" the following:

- *Process*, in the sense that when someone is informed, then what this person knows is changed (*dynamics of information*);

- *Knowledge*, when information used to denote that what is perceived in information-as-process;
- *Thing*, when the term "information" is used attributively for objects, such as data and documents that are referred to as information because they are regarded as being informative.

MacKay (1969) defines information as *that which determines form*, and introduces three forms of information:

- *Structural information* is measured by "logons" (the frame of relevance used to construct a mapping between messages and referents).
- *Metrical information* is measured by "metrons" (the degree of resolution such a mapping is capable of capturing).
- *Selective information* is measured by "bits" (Shannon's original definition).

The first two types are called *constructive information.*

Note that MacKay's definition of information perfectly correlates with the principles of the general theory of information, because form is a kind of structure and, according to Ontological Principle O2a, information operates in the structural world as energy acts in the physical world (Burgin, 2011b).

A different approach is taken by Schroeder (2003) who makes the distinction between two fundamental and complementary types of information, unlike the MacKay tripartite scheme. Schroeder provides the following distinction based on the categorical opposition of *one* and *many*:

- *Structural information* as the structural property of variety;
- *Selective information* as the result of selection of one out of a variety.

The duality of selective/structural information manifests itself in the duality of function and form.

> *Functions of the elements of a system give them identity by distinguishing them from, and giving them their place in, the differentiated variety. On the other hand, this distinction is a consequence of the specific structural characteristics that they possess, their internal structure allows them to play specific roles in the system* (Schroeder, 2003).

Another manifestation of the same dualism is between *structure* and *process*. Dodig-Crnković (2006, 2016, 2017) and Dodig-Crnković and von Haugwitz (2017) analyze information as a structure and its dynamics/process as computation. Information and computation have been objects of extensive study in Dodig-Crnković and Burgin (2011). Parallel to the taxonomy of information, taxonomy of computation has been developed in Burgin and Dodig-Crnković (2015) to describe classification of the dynamical phenomena involved in information processes.

31. Conclusion

All constructed taxonomies of information together form a unified hierarchical multiscale information taxonomy, which gives a systematic picture of information, delineating its properties and relations. The hierarchical multiscale information taxonomy represents information in a multidimensional parametric space. The aim of this taxonomy is more efficient organization of information exploration, processing and utilization. To make this representation more specific, it is necessary to determine numerical characteristics for dimensions in the parametric space. This is an important task for the further research.

Acknowledgment

The core ideas and initial results of this chapter were presented at the International Summit on the Studies of Information held in Gothenburg, Sweden, 12–16 June 2017, by Burgin and Dodig-Crnković (2017). The authors would like to express sincere thanks to colleagues for their encouraging comments and discussions.

References

Arrow, K. J., Marschak, J. and Harris, T. (1951). Economics of information and organization: Optimal inventory policy, *Econometrica* 19(3), 250–272.

Banathy, B. A. (1995). The 21st century Janus: The three faces of information, *Syst. Res.* 12(4), 319–320.

Bar-Hillel, Y. and Carnap, R. (1952). An Outline of a Theory of Semantic Information, Technical Report No. 247, Research Laboratory of Electronics, MIT, Cambridge, MA (reproduced in Bar-Hillel (1964) (ed.), *Language and Information*, Addison-Wesley, Reading, MA, pp. 221–274.)

Bloch, M. (1949). *Apologie pour l'Histoire ou Métier d'Historien*, Armand Colin, Paris.

Bosmans, A. and Baumgartner, H. (2005). Goal-relevant emotional information: When extraneous affect leads to persuasion and when it does not, *J. Consumer Res.* 32, 424–434.

Braman, S. (1989). Defining information: An approach for policymakers, *Telecommuni Policy* 13(1), 233–242.
Buckland, M. K. (1991). Information as thing, *Journal Am. Soc. Inform. Sci.* 42(5), 351–360.
Burgin, M. (2003). Information: Problems, paradoxes, and solutions, *TripleC* 1(1), 53–70. Available online: http://triplec.uti.at.
Burgin, M. (2004). Data, information, and knowledge, *Information* 7(1), 47–57.
Burgin, M. (2005). *Super-recursive Algorithms*, Springer, New York.
Burgin, M. (2009). Extended probabilities: Mathematical foundations, preprint in *Physics*, math-ph/0912.4767. Available online: http://arXiv.org.
Burgin, M. (2010a). *Theory of Information: Fundamentality, Diversity and Unification*, World Scientific, New York.
Burgin, M. (2010b). Interpretations of negative probabilities, preprint in *Quantum Physics*, quant-ph/1008.1287, p. 17. Available online: http://arXiv.org.
Burgin, M. (2011a). Epistemic information in stratified M-spaces, *Information* 2(2), 697–726.
Burgin, M. (2011b). Information in the structure of the world, *Inform.: Theories Appl.* 18(1), 16–32.
Burgin, M. (2012). *Structural Reality*, Nova Science Publishers, New York.
Burgin, M. (2015). Picturesque diversity of probability, in *Functional Analysis and Probability*, Nova Science Publishers, New York, 2015, pp. 301–354.
Burgin, M. (2016a). *Theory of Knowledge: Structures and Processes*, World Scientific, New York.
Burgin, M. (2016b). Axiomatizing negative probability, *J. Adv. Res. Appl. Math. Statist.* 1(1), 1–17.
Burgin, M. (2017). An introduction to symmetric inflated probabilities, in *Quantum Interaction, Theoretical Computer Science and General Issues*, Vol. 10106, Springer, pp. 206–223.
Burgin, M. and Dodig-Crnković, G. (2011). Information and computation — Omnipresent and pervasive, in *Information and Computation*, World Scientific, New York, pp. vii–xxxii.
Burgin, M. and Dodig-Crnković, G. (2015). A taxonomy of computation and information architecture. ECSA 2015 ASDS Workshop, in *Proc. 2015 European Conf. Software Architecture Workshops* (*ECSAW '15*), ACM, New York, NY, USA, doi:10.1145/2797433.2797440.
Burgin, M. and Dodig-Crnković, G. (2017). Prolegomena to information taxonomy, *Proceedings* 1(3), 210, doi:10.3390/IS4SI-2017-04035.
Burgin, M. and Kuznetsov, V. (1993). *Nomological Structures in Scientific Theories*, Naukova Dumka, Kiev (in Russian).
Burgin, M. and Kuznetsov, V. (1994). *Introduction to Modern Exact Methodology of Science*, International Science Foundation, Moscow (in Russian).
Burgin, M. and Meissner, G. (2012). Negative probabilities in financial modeling, *Wilmott Magazine*, March, 60–65.
Camandola, S. and Mattson, M. P. (2017). Brain metabolism in health, aging, and neurodegeneration, *The EMBO J.* 36(11), 1474–1492.

Csanyi, V. (1989). *Evolutionary Systems and Society*, Duke University Press, Durham and London.
Dodig-Crnković, G. (2006). *Investigations into Information Semantics and Ethics of Computing*, Mälardalen University Press.
Dodig-Crnković, G. (2016). Information, computation, cognition. Agency-based hierarchies of levels, in *Fundamental Issues of Artificial Intelligence*, Müller V. C. (ed.), Synthese Library 377, Springer International Publishing, Switzerland, pp. 139–159, doi:10.1007/978-3-319-26485-1_10.
Dodig-Crnković, G. (2017). Nature as a network of morphological infocomputational processes for cognitive agents, *The Eur. Phys. J. Special Topics*, doi:10.1140/epjst/e2016-60362-9; *Eur. Phys. J.* 226, 181–195.
Dodig-Crnković, G. and Burgin, M. (2011). *Information and Computation*, World Scientific, Singapore.
Dodig-Crnković, G. and Giovagnoli, R. (eds.) (2017). *Representation and Reality in Humans, Other Living Organisms and Intelligent Machines*, Springer Nature Switzerland AG, ISBN: 978-3319437828, doi:10.1007/978-3-319-43784-2.
Dodig-Crnković, G. and von Haugwitz, R. (2017). Reality construction in cognitive agents through processes of info-computation, in Dodig-Crnković G. and Giovagnoli R. (eds.) *Representation and Reality in Humans, Other Living Organisms and Intelligent Machines*, Springer, pp. 211–235.
Fisher, R. A. (1922). On the mathematical foundations of theoretical statistics, *Philos. Trans. Roy. Soc., London, Sec. A* 222, 309–368.
Gailliot, M. T. *et al.* (2007). Self-control relies on glucose as a limited energy source: willpower is more than a metaphor, *J. PersSoc. Psychol.* 92(2), 325–336.
Goleman, D. (1995). *Emotional Intelligence*, Bantam Books, New York.
Gödel, K. (1990). *Collected Works*, Oxford University Press, Oxford.
Hofmann, H. F. (2009). How to simulate a universal quantum computer using negative probabilities, *J. Phys. A: Math. Theor.* 42, 275304 (9pp).
Kampis, G. (1991). *Self-modifying Systems in Biology and Cognitive Science*, Pergamon Press, Oxford.
Landauer, R. (1996). The Physical Nature of Information, *Phys. Lett. A* 217, 188.
Linnaeus, C. (1735). *Systema Naturae, Sive Regna Tria Naturae Systematice Proposita Per Classes, Ordines, Genera, & Species*, Johann Wilhelm de Groot for Theodor Haak, Leiden.
MacKay, D. M. (1969). *Information, Mechanism, and Meaning*, MIT Press, Cambridge, MA.
MacLean, P. D. (1973). *A Triune Concept of the Brain and Behavior*, University of Toronto Press, Toronto.
Marschak, J. (1954). Towards an economic theory of organization and information, in *Decision Processes*, New York, pp. 187–220.
Mowshowitz, A. and Kumar, N. (2009). And then there were three, *Computer* 42(2), 106–108.
Peirce, C. S. (1931–1935). *Collected Papers*, Vols. 1–6, Cambridge University Press, Cambridge, England.

Plato (1961). *The Collected Dialogues of Plato*, Princeton University Press, Princeton.

Scerri, E. (2011). *The Periodic Table: A Very Short Introduction*, Oxford University Press, Oxford.

Schroeder, M. J. (2013). Dualism of selective and structural manifestations of information in modelling of information dynamics, in *Computing Nature*, Dodig-Crnković G. and Giovagnoli R. (eds.), Studies in Applied Philosophy, Epistemology and Rational Ethics, Vol. 7, Springer, Berlin, Heidelberg.

Shannon, C. E. (1948). The Mathematical Theory of Communication, *Bell Syst. Tech. J.* 27(1), 379–423; No. 3, pp. 623–656.

Vayenas, C. G. and Souentie, S. N. A. (2012). The world of particles and the standard model, in *Gravity, Special Relativity, and the Strong Force*, Springer, Boston, MA, pp. 31–49.

von Weizsäcker, C. F. (1974). *Die Einheit der Natur*, Deutscher TaschenbuchVerlag, Munich, Germany.

von Weizsäcker, C. F. (2014). Major texts in physics, in *Springer Briefs on Pioneers in Science and Practice*, Drieschner M. (ed.), Vol. 22, Springer, New York.

Chapter 2

Regeneration of Information as Forced Expansion

Meir Buzaglo

The Hebrew University of Jerusalem, Israel

meir.buzaglo@mail.huji.ac.il

Regeneration is an operation whereby an organism restores a segment that has been severed from one of its limbs. Analogs to this process can be found in non-biological contexts, such as restoration. When an area in a painting has faded or been damaged, experts apply a variety of techniques in their efforts to restore the original work. Image processing provides yet another context; faced with a photograph from which a portion is missing, we can use a number of methods for its restoration. Another example, very relevant to vision, can be found in the eye's "filling in," or completion, of missing visual information. In order to describe what takes place in regeneration, we use words such as "extrapolation" or "interpolation" to indicate what we mean. However, this has neither yet approached formal analysis nor any methodical investigation of the operation.

1. Introduction

Regeneration is at the core of several important research projects, with the emergence of regenerative medicine as a most promising endeavor (Stocum, 2012). Yet, except for a few synonyms and general words ("restoration", "extrapolation", etc.), there is no systematic and general analysis of the concept.

This lack of definition is similar to what used to be the case with the concept of proof. Although it was in use and intuitively well understood for a long time, it was not until Frege that significant progress was made in defining it precisely. Just as in the case of proof, we expect an account of regeneration to apply to an idealization. We can also expect that there will be several concepts of regeneration, although here we will presume to describe the treatment of only one type of regeneration. In this chapter, I wish to apply an analysis of the process of non-arbitrary expansion of functions, that is found in mathematics, to the study of regeneration.

I will next propose a definition of one sense of regeneration in model-theoretic terms. More specifically, I wish to apply an analysis of the expansion of functions in mathematics, and of definitions of the process of natural expansions, to the study of regeneration in general. I will propose several requisite concepts and point out a number of basic characteristics and questions.

2. Informal Motivation

The type of regeneration that I intend to treat can be understood from the following example. Consider the natural numbers with addition and multiplication. It is important that they be given as a set on which functions are defined and that they conform to the usual commutative, associative and distributive laws. Now, "erase" an equation, by declaring, e.g., that addition is undefined for the ordered pair (1, 7). In the new situation, $1+7$ does not equal 8 and becomes, in effect, undefined. We have obtained a model in which the function of addition is partially defined. Something in this model is "flawed" and yet, it is, nevertheless, straightforward to repair. Should we wish to retain the commutativity of addition, the only way is to expand the addition function in such a way that $1+7$ will equal 8.[1]

Note that this is not a proof of the fact that $1+7$ equals 8, but an expansion of the function's domain of definition. The distinction should be clear: while in the case of the proof, we seek to determine the truth value of a claim, here an arithmetic equation, in the case of expansion of the equation, we had no truth value prior to the expansion.

Even if we erase the equation $7+1=8$, it will still be possible to repair the model, only that in this case, the law of commutativity cannot help us, and we need to rely on the law of associativity: $6+2=8$ and $6+2=6+(1+1)$, so $6+(1+1)=8$; by associativity, we have $6+(1+1)=(6+1)+1$,

[1]Note that in this case, we recover information, while in biological cases, we regenerate new cells, i.e., objects. In the following, Definition 2.2., I shall define regeneration of objects.

so by transitivity of identity, together with $6+1=7$, we are forced to define $7+1$ equals 8.

The likeness analogy to regeneration is clear. Both in the case of biological cells and in the case of information in a model, a part of the object is destroyed, but the organism and the model have the ability to restore themselves and develop back to the original state.

When we erased the second equation, we needed to use a different route for regeneration from the one we used in the first equation. It is likely that if we continue to erase information from the model, it will still be reparable and restorable up to a certain point. But if we erase a sufficiently large part, it will be impossible to restore the model to its previous state. This is what we find in the case of biological regeneration as well. We know that if we sever a small piece of a lizard's tail, the lizard will succeed in restoring it, but sever a large enough piece and the restoration will fail. In addition, there are different degrees of the capacity to regenerate: although some organisms have a wondrous capacity to restore different limbs, mammals, for example, have very limited abilities of regeneration. The same goes for the restoration of images: If all we have is the beginning of a drawing of a hand and some additional parts of the image, we will be able to complete it in several different ways; if all we have removed is a small spot of color, there will probably not be too many ways of restoring it. We find the same in music: if we only omit a single note, there are not many ways to replace it that make sense. In any case, this is so in certain musical styles.

We should note that in the case of restoring an area in a painting, as in the case of completing the definition of functions, it is the immediate surroundings that provide the missing information. However, there may be cases in which the immediate surroundings are of little help, and the scope in which the behavior of a function must be examined in order to complete the picture is a broad one. Remove a dot of color from a butterfly's wing and symmetry will oblige you to complete the wing according to its reflection, which is likely to be distant.

3. Analysis

These analogies between the mathematical case of the restoration of functions and the restoration of an image suggest a unified treatment. We begin with a more general definition of the regeneration of a function. The main challenge is a precise understanding of what we mean when we say that commutativity "forces us" to define $7+1$ equals 8 and that associativity, but not commutativity, "forces us" to define it so once we remove (1,7) from the domain of definition of addition.

The removal of an object from the domain of a function requires no special effort. We need but to accept the existence of models with partial functions. The more important question is that of the idea that a model "knows" how to regenerate itself. The model looks into itself, as it were, and finds a certain constancy which enables it to restore itself. Here, I would like to implement the concept of a natural expansion, or a forced expansion, that is studied in Buzaglo (2002).

3.1. *Regenerating information*

Take the expansion of the exponentiation function to the argument 0:

$$\text{A.} \quad 2^0 = 1.$$

This is forced by the rule:

$$\text{B.} \quad 2^x : 2^y = 2^{x-y}.$$

To make it fully rigorous, we have to write B in a way that takes the possibility that one of the terms in B is not well defined. But, this is straightforward (Mikenberg, 1977). Introduce a new symbol to the language "X" to indicate that a term is undefined and then replace B with the following law that guarantees that each term is well defined:

$$(x)(y)(2^x \neq X \wedge 2^y \neq X \wedge 2^{x-y} \neq X \rightarrow 2^x : 2^y = 2^{x-y}).$$

Each time I consider a law, I assume that it is written in a way that sees that the points that are defined are taken care of in a similar way.

This is a case of concept expansion. But — and this is simple though significant — this is exactly what we are doing when we are restoring the equation $7+1 = 8$, where the commutative law C is now doing the imposing:

$$\text{C.} \quad x + y = y + x.$$

We introduce the following notation: "$F(L, h(a) = b)$", which means that the law L forces the expansion $h(a) = b$. Thus, "$F(B2^0 = 1)$" is true in the expanded model of the former example, and "$F(C, 7 + 1 = 8)$" is true in the regenerated model of the latter example. Generally, given a model M for a first-order language L that includes functions and identities, we can expand L with sentences of the form "$F(K, h(t) = s)$" where "h" is a function symbol and "t" and "s" are terms. A precise definition of truth in M for these sentences is given in Buzaglo (2002, p. 42). Here is the definition of an internal expansion, i.e., an expansion that does not add new objects to the domain of the model, but does expand the domain of at least one function:

$$M \models F(K, h(a) = b)$$

if

1. $M \models K$;
2. $h(a)$ is undefined in M;
3. there is a model $N \supset M$, where $N \models h(a) = b$ and $N \models K$;
4. for every model T such that $T \supset M$ and $T \models K$ and $h(a)$ is well defined, $T \models h(a) = b$.

In other words, there is an expansion of the domain of h that satisfies K, and any manner of expanding the model so that it will retain K must necessarily be consistent with $h(a) = b$.

We are ready to define the main concept.

Definition 3.1 (Partially Regenerative at $h(a) = b$). Let M be a model for language L, which includes the expression "$h(x)$" such that $h(a) = b$. We will say that M is partially regenerative at point a if, following the removal of a from the domain of the definition of h, there is a law C that imposes $h(a) = b$ anew. Formally, the condition is that $M \models h(a) = b$ and that, where M' is the model that results from erasing a from the domain of the function $h(x)$, there is a sentence C such that

$$M' \models F(C, h(a) = b).$$

It is straightforward to extend this definition so that it applies to a subset of the universe of M. One may limit, in advance, those statements that impose expansion and say that C is a universal law. From here onward, we will simplify and assume this unless otherwise stated.

Another useful definition is as follows.

Definition 3.2 (Universally Regenerative at $h(a) = b$). M is universally regenerative at $h(a) = b$ if M is partially regenerative at $h(a) = b$, and all the possible ways to restore $h(a)$ by a universal law are compatible with $h(a) = b$.

Theorem 3.3. *If a Model M is partially regenerative at $h(a)$, and the removal of a from the domain of definition of $h(x)$ did not create new equations, then M is universally regenerative at $h(x)$.*

Proof. Straightforward. □

Examples

1. Define $f(x, y)$ on ordered pairs of natural numbers in the following way $(i, j \in N)$:

$$f(2, j) = 2 \quad \text{if } j \neq 5,$$

$$f(i, 5) = 3 \quad \text{if } i \neq 2,$$

$$f(2, 5) = X,$$

$$f(x, y) = 0, \text{ otherwise.}$$

   Note that if $f(2, 5) = 3$, then f is partially regenerative at $f(2, 5)$. After one erases $(2, 5)$ from the domain of definition of f, the law (x) $f(x, 5) = 3$ forces $f(2, 5) = 3$.
2. Another example is the symmetric function g. It is easy to prove that if $g(i, j) = 5$, then the function is universally regenerative at $g(2, 5)$. If $g'(2, 5) = 100$ and $g(i, j) = g(i, j)$, then $g'(2, 5) = 100$ is removable.

3.2. *Regenerating an object*

In the above examples, we have restored information, a value of a function, through non-arbitrary expansions. Many kinds of external expansions are studied in model theory. Of special interest are the examples of an external expansion that is constructed from internal expansions.

Here is an example. Assume that we eliminate 7 from N. It is not difficult to see that N' without 7 can be expanded in a natural way (that can be defined) to the original model N. We note that many functions will be undefined at points that would yield $7(6+1, 35 : 5, \sqrt{49}$, etc.), a situation we can overcome by "recreating" 7 again. To create number 7, we add a "multipotent stem number" t with $6 + 1 = t$ and then ascribe it all the properties of 7 by forced expansions.

Almost all mathematical functions and models are regenerative in some way. Regeneration points to a specific relation between parts and wholes in mathematics: If one "cuts" a part of a function or of a common mathematical structure, the rest can bring it back to life. One wonders, however, how philosophically relevant this phenomenon is.

Definition 3.4 (Removable Subset). Given a subset A of M, and a function $h(x)$, we say that A is removable from M if M is not partially regenerative at A, but it turns out to be regenerative after a redefinition of $h(x)$ at A.

4. Implications

Let me conclude this programmatic chapter with two remarks that should guide further research on the subject of regeneration.

4.1. *Symmetry and regeneration*

The study of regeneration can benefit from looking for possible applications. In Example 2.4, we gave a symmetric matrix as an example of a case of a universally regenerative point. This is no accident. The following picture is regenerative for many subsets. If one removes a subset B that does not include symmetric points, i.e., there is no pair of a and $R(a)$, where R is the reflection along the axis of symmetry in B, and the remaining picture is symmetric, then the picture is B-regenerative (Figure 1).

On the other hand, in Figure 2, the set of points on the scratch is removable.

4.2. *Generalizations*

Regeneration is a fundamental concept that can be defined in general terms: it is the phenomenon where a whole loses some of its parts and is regenerated from its remaining parts. Model theory is perhaps the best candidate for a study of this phenomenon in general terms. From the above discussion, the following generalization emerges: there are several ways to expand

Figure 1. Symmetric pictures are regenerative.

Figure 2. Repairing a scratch.

a model M to a model N and each type of expansion gives rise to a specific notion of regeneration. To put it in a non-rigorous way, suppose that E is a type of expansion of models: Given a model M, there is a way to expand it by a certain kind of expansion, E, to a larger model N. In that case, we may define "K is E regenerative relative to a subset of models T" if each M in T can be expanded by E to K.

References

Buzaglo, M. (2002). *The Logic of Concept Expansion*, Cambridge University Press, UK.

Mikenberg, I. (1977). From total to partial algebras, in *Mathematical Logic, Proceedings of the First Brazilian Conference*, Arruda A. (ed.), N.C.A. de Costa and R. Chuaqui. Dekker, pp. 203–223.

Stocum, D. (2012). *Regenerative Biology and Medicine*, 2nd edition, Academic Press, USA.

Chapter 3

Why is Entropy not Enough? — Good Emerges from the Relaxation between Order and Disorder

Wu Kun* and Wang Jian†

School of Humanities and Social Science,
Xi'an Jiaotong University, China
**wukun@xjtu.edu.cn*
†wangjian0501@163.com

The moral principle that Luciano Floridi merely judges Good and Evil by the amount of entropy is based on a simplistic and unipolar way of thinking. If one were to practice in accordance with his principle of Goodness that absolutely excludes entropy, then, in the field of nature and biology, that can only lead to the end of the dynamic changes, in the field of human mind and scientific development it can only lead to rigidity and stagnation, and in the social field it can only lead to a fascist autocratic centralized system. The orderly and disorderly development of things has those limits; entropy and its increase are not absolutely "Evil", while information and entropy decrease are not absolutely "Good". In the evolution of the universe and things that are embodied in it, there is no evil in eternal increase of entropy and no good in eternal decrease of entropy. When the evolution of total entropy increases to a certain extent, it will naturally turn into the evolution of the decrease of the total entropy, and vice versa, when the evolution of the total entropy decreases to a certain limit, it will naturally turn into the increase in the total entropy of the evolution process. A reasonable ethical principle should reconcile opposing factors, such as information and entropy, order and disorder, integrity and reducibility, certainty and uncertainty, determinism and non-determinism, purpose and randomness, inevitability and contingency, and maintain a reasonable tension between these opposing factors. A reasonable conclusion is as follows: Good emerges from the relaxation between order and disorder. In addition, the entropy theory (whether it is physical entropy or information entropy) that Floridi borrowed corresponding to order and disorder only deals with grammatical information. The problem of information directly related to ethics and value is mainly in semantics and pragmatics, rather than simply in its grammar, and it is impossible to derive it simply and directly from the size of this formalized entropy. The lagging way of thinking, the deviation of the entropy, and the understanding of information theory led to the inefficiency of Floridi's information ethical framework in supporting his ambition to establish macroethics. In the same way, his information ethics has also proved insufficient to be the philosophical foundation of his commitment to the construction of human ecological civilization. In fact, as early as the 1990s, Chinese scholars have put forward a general philosophy of value that transcended the human-centered narrow position in the name of natural ontology. This philosophy of value is not only compatible with natural values and human values but also compatible with material values and information values. It is the value and ethical paradigm put forward by such a philosophy of value that has laid the foundation for the construction of the human information ecological civilization and the general philosophical basis of sustainable development theory and practice.

1. Floridi's Macroethics Model

In 2013, Oxford University Press published *The Ethics of Information* (Floridi, 2013) by the British philosopher Luciano Floridi. In this academic work, Floridi is ambitious to establish macroethics. He also believes that the information ethics he founded, which transcends the perspective of pure bioethics, is based on "existence" compared with the previous computer ethics based on "behavior". It is a more universal and unbiased digital-environmental ethics that is involved in the information sphere and can thus become true macroethics.

To evaluate the rationality of a doctrine is not just to see how many gorgeous laurels he or she has put on the doctrine, but is, more importantly, about the kind of content explained in the doctrine, and the kinds of methods it adopts and their theoretical basis.

Floridi's macroethics is based on the "entropy" theory in science. He first established the following basic ethical principles:

(0) entropy ought not to be caused in the infosphere (null law).
(1) entropy ought to be prevented in the infosphere.
(2) entropy ought to be removed from the infosphere.

(3) the flourishing of informational entities as well as of the whole infosphere ought to be promoted by preserving, cultivating, and enriching their well-being (Floridi, 2013, p. 71).

I really don't understand it. Why do you have four? Is this a mystery, or a cumbersome philosophy?

In fact, the above four principles can be summed up in one way: to eliminate the existing entropy and to prevent the generation of new entropy.

In a specific explanation, Floridi directly relates entropy increase or entropy decrease to evil or good in ethics. He believes that unnecessary increase in entropy is a kind of evil behavior because the increase of entropy leads to system disorder; on the contrary, any behavior that maintains and promotes the increase of information is good, because the increase of information contributes to the stability of the system, to make the system more orderly. Thus, any object or structure having the latter behavioral characteristic has minimal intrinsic value: "an action is said to be morally qualifiable if and only if it can cause moral good or evil, if it decreases or increases the degree of metaphysical entropy in the infosphere" (Floridi, 2013, p. 147).

Here, Floridi follows the method of quantifying the degree of system structuring in the usual entropy and information theory, unifying entropy and disorder, information and order, and then directly uniting entropy, entropy increase, and disorder with evil, and uniting information, entropy decrease, and order with good simply.

Here, there are three real Floridi innovation points:

(1) entropy (disorder) = evil, information (ordered) = good;
(2) entropy (disorder) = evil (ordering), entropy decrease = good;
(3) the highest good behavior strategy: absolute rejection of entropy — remove the existing entropy and prevent the emergence of new entropy.

According to this standard, all systems (things), whether animate or inanimate, in the static state, as long as they are in accordance with the degree of structure size (ordering degree) can directly determine the extent of good or evil; in the dynamic state, as long as they are in accordance with the increase or decrease of entropy, they can directly determine the degree of conversion to evil or good.

This innovation is indeed bold and novel, giving us a great deal of information (because its unexpected degree is unprecedented). However, in the face of such a bold innovation, we should also be careful to verify it.

2. Back to Physical Entropy

The entropy theory of good and evil discussed by Floridi originated from the related research of physics.

In 1850, the German physicist Rudolf Clausius proposed the second law of thermodynamics. In 1865, Clausius proposed the concept of "entropy" and correspondingly expressed the second law of thermodynamics as the "Entropy Increasing Principle": the actual process in the isolated system always increases the value of the entropy of the whole system. Because Clausius only grasps the concept of entropy in the sense of "transformation" and "change", it only points out that the entropy increase phenomenon, the absolute value of the entropy of a physical system itself, and the more general creative meaning and value of entropy are not clearly defined and interpreted by Clausius. This makes Clausius's "entropy" have some mysterious and speculative colors. In 1877, from the perspective of molecular motion theory, Boltzmann made a probabilistic interpretation of the physical meaning of entropy and the principle of entropy increase using statistical methods. He pointed out that the isolated system must evolve from a macroscopic state containing a small number of microscopic states to a macroscopic state containing a large number of microscopic states; it must evolve from a state in which the probability distribution of each microscopic state is uneven to a state in which the probability distribution of each microscopic state is uniform. He gives the statistical formula of the equilibrium state entropy as

$$S = k \ln \Omega_o.$$

This formula corresponds to the equilibrium state system in which the microscopic states appear, where k is the Boltzmann constant and Ω is the number of microscopic states of the system.

If the equilibrium entropy is generalized to the calculation of the system entropy of the non-equilibrium state, a statistical entropy formula more general than the above formula can be obtained, as follows:

$$S = -K \sum_{i=1}^{n} p_i \log p_i,$$

where p_i is the probability that the ith microstate in n microstates may occur.

The significance of Boltzmann's work is not only that it introduces a probabilistic method but also that it provides a feasible solution for the calculation of the absolute value of the entropy of the system, and more precisely reveals the general creativity of the concept of entropy through this calculation, that is, what entropy describes is not the general quality of the system or the way and state of energy, but the organization, matching, distribution, and state of these masses or energies. This links entropy with the structure and order structure of the system elements. For thermodynamic systems, entropy is a measure of the distribution of thermodynamic molecules in the system and the order of the organization and the state of the arrangement. Consistent with this, the change in entropy (entropy increase or entropy decrease) means the organization, matching, distribution, and state of the system elements. Boltzmann's work not only unveils the mysteriousness veil of the entropy concept and the entropy-increasing principle, but also reveals that it is from the introduction of the concept of entropy that the scientific vision begins to shift from the study of the quality and energy of the general object into the study of the structure and relationship of the general object. In addition, Boltzmann's work provides a scientific basis for the development of the entropy concept and the generalization of the entropy theory, because if the system elements that are faced are no longer molecules in the thermodynamic sense, then the structure and relationship of the system elements, as well as the entropy describing the structure and relationships, can achieve a more universal character. This is why the concept of entropy has not lost its charm for more than a hundred years, and it has been inseparable from the concepts of "information" and "negative entropy". It has penetrated a wide range of disciplines and triggered a series of emerging, marginal, intersecting, and comprehensive disciplines.

In thermodynamics, the third law of thermodynamics (the energy theorem) is also related to the entropy theory. In 1912, German physical chemist Nernst proposed the third law of thermodynamics. This law has a way of saying that when the temperature is absolute zero, the entropy of the system is zero, and it has nothing to do with other properties such as pressure.

We know that heat is the macroscopic effect produced by a large number of microscopic particles in an irregular movement. The thermodynamic temperature (absolute temperature) measures the extent to which a large number of microscopic particles interact in motion. The so-called absolute zero corresponds to the limit state wherein all microscopic particles constituting the system do not have any form of motion (absolutely stationary), that is, the "absolutely non-thermal state".

3. Information Entropy and Negative Entropy Theory in Communication and Control Theory

The Shannon information entropy theory developed in the late 1940s and the negative entropy theory of Wiener are an extension of the theory of physical entropy.

An obvious fact is that the information entropy explained by Shannon and the statistical physical entropy explained by Boltzmann are intrinsically identical. From the process of production, Shannon's information theory was founded under the direct inspiration of Boltzmann's statistical entropy theory. Shannon used two aspects of Boltzmann's statistical entropy theory, one being the statistical method and the other being the entropy formula. In Shannon's view, the source is a set system that can generate a set of random messages with their respective probability of occurrence. Based on this, a mathematical formula for measuring the amount of information generated by the source, "entropy of the information source", is proposed. The general form of its formula is

$$H = -K \sum_{i=1}^{n} p_i \log_2 p_i,$$

where H is the average amount of information, p_i is the probability of the occurrence of the ith message, n is the total number of messages in the message set, and K is the normal number. According to Shannon, the formula of H is the same as the formula of the so-called entropy in statistical mechanics, where p_i represents the probability that a system is in the ith element of its phase space. Therefore, H here is H in the famous Boltzmann's H theorem. We will refer to $H = -\Sigma p_i \log p_i$ as the entropy of the probability set $p_1, \ldots, p_n$ (Shannon, 1965). From this point of view, Shannon's information theory is actually about the theory of information entropy, that is, the theory of entropy.

From the perspective of the correspondence between macroscopic and microscopic representations, the calculation of Shannon's amount of information also reveals the mutual regulation and connection between certain macroscopic states and microscopic states as well as the calculation of physical entropy. If we consider the total set of messages that the source may emit as a macroscopic representation of the source characteristics, then each message event sent by the source constitutes a microscopic representation corresponding to this macroscopic performance. The combined effect of microscopic performance reflects the macroscopic characteristics of the source. If each distribution of molecules in a physical system is considered to be a possible message event, then entropy becomes a measure of information entropy.

The inherent unity of physical entropy and information entropy illustrates the fact that both entropy and information are measures of a certain state of a material system itself, and the concept of physical entropy is promoted in the field of communication as the uncertainty of the source. The concept of source uncertainty in the molecular physics system is physical entropy. It can also be seen from the comparison of the above formulas that Boltzmann's statistical physical entropy formula is only a special case of Shannon's information entropy formula in the case of equalization. In fact, as early as 1927, the Hungarian mathematician von Neumann extended the Boltzmann statistical physics entropy formula to a system with unequal distribution of microscopic states and obtained $S = -K\Sigma p_i \log p_i$, which is completely consistent with the Shannon information entropy formula.

At the same time that Shannon founded the information entropy theory of communication, American mathematician and founder of cybernetics Wiener also proposed the theory of negative entropy of information in the process of integrating communication and automatic control theory and creating cybernetics. In the book *Control Theory* published in 1948, he independently provided the Wiener information formula that lacks the negative sign of the Shannon information formula.

What the Wiener formula derives is not the amount of information contained in the message (a certain amount) itself, but how much new information the message brings to the receiver. The measure of his formula is obviously not the information entropy in Shannon's sense, that is, it is obviously not the uncertainty of the source to generate the message. In Wiener's explanation of the meaning of his formula, the concept of information quantity is very naturally subordinated to a classical concept of statistical mechanics — entropy. Just as the amount of information in a system is a measure of its degree of organization, the entropy of a system is a measure of its unorganized degree; one happens to be the negative of the other. The amount of information is a negative of the logarithm of a quantity that can be seen as a probability, which is essentially negative entropy (Wiener, 1963).

Perhaps the analysis from the perspective of the difference in the way of understanding can help us find the root cause of the difference between Shannon's information and Weiner's information.

We have already pointed out that Shannon regards the source as a probability set system that can generate a set of messages with their own probability of occurrence, and the amount of information is the uncertainty of what the system itself will produce. For Wiener, what he wants to derive is not the amount of information contained in the message itself, but the amount of new information that the message brings to the recipient. Here, the amount of information is not specified as a feature metric of the probability statistical uncertainty of the source information itself, but is defined as a measure of the probability of the statistical feature of the "knowledge" change that causes the recipient to receive the source information.

In the field of mechanical communication, the number of primitives for sending messages by the source and the probability of sending each message are predefined, and the recipient is completely clear about this rule. The *a priori* estimate of the uncertainty of the message sent by the source to the recipient is also derived from this preregulation. In this way, the uncertainty of what kind of message is sent by the source can be regarded as the characteristic of the recipient's estimation of the source state, and can also be regarded as the characteristic of the source itself. The difference between the negative signs of the Shannon and Wiener information formulas can be seen as the result of their separate observations from these two different perspectives.

The amount of information can be derived from the principle of relativity between source and recipient's interactions and mutual stipulations. This leads to the provisions of "prior probability" and "posterior probability".

According to Shannon's principle, the information amount formula is derived from the perspective of the state characteristics of the source itself, and then the contribution of the prior probability to the information amount is reversed because it provides the "uncertainty of the sink's estimate of the source"; its direction is opposite to the direction of the source's own characteristics. The contribution of the posterior probability to the amount of information is positive because it provides the information state itself wherein the source itself actually occurs at the moment, and its direction is consistent with the direction of the state characteristics of the source itself.

If you use a logarithmic function, it is as follows:

$$\text{Information amount}_{\text{Shannon}} = \log[(\text{posterior probability})/(\text{prior probability})].$$

If, like Wiener, the information formula is derived from the perspective of the source of knowledge, then the contribution of the prior probability to the amount of information is positive. On the contrary, the contribution of the posterior probability to the amount of information is reversed. So in Wiener, the information amount formula will lack a negative sign found in the Shannon formula:

$$\begin{aligned}\text{Information quantity}_{\text{Wiener}} &= \log[(\text{prior probability})/(\text{posterior probability})]\\ &= -\log[(\text{posterior probability})/(\text{prior probability})].\end{aligned}$$

From the two opposite angles and directions of source and recipient's interaction and mutual definition, the facts of Shannon and Wiener's information formula can be derived, respectively, indicating that the difference between the negative signs of these two formulas has a profound epistemological source. This reflects the difference and unity of the philosophical, ontological, and epistemological methods to a certain extent and meaning.

Before and after the negative entropy theory of Wiener's information, the more famous negative entropy theories of Schrödinger's negative entropy theory of life and Prigogine's negative entropy theory of dissipative structure, respectively.

In fact, no matter how many kinds of negative entropy theories have been proposed by scholars, their meanings are only of two kinds: one is the degree to which its structural organization deviates from the standard value (maximum entropy) relative to the same system; the other is the degree to which the entropy value of a system is reduced in the process of structural changes.

4. Unification of the Scientific Meanings of Entropy, Negative Entropy, Information, Entropy Increase, and Entropy Decrease

Although concepts such as entropy, negative entropy, information, entropy increase, and entropy decrease often have very different specific meanings in different subject areas and among different scientists, these concepts are essentially consistent. They all study the same kinds of phenomena in a unified sense, and the differences between them are produced when the same kinds of concepts are applied to the research of different directions of the same kinds of phenomena.

We have reason to divide the provisions of these concepts into two categories: one is the provision given in the static sense and the other is the provision given in the dynamic sense.

Boltzmann's statistical entropy, Shannon's information entropy, and the negative entropy of the degree of deviation from the system's standard entropy values, in essence, are the quantities of entropy (information entropy) and negative entropy (information). In the static, systematic absolute value sense, the basic meaning of this provision is the degree of uncertainty in the microscopic state of the computing system and the extent to which this uncertainty deviates from the maximum possible uncertainty.

The dynamic definition of entropy and information concept is developed in two directions: one is the direction of entropy increase based on the second law of thermodynamics and the other is the negative entropy constructed in the sense of entropy increase of the resistance system on the entropy decrease direction of the framework.

The very interesting fact is that the studies of the dynamics of entropy and information were done earlier than the studies of its statics. While people don't really understand what entropy is, Clausius is already quantifying its changes.

The various forms of negative entropy theory are dynamic measures of the change of information (entropy) from another direction opposite to the second law of thermodynamics. Schrödinger, Wiener, and Prigogine all share a common idea that the system can input an external entropy (information) flow from the environment to resist the entropy increase trend within the system. They measure the amount of this external entropy (information) stream by the amount of change in the entropy (information) within the system caused by the outer entropy (information) stream. Because the external entropy (information) flow may cause the entropy decrease effect inside the system, the amount of this external entropy (information) flow can be measured by the degree of entropy decrease effect inside the system caused by it, and it is simultaneously determined as negative entropy.

It seems that the function $\Sigma p_i \log p_i$ has some unique value. Statically, its absolute value indicates the uncertainty of the microscopic state of the system. Dynamically, the amount of change in the value of this function indicates the change in the uncertainty of the microscopic state of the system. This change is caused by a change in the value of n indicating the number of microscopic states of the system and a change in the probability distribution of p_i. In general, the increase in the value of n and the convergence of the p_i value result in an entropy increase process, while the decrease in the value of n and the differentiation of the p_i value are caused by the entropy decrease process. As for the general opinion that regards information and entropy as opposites, it has more artificial colors. It is said that the external information causes the entropy decrease effect of the system, and the entropy flow introduced by the outside world causes the entropy decrease effect of the system to be equivalent. Prigogine uses external entropy flow, Schrödinger uses negative entropy, and Wiener uses both information and negative entropy, and in fact they all measure the same type of change in the same process. We have every reason to regard the various theories of entropy, information, and negative entropy as theories about entropy, and there is also a good reason to regard them as theories about information. Based on this, we can establish a general entropy theory or a generalized information theory, and unify the discussions about entropy and information that have been carried out in different subject areas to date.

When entropy is discussed in relation to the degree of order or disorder of the system, the following scenarios occur: In the general entropy theory, a system with a large entropy value corresponds to a state in which the number of microscopic states of the dominant system is large and its probability distribution is uniform, that is, a state in which the degree of disorder is high and randomness and uncertainty are large. The system with small entropy value corresponds to a state in which the number of microscopic states of the dominant system is small and the probability distribution is non-uniform, that is, a state with a high degree of order, large decisiveness, and large certainty. There are two opposite limit states: the entropy maximum state and the entropy minimum (zero) state. The former refers to the state in which the microscopic state of the dominant system is more than one and its probability distribution is similar, which is also dominated by the same microscopic state number, the most disordered state in the system; the latter refers to the dominant state of the microscopic state of the system, which is the most ordered system with an entropy value $S = k \ln 1 = 0$. Entropy increase or entropy decrease is a situation in which the degree of ordering and disordering changes with the evolution of the system.

The enhancement of the degree of disorder corresponds to the entropy increase, and the specific mechanism is to increase the number of microscopic states of the dominant system or to equalize the probability distribution of the microstate; the degree of order corresponds to the entropy decrease, and the specific mechanism is to reduce the number of microscopic states of the dominant system or the non-equalization of the probability distribution of the microstate.

5. Entropy and Information, Order and Disorder, Good and Evil

There is no direct relationship between entropy and information, ordered and disordered theory in science, and good and evil in ethics or moral science. However, after Floridi's creative simple correspondence grafting, they have a relationship. As disclosed earlier, Floridi has three creations:

(1) entropy (disorder) = evil, information (ordered) = good;
(2) entropy (disorder) = evil, entropy decrease (ordering) = good;
(3) the highest good behavior strategy: the absolute rejection of entropy — remove the existing entropy and prevent the emergence of new entropy.

If such a correspondence is followed, then what is the kind of situation that is arrived at when Floridi's desired good strategy of removing entropy and suppressing new entropy is realized?

In Nature: all systems, and even the entire universe, will approach a state of extremes: all microscopic motion ceases (according to the third law of thermodynamics); all macroscopic states will be governed by a single, fixed pattern (according to the calculation formula of the system entropy, $S = k \ln 1 = 0$).

In the ecosystem: all species that are different will eventually become extinct, and the ecosystem can only be dominated by a single species that is fixed. Moreover, the individual degrees of freedom of this single species must also be strictly limited to the overall standardized order dictated by a certain will. Because only such a state of the entire ecosystem can reach the state of entropy minimally ordered.

At the level of personal consciousness: all possible whimsy is eliminated, and everyone's consciousness will be strictly limited to a rigid and fixed cognitive paradigm. The content of knowledge and thought is stubborn and obsolete, and no novelty and creativity are possible, because any novelty and creativity will be considered as entropy-increasing behaviors relative to the original cognitive paradigm, and be strictly classified as aspects that must be prevented and banned.

At the level of social order: the entire society should be strictly regulated under a singular standardization system, subject to the domination of a single leader, a single idea, and a unitary system. Any opinions, ideas, and behaviors that deviate from this unity should be removed or strangled. It is not only necessary to obey this singular heresy, but it is more necessary to eliminate the deviations physically. Such an orderly social system, with small entropy and the state of social consciousness, is more reminiscent of the centralized system of the underworld, of fascism, and of authoritarianism. Correspondingly, it is full of oppression and slavery, and absolutely kills human rights, democracy, and freedom.

We note that Floridi and Wiener's information ethics views have great similarities: they both consider that the universe is essentially information; they regard humans as information objects; and they regard entropy as a kind of extremely important evil (Bynum, 2010). However, Floridi did not know that Wiener did not regard the same relationship between entropy and evil as absolute. He did not think that it would be effective to extend such a viewpoint to a wider field. On the contrary, in Wiener's view, in some fields, such as human thinking, social order, and scientific development, orders that pursue exclusionary entropy, exclusion of contingency, and randomness may be dangerous, even evil.

Here, we need to refer to several relevant statements of Wiener:

> "In physics, the idea of progress opposes that of entropy, although there is no absolute contradiction between the two" (Wiener, 1989, p. 38).
>
> "Progress imposes not only new possibilities for the future but new restrictions. It seems almost as if progress itself and our fight against the increase of entropy intrinsically must end in the downhill path from which we are trying to escape" (Wiener, 1989, p. 46).
>
> "In the ant community, each worker performs its proper functions. There may be a separate caste of soldiers. Certain highly specialized individuals perform the functions of king and queen. If man were to adopt this community as a pattern, he would live in a fascist state, in which ideally each individual is conditioned from birth for his proper occupation: in which rulers are perpetually rulers, soldiers perpetually soldiers, the peasant is never more than a peasant, and the worker is doomed to be a worker" (Wiener, 1989, p. 51).
>
> "When human atoms are knit into an organization in which they are used, not in their full right as responsible human beings, but as cogs and levers and rods, it matters little that their raw material is flesh and blood.

What is used as an element in a machine is, in fact, an element in the machine. Whether we entrust our decisions to machines of metal, or to those machines of flesh and blood which are bureaus and vast laboratories and armies and corporations, we shall never receive the right answers to our questions unless we ask the right questions" (Wiener, 1989, p. 185).

"Those who would organize us according to permanent individual functions and permanent individual restrictions condemn the human race to move at much less than half-steam. They throw away nearly all our human possibilities and by limiting the modes in which we may adapt ourselves to future contingencies, they reduce our chances for a reasonably long existence on this earth" (Wiener, 1989, p. 52).

Wiener once expressed "anger", "disappointment and sorrow" on our universities' preferring imitation over originality, favoring vulgarity, superficiality, mass reproduction rather than new life, preferring unhelpful precision, shortsightedness, and methodological limitations, and rejecting novelty and creativity. Wiener also opposes the formation of military and religious organizations. To organize scientific research, he said, "such order or organization is intrinsically unsuited to encourage an Augustinian attitude within the scientist; and furthermore, it does not tend to rate limpid intellectual honesty very high in its scale of virtues. Against an insidious enemy who himself plays tricks, military strategies are permissible. Thus a religious military order is almost bound to set a great value on obedience, confessions of faith" (Wiener, 1989, p. 191).

"Without faith that nature is subject to the law, there can be no science. No amount of demonstration can ever prove that nature is subject to the law. For all we know, the world from the next moment on might be something like the croquet game in *Alice in Wonderland*, where the balls are hedgehogs which walk off, the hoops are soldiers who march to other parts of the field, and the rules of the game are made from instant to instant by the arbitrary decree of the Queen" (Wiener, 1989, p. 193).

"Science is a way of life which can only flourish when men are free to have faith. A faith which we follow upon orders imposed from outside is no faith, and a community which puts its dependence upon such a pseudo-faith is ultimately bound to ruin itself because of the paralysis which the lack of a healthily growing science imposes upon it" (Wiener, 1989, p. 194).

Obviously, Wiener does not give an ethical standard that simply determines good or evil by the amount of entropy. On the contrary, he resolutely opposes the principle of absolute exclusion of uncertainty, randomness, contingency, and the individual will, thus stifling novelty and creativity, and pursuing the so-called entropy decrease effect of overall order. He particularly emphasized the so-called orderly absurdity and evil of using this principle to regulate human thinking, scientific research, and social systems.

According to the ethical principles that Floridi gave us, what it may bring us is definitely not "the prosperity of the information entity and the entire information circle", and it is definitely not "by preserving, nurturing and increasing their welfare". If you follow the ethical principles set out by Floridi, the result is only one, that is, the infosphere, the ecological circle, the human society, and the individual's living state deteriorate step by step until catastrophic destruction.

The value ethics of human information civilization should definitely not be the theory of serving the underworld, fascism, and totalitarian autocracy. On the contrary, it should be the theory of serving a new human civilization system that is more democratic, free, pluralistic, and harmonious. In addition, it should provide the philosophical foundation of value ethics for the harmony of the Earth's ecology and sustainable human development.

6. The Perspective of Complexity: Maintaining Some Moderate Tension between Order and Disorder

Why did the ethical principles that Floridi gave us produce such a bad result? This is because Floridi's ethical principle of judging good and evil solely according to the single dimension of entropy is based on simplicity and unipolar thinking. The complexity of ethical issues itself determines that the establishment of reasonable ethical principles must go beyond simple, unipolar thinking.

An obvious fact is that no matter how ambitious Floridi has been, no matter how he advertises his ethical principles as true macroethics with transcendence, in terms of his way of thinking to establish ethical principles, he still stays in the early stages of entropy and information system theory research. This stage extended from the physical entropy theory in the middle of the 19th century to the information communication and control theory in the middle of the 20th century and further extended to the self-organization theory of the 1960s and 1970s. The development of related theories at this stage advocates the pessimistic hot silence theory derived from the evolution of the second law of thermodynamics, the entropy-increasing principle, and the absolute superiority of the integrity constructed by the orderly way and the eternity of the entropy decrease process of self-organizing evolution. This is the theoretical basis for the emergence of Floridi's information ethics of absolutely excluding entropy and disorder and absolutely respecting information and order.

In fact, the principle of entropy decrease and orderly development is admired as the ultimate principle. Both in theory and in practice, it has been proved that it is not always reasonable or "good". We have already described and discussed in the past Wiener's concern and criticism of the application of the cybernetic method to control human beings, the norms of scientific research, and the negative effects of the construction of social order. In fact, the former Soviet Union as the core, including some Eastern European countries as its allies, has established some theory of social cybernetics to maintain its long-term stability of centralized politics and planned economic systems. It is hoped that through a single planning approach and a reasonable design of the way and path of social and economic development, the implementation of a highly ordered process will effectively eliminate the factors of heterogeneous entropy, thereby leading to the rapid development of society and the economy. However, the results of social practice under the control of such a theory are contrary to the wishes of the theoretical designers. They not only fail to bring them the expected prosperity but also bring about the deviation of the human heart and the decline of the economy and the destruction of the system.

Why did it turn out this way? This is because the orderly and disorderly development of things has limits. Entropy and entropy increase are not absolutely "evil". Information and entropy decrease are not absolutely "good". Entropy and information are a pair of twin brothers who must maintain a certain entropy and information ratio in any realistic, dynamic system. When one party destroys the other, that is, when the entropy reaches a great or very small level, the system loses its vitality of development, and thus goes to the opposite side of "goodness".

A reasonable ethical principle should be compatible with seemingly opposing factors and maintain some reasonable tension between these opposing factors, for example, information and entropy, order and disorder, integrity and reductivity, certainty and non-determinism, determinism and non-determinism, purpose and randomness, necessity and contingency, etc.

Ordering is developed in disorder by defining constraints. If the disorder is eliminated, the order is also rigid and stagnant. Faced with this scenario, the introduction of heresy and novelty through the entropy increase effect will break the rigidity and stagnation, and make things regain the vitality of the movement change. At this time, "entropy increase" is not only evil but must be "good".

Since complex systems have multiple possibilities in the direction of their evolution, when the system develops its integrity in a certain direction through the evolution of entropy decrease, it also becomes part of it. The constraint of restricting freedom is implemented. At the same time, it also means the shielding of other developmental possibilities. Thus, in the process of entropy decrease of the same order, multiple complex effects of "good" or multiple "evil" are generated.

At this point, it is necessary for us to introduce and evaluate the discussion of the complex relationship between the whole and the parts, the order and the disorder, by the famous French scholar Edgar Morin.

In the discussion of the relationship between the whole and the parts, Morin concentrated on criticizing the one-sided and mechanical features of reductionism and holism. This reveals the multifaceted and complex nature of the overall and partial relationships.

In response to the basic idea of popular holism, "the whole is greater than the sum of the parts," Morin proposed that the system is greater than, less than, and different from the sum of the parts. The part itself will also have the difference of becoming smaller or larger. In short, it is different inside and outside the system.

Morin stressed the following: "A system has some new features and new attributes that exist alone or are housed in another type of system, we can call these new features and new attributes emergences. All overall states have an emerging feature when there are more things in the whole than when the parts are superimposed: organization, overall unity (constituting a 'whole'), new features and new attributes emerging from the organization and overall unity" (Morin, 2002, p. 99).

"In addition to the emergence principle, there are constraints." "All unions necessarily lead to constraints: constraints between the various parts of the interdependence, partial constraints on the whole, and overall constraints on the part." "Constraints are restrictions and limitations on parts." These constraints, restrictions, and limitations deprive and repress the advantages or attributes of the various parts. In this sense, the whole is less than the sum of the parts." "The whole is less than the sum of the parts: it means all the good quality when partially isolated will disappear in the system." He also cites, for example, "All organizations that define and develop tiered and divisional labor will identify and develop constraints, oppression, and slavery. We also know today that every cell in the body stores the entire body's information. However, most of the information in the cell has been suffocated, and only a small part of the information related to the specialized division of labor can be expressed" (Morin, 2002, p. 106).

Based on the above aspects, Morin wrote, "When we face the system, we must not only see its gains in emergency, but also the losses it suffers under constraints, wills and restrictions, which means not only being richer, but also more impoverished, and the degree of impoverishment may be greater than the degree of enrichment." "The development of certain systems comes at the cost of shrinking certain possibilities in the system." "The system is such a whole, it transforms its ingredients in the process of formation," he said and also accused, "It is absurd to simplify the description of the system to a quantitative one. He also shows that the description of the system should not only be qualitative, but also complex" (Morin, 2002, p. 108).

In fact, whether "the whole is greater than the sum of the parts" or "the whole is less than the sum of the parts," or even as some people put forward "the whole is equal to the sum of the parts" is only a symbolic and metaphorical statement. The main purpose of the argument is to emphasize the nature of the system as a whole and the nature of the elements of the system in an isolated state that cannot be simply attributed, and that they are not directly comparable. This means that it is generally impossible to attempt to logically derive the overall nature from a partial nature through purely mathematical methods.

On the important issue of order and disorder, Morin proposed a theory on the relationship between the two, which is compatible and complementary. He wrote, "It is necessary to confirm the composition of complexity: it not only contains diversity, disorder, randomness but also its laws, its order and its organization." The relationship with organizational phenomena is not necessarily mutually exclusive. It is likely that there is a complementary relationship. "Anomalies, disorders, dissipations, etc. will also produce 'structures', that is, they can generate organization and order." "The order and organization of the universe are developed in disorder, uncertainty, abnormality, impossibility, and energy dissipation" (Morin, 2001, p. 24). Morin also emphasized the complementary role of order and disorder in the two directions of evolution: "The information that disorder brings us is to command us to find complexity. Evolution cannot be a simple concept: upward evolution must contain construction and degradation, concentration and dispersion.... Order, disorder, and potential organizational strength should be considered together. We should consider both their opposite characteristics and their mutual complementarity characteristics. The above concepts are interdependent to form a moving circle." "Disorder is an integral part of all processes of the universe. It becomes a very rich concept: there is not only one disorder in the universe, but a variety of disorders (orders as well): imbalance, disorder, eddy current, random encounter, fracture, catastrophe, fluctuation, instability, scattering, dispersion, positive feedback, runaway, and explosion" (Morin, 2002, p. 29).

Morin's discussion of the complexity of the whole and partial, ordered, and disordered relations gives us a clear hint: trying to directly judge the relationship between the whole and partials, order and disorder by directly specifying the effects of entropy increase or entropy decrease, and making them simply and more directly "good" or "evil," is absurd and ridiculous.

If we must use the measure of entropy increase or entropy decrease, order and disorder, to judge good or evil, then we may only give a qualitative macro statement: to truly achieve good, it is necessary to have maintained some moderate tension between order and disorder.

7. What Kind of Philosophical Foundation of Value Does Ecological Civilization Need?

The lag of the thinking mode, the deviation of entropy, and the understanding of information theory make it difficult for the information ethics framework proposed by Floridi to support his ambition to establish a macroethics. Similarly, his information ethics is inconducive to become the ethical foundation of value he promised on which to construct human ecological civilization.

In fact, as early as the 1990s, Chinese scholars proposed a general value philosophy that transcends the narrow position of humanity, which was explained in the name of natural ontology. This philosophy of value is not only compatible with natural values and human values but also compatible with material and informational values. It is this kind of value philosophy that proposes the ethics paradigm of value that lays the general philosophical foundation for the theory and practice of constructing human information ecological civilization and sustainable development.

In 1997, Wu Kun published a paper on the interpretation of natural values. In this paper, Wu Kun takes a general rule from the perspective of natural ontology: "From a philosophical level, value is the effect that things (matter, information, including the subjective form of information — the spirit) achieve through internal or external interactions." This definition contains at least some of the following key points (Kun, 1997):

(1) The phenomenon of value exists not only in the relationship between subject and object in the human world but also in the internal or external interaction of all things.
(2) The things here refer to the existences in the broad sense. It is the reference to all phenomena in the universe, including all material phenomena, information phenomena, and spiritual phenomena as the advanced form of information activities. In this way, whether it is the effect achieved in the interaction between the material system, and the information system, the spiritual system, or the effect achieved in the interaction between matter and matter, information and information, spirit and spirit, or the effect achieved in the interaction of matter and information, matter and spirit, information and spirit, they are all values.
(3) The role of value is by no means one way because the interaction between things is mutual, so the value realized in this interaction must be two ways or multidirectional.
(4) Only the interaction is not a value. The effects of the system itself or the changes of both sides and parties caused by the interaction are the value.
(5) For the existence and development of something, the effects of interaction may be beneficial or harmful, either positive or negative. But the effect of whatever nature or direction of action is a value relationship. This makes it possible to distinguish between positive value, negative value, neutral value, and so on. It is not what some scholars believe, that only beneficial effects or effects that promote the existence and development of things are values, or else they do not constitute value.

If we want to discuss the good or evil in ethics with the positive, negative, and neutral values proposed by Wu Kun, then we have reason to regard positive values as good and negative values as evil. As for neutral value, it is neither good nor evil. However, the positive or negative value of the judgment here must not be simply based on the size of the entropy, or the degree of entropy increase or entropy decrease. Because when the system is overordered, it is beneficial to increase its entropy value for its existence and development. When the system is excessively disordered, appropriately reducing its entropy may be beneficial to its existence and development.

In addition, the positive or negative value here is a concept of relativity, which is only relative to things that endure the effects of realistic interactions. However, the effects of the same incident may be multifaceted and multilayered. From different aspects or levels, the nature of its value, its positive or negative degree, may be diverse or even distinct.

We can give a simple example: the wolf eats sheep. In general, the effect of this event on the wolf is positive because it continues the life and normal development of the wolf; the effect on the sheep is negative because it ends the life and normal development of the sheep. However, its effect on the grassland is positive, because the end of the life of a herbivorous sheep protects the pasture from minor damage. In addition, if the sheep that are eaten are carriers of bad genes (as well as old and weak), then this event is of positive value to the development of the entire flock; of course, in the opposite case (e.g., the sheep that are eaten happen to be the fine breeds in the flock or the ewes that are breastfeeding), it may have other negative values at that level. However, such positive or negative values are difficult to determine in a one-to-one correspondence using a single evaluation criterion of entropy decrease or entropy increase.

Here, we must also emphasize that as far as the evolution of nature and the universe is concerned, as long as it is a pattern or a way of a motion that conforms to its basic evolutionary law, it should be good.

According to the relevant theories of contemporary cosmology, the overall evolutionary model of the universe depends on the total mass of the universe. The basic evolution is the result of the interaction between the self-gravity determined by its total mass and the repulsive force of spontaneous instability. There is a threshold for the mass density of the universe. If the mass density of the universe in which we live is greater than this threshold, then the self-gravity of the universe will be greater than the outward expansion force, and it will be a "closed universe". If this is the case, the universe in which we are currently in the expansion phase can only continue to expand for a limited time, then it will stop and contract to the center of the original explosion under self-gravity.

The closed universe is an oscillating universe that unfolds in the cyclical motion of expansion and contraction. Perhaps the universe we are in has oscillated many times between very high density and very low density states. The universe is likely to exhibit its evolution in an infinite loop of expansion, contraction, reexpansion, and then contraction, and so on.

According to the model of the closed universe as illustrated in cosmology, we have drawn a "general pattern diagram of the evolution of the closed universe":

The cyclical evolution of the universe is highly self-organizing. The universe does not need to introduce patterns from outside itself: the mass-energy of the universe itself; the space-time dimension of the universe itself; the mass form of the universe itself; the form of energy; the diversity and complexity of information patterns; the intertransition and transformation between various quality forms, between energy forms, and between information patterns; the complex interweaving interactions, mutual support, mutual transition, and transformation; and synergy between quality forms, energy forms, and information patterns, and the synchronous generation, evolution, and degradation all fully demonstrate the endogeneity of specific patterns in different stages, different fields, and levels of the evolution of the universe, and the spontaneous construction of mutual transition, transformation, and transition between various modes.

This also shows the holographic complexity of the self-organizing evolution of the universe, its own regulation, its own network, its own coding, its own decoding, its own isomorphism, its own settings, and its own display.

Obviously, in this process, there is no eternal entropy increase, and there is no eternal entropy decrease. When the evolution of the overall entropy increases to a certain limit, it naturally shifts to the evolution of the overall entropy decrease; vice versa, when the evolution of the overall entropy decrease reaches a certain limit, it naturally shifts to the whole evolution process of entropy increase. In the process of this cycle of evolution, whether it is the process of entropy increase and evolution, or the process of entropy decrease and evolution, whether it is the process of transformation and evolution from entropy increase to entropy decrease or the process of transformation and evolution from entropy decrease to entropy increase, they are all good.

According to this, we can also draw the following conclusion: Good emerges from the relaxation between order and disorder. This is also the true charm of the title of this chapter.

References

Bynum, T. W. (2010). Philosophy in the Information Age. In *Putting Information First: Luciano Floridi and the Philosophy of Information*. Allo, P. (ed.), Wiley-Blackwell, pp. 171–193.

Floridi, L. (2013). *The Ethics of Information.* Oxford: Oxford University Press.

Kun, W. (1997). An Outline of General Value Philosophy in the Name of Natural Ontology. *Humanities Journal,* (2).

Morin, E. (2001). *Complexity Thought: Conscious Science.* Translated by Chen Yizhuang. Beijing: Peking University Press.

Morin, E. (2002). *Method: Nature as Nature.* Translated by Feng Xuejun. Beijing: Peking University Press.

Shannon, C. E. (1965). Mathematical Theory of Communication. In *Elements of Information Theory* Shanghai: Shanghai Science and Technology Compilation Hall.

Wiener, N. (1963). *Cybernetics.* Translated by Hao Jiren. Beijing: Science Press.

Wiener, N. (1989). *The Human Use of Human Beings: Cybernetics and Society.* London: Free Association Books.

Chapter 4

Qualitative Stance, Effective Procedures and Common Sense Computation

Gianmarco Tuccini[*,‡], Roberta Lanfredini[*,§], and Luca Baronti[†,¶]

[*] *Università degli Studi di Firenze, Italy*

[†] *School of Computer Science, University of Birmingham, UK*

[‡] *gianmarco.tuccini@stud.unifi.it*

[§] *roberta.lanfredini@unifi.it*

[¶] *l.baronti@bham.ac.uk*

The philosophical analysis of the qualitative may serve as the basis for carrying out qualitative applied ontology and, consequently, it would permit a more accurate representation of human actions and natural language, which would be useful in performing complex tasks. However, in the debate within contemporary philosophy of mind, the dimension of the qualitative states (qualia) represents a hard problem and is reputed to be in principle insoluble. The problem is closely connected to the irreducibility of the qualitative dimension to physical-neuronal or functional states, and to the consequent establishment of a "logic" which sees the qualitative dimension as inevitably residual in relation to the formal and computational dimension. This logic is defined here as a "logic of the residue". This chapter is intended as a reconsideration of the relationship between the computational theory of the mind and its qualitative dimension by analyzing the notion of effective procedure and the more general notion of information. The notion of procedure is a complex one, capable of incorporating closed or computational procedures, open quantitative procedures and open qualitative procedures.

The perspective of computational mind is here replaced by an adaptive and common sense computational model, able to incorporate the performative aspect of linguistic use and, in a consistent way, to take account of social context, emotional intelligence, flexibility, openness and adaptability to the environment.

1. Church–Turing Thesis

The problem of the possible existence of an *effective procedure* able to determine if any formula of the functional calculus is provable was first posed by Ackermann and Hilbert (1928). The first step to provide a solution for this *Entscheidungsproblem* was defining exactly what an "effective procedure" was in this context. Two distinct definitions, and solutions to the *Entscheidungsproblem*, were independently presented by Church (1936a,b) and, only a few months later, by Turing (1936).

Church (1936a,b, p. 356) defined the notion of effectively calculable functions in terms of λ-definable functions. This class of functions is defined in the formalism of the λ-calculus, a basic formal system defined by Church (1932) a few years earlier, and it's equivalent (Kleene, 1936) to the class of the *recursive functions.* In the λ-calculus formalism, a function in the natural numbers is λ-definable if (Copeland, 2007, p. 103) the values of the function can be calculated by a process of repeated substitution. Church (1936a,b, p. 356) then identified the class of "effectively computable" functions, in the natural numbers, as the class of recursive functions.

Turing's vision of an effective procedure, in the context of the *Entscheidungsproblem,* was a procedure computable by what he (Turing, 1948, p. 7) called *logical computing machine* (LCM). An LCM, commonly referred to as simply Turing machine, is an ideal machine that is composed of a *state machine* and a *head* able to read and write symbols present on a *tape* that can be moved forward and backward. A finite set S contains all the possible symbols present on each *cell* of the tape, which is assumed to be either infinite or long, "as required". At each step, the machine is able to carry out a specific action out of a limited set (e.g., read or write a symbol, move the tape to the next or previous cell) based on the currently read symbol and the current state of its state machine. Turing (1948) then identified the functions that can be computed by his machines as the class of "effectively computable" functions.

The negative answer to the *Entscheidungsproblem* came independently from both these formalisms. Church (1936a,b) proved the general case of the *Entscheidungsproblem* to be unsolvable in the λ-definable functions, on the natural numbers. Turing proved that a Turing machine able to determine in finite time if any formula of the predicate calculus is a theorem of the calculus cannot exist.

The equivalence of the Church and Turing thesis, thus commonly referred together as the Church–Turing thesis (Kleene, 1967, p. 232), is based on the grounds that, on the natural numbers, the class of the λ-definable functions and the class of the functions that can be computed by a Turing (1937) machine are equivalent.

Despite their equivalence, the formalisms used in the Church and Turing theses are expressed in quite different terms. In case of Turing, he expressed his formalism in terms of objects (e.g., a tape, a head, etc.) that are ascribable to a physical connotation and this, perhaps, contributes to many (Copeland, 2007, p. 104) misunderstandings of his thesis.

In some cases, the premises under which the Turing machine was formulated were bent, providing it a context more general than what Turing originally considered. Examples can be found in Dennett (1991, p. 215) where the Universal Turing machine "can compute any function that any computer, with any architecture, can compute" or in Churchland and Churchland (1990, p. 26) where it's claimed that the machine "can compute any rule-governed input–output function".

In other cases, a different, *strong* version of the Turing thesis was considered. This variant assumes (Smolensky, 1988; Newell, 1980, pp. 3, 150) that a Turing machine is able to compute what any machine working on finite data under a finite set of instructions is able to compute. Deutsch (1985) went one step further, formalizing (Deutsch, 1985, p. 99) a "physical version of the Church–Turing" thesis where "Every finitely realizable physical system can be perfectly simulated by a universal model computing machine operating by finite means". Copeland (2007), borrowing (Gandy, 1980) notation, expressed the similar principle in a *Thesis M* (Copeland, 2007, p. 104), "Whatever can be calculated by a machine", under certain assumptions, "is Turing-machine-computable". The Thesis M is more general (Copeland, 2000) than the Church–Turing one, yet is less general with respect to the thesis suggested by Deutsch (1985), being restricted only to what can "be calculated by a machine" rather than a computing machine able to simulate every *physical system*.

Similar formulations of the Thesis M can be found in many different fields, like theory of the mind or psychology (Fodor, 1981; Boden, 1988) where the mind or consciousness is expressed in terms of the Turing machine. For instance, in Guttenplan (1994, p. 595) it is claimed that "we can depend on [...] a Turing machine that captures the functional relations of the brain", while in Searle (1992, p. 200) Church's thesis is used to suggest that "the operations in the brain" can "be simulated by a computer".

However, the Turing machine was not formulated as a model to describe — in part or in total — physical events, let alone the human mind and consciousness. Rather, it was initially (Turing, 1936) formulated as a machine designed to perform effective computations following the same logical steps Turing identified a human would follow. This formulation has been also described in reverse, where the Turing machine is compared (Wittgenstein, 1980, p. 1096) to a human, freed from certain physical constrains, who computes an effective procedure. In fact, Turing (1948, p. 9) stated that "A man provided with paper, pencil and rubber, and subject to strict discipline, is in effect a universal machine", confirming the equivalence between the two formulations. As for this definition, the set of actions and, more generally, the expressive power of a Turing machine are limited (Copeland, 2007, p. 105) to the set of actions that can be performed by an *unaided* human. Assuming that a set of actions that cannot be carried out by a human but that can be executed by a machine does exist, it's possible that some of those actions cannot be computed by a Turing machine.

As a formal system, the Turing machine is defined under a certain set of assumptions (i.e., having an infinite tape, read/write operations with perfect accuracy) that, generally, cannot be applied under a physical environment. Moreover, the class of functions (i.e., the effective procedures) that can be implemented with a Turing machine is a subclass of the possible functions,[1] therefore the possibility to simulate every physical event using only — concrete or formal — Turing machines is open to debate.

2. The Computational Theory of the Mind and the Qualitative *Hard Problem*

The qualitative dimension is as essential to everyday experience as it is problematic from the viewpoint of scientific treatment. The apparent obviousness and naturalness of consciousness, understood as *what it is like to be* (Nagel, 1974), contrasts with certain characteristics which make it hard to "attack" from the theoretical viewpoint, such as subjectivity, inexpressibility, incommunicability and ineffability. It does not seem possible to construct any science out of the content of what is experienced or felt — whether it be sensation, perception or emotion. In philosophy of mind, this problem takes the name of the *qualia*. The qualia are all those qualitative

[1] Turing identified in the *halting function* an example of a function that cannot be computed by a Turing machine.

determinations (colors, sounds, tactile sensations, emotions, passions, etc.) which refer back to the effect of being made conscious of something. The nature of the qualia is therefore essentially subjective (access to them is privileged) and qualitative (to have them creates a certain effect). The debate in contemporary philosophy of mind has been dominated by two main theoretical positions: physicalism and functionalism. From the physicalist standpoint, mental states are physical entities, more specifically, states of the central nervous system. The theory that mental states and cerebral states are identical leads to the impossibility of taking qualitative states into account. For Thomas Nagel, even if we could understand in detail how a bat's perceptual system functions, we could not in any way imagine the effect of perceiving the world by means of echolocation. The thought experiment of the scientist Mary, elaborated by Jackson (1982, 1986), shows how the possession of a complete explanation of the physiological processes which enable color to be seen does not by any means entail a real comprehension of the experience of seeing color. Having a conscious experience of color cannot be a physical property of the brain. The central idea of Jackson's argument is that the knowledge of physical properties cannot provide knowledge of what it is like to have a certain experience of color.

For functionalism, which stands in explicit contrast to physicalism, mental states are characterized by their functional role. According to the functionalist picture, computational processes are not material processes, but are implemented by or realized in different material systems. In this sense, computational processes are multiply realizable: a computing machine is a device that operates in terms of formal operations on symbols or, more simply, symbol manipulation. Such a device could be made from any number of materials and organized in any number of ways: in other words, the actual or material makeup is totally non-essential. Just as you abstract from the size, color and spatial location of a geometrical figure when it's the subject of a geometrical proof, so you abstract from a computing machine's material composition when you consider it as a computational device. The thesis of multiple realization, originally proposed by Putnam (1960), endorses the total irrelevance of physico-material properties for identifying mental states. In support of this thesis, functionalism speaks of functional isomorphism: different material structures can turn out to be equivalent from the mental perspective employing the same program. For Fodor (1981) neurological descriptions are like those of the hardware of a computer, but the logical operations of a computer are independent of any hardware.

As stated by one of the commonplaces of functionalism, the mind is to the brain as software (the program) is to the hardware is to a computer. If the mind can be likened to a program for a calculator, and if such a program can be executed by any sufficiently stable and powerful material support, the essential characteristics of the mind can be described and understood without knowing how the brain functions. Therefore, from the standpoint of computational functionalism, the human mind is an instrument for the formal manipulation of symbols; on the other hand, a Turing machine is an abstract device, physically realizable in an almost infinite number of ways, of which the human brain is only one example. What matters is the functional relationships between its parts, and not the material substrate (mechanical, electronic, biological) which realizes it. Thinking formally means manipulating sequences of representational signs, thereby effecting trains of thought about reality even when there is no effective comprehension of what the sequences themselves mean. Such manipulation enables knowledge to be inferred about reality since the signs stand, to some degree, for things. The notion of isomorphism, or of structural identity, is based on the presupposition that complex entities (facts, thoughts and expressions) are constructed out of simple entities — objects, ideas and words, respectively — similarly and in parallel in all three worlds by means of syntactical rules. The analytical or compositional hypothesis is that the interpretation, representation and meaning of the whole are univocally determined by the interpretation, representation and meaning of its parts. The analytical and compositional idea, united to the functional paradigm, constitutes the foundation of the computational hypothesis of the mind. The knowledge of a phenomenon depends not on the intuitive significance of a datum, but on the possibility of its being inserted into a system of calculable connections.

Conceiving of mental states as functional or computational nevertheless presents certain substantial limitations. First of all, the computational thesis about the mind takes no account of the fact that cognitive activity cannot be separated from living and acting within a world. What are absent from the computational theory of mind are the categories of being-in-the-world (*Dasein*) and openness to the world: that is, those intuitive, preliminary, tacit capacities for pre-comprehension which seem to characterize the human mind (Dreyfus, 1992; Dreyfus *et al.*, 1987). Further, the computational thesis about the mind takes no account of the fact that manipulating abstract symbols does not mean having a real understanding of what those symbols mean. Computers behave *as if* they had an effective understanding, but they actually execute mechanical procedures void of meaning. This occurs for the simple reason that the machine lacks fundamental, primitive concepts such as those of consciousness, intentionality, interest in social context, and so on. Searle's (1980) argument about the Chinese room shows how a system can simulate a human cognitive capacity to perfection in the total absence of effective comprehension. Finally, the computational thesis about the mind, like functionalism, is unable to take account of qualitative states, or qualia. When a red object is seen or a pain is felt, qualitative experiences are in fact involved which remain excluded from a purely functional description.[2] The objections to the computational thesis about the mind converge in pointing out the closed, non-interactive, non-environmental character, the so to speak "extractability", of computational procedures. As Putnam (1991) claims in critiquing the functionalism he had himself previously maintained, mental states are not just *compositionally plastic* (the same mental state can in principle be a property of systems which do not possess the same physical constitution) but are also *computationally plastic* (the same mental state can in principle be a property of systems that do not possess the same computational structure). Therefore, to put it in epistemological terms, there is an overdetermination which is not only physical (thesis of multiple realization) but also computational (thesis of multiple functionality). This renders both physicalism and functionalism inadequate to take account of the mental dimension. Furthermore, both are inadequate to take account of a qualitative and phenomenal dimension, that of consciousness.

Although they are complementary, functionalism and physicalism share a common structure or logic. We shall call this structure a residual theory of the qualitative, and its underlying logic a *logic of the residue*. We are concerned with an approach that, generally speaking, considers the experience of the qualitative as *something more* (and inaccessible) with respect to the physical–material dimension (in the case of physicalism) or the formal-computational dimension (in the case of functionalism).

[2]Thought experiments with the inversion of the color spectrum and Block's so-called absent qualia (Block, 1980a, 1980b) exactly show this type of insufficiency.

Take the case of pain as an example. As is well known, procedures exist for "measuring" the experience of pain, which usually resort to sensory, affective, evaluative parameters (Melzack, 1987). The theoretical problem lies nevertheless in understanding what may be entailed in having to rely on a questionnaire which enables the patient to express their own internal experience in order to go on to operate from a clinical viewpoint. There are two routes that can be followed. The first is the reductionist path, which attributes the subjective experience of pain entirely to neurological events. The second is the functionalist path, which puts the pain that is felt, for example, in close relation to its linguistic connotation. However, both paths seem unable to catch the experience of pain as distinct from simply having a pain, in the sense of having a painful part. There is a surplus in feeling pain or having an experience of pain, which cannot be reduced either to a physical state or to a computational procedure.

With reference to the analysis of what is meant by "mentalistic" or "folk psychology" terms like "desire", "expect", "decide" and "intend", traditionally attributed to the sphere of the mental, Dennett (2009) elaborates an Intentional System theory which allows for three "stances": the physical stance (relating to physical things), the design stance (relating to artefacts) and the intentional stance (the strategy of interpreting the behavior of an entity — person, animal, artifact — by treating it *as if* it were a rational agent who governed its actions by a consideration of mental terms). According to Dennett, "the intentional stance is thus a theory-neutral way of capturing the cognitive competences of different organisms (or other agents) without committing the investigator to overspecific hypotheses about internal structures that underlie the competences" (Dennett, 2009, p. 344). The absence of an ontological engagement on the "internal" character of the mental derives from the private, ineffable, subjective character of the qualia.

We think it appropriate, *contra* Dennett, to add a fourth to his three stances, one which we will call the Qualitative Stance since it is the only one able to take into account certain aspects of the mind (perceptual, emotional, social, etc.) in relation to which the standard computational model falls short. However, our challenge will be that of interpreting the Qualitative Stance without exclusive recourse to "internal" or "purely interior structures" which would be covered by that logic of the residue to which we referred a little earlier. Our attempt will, on the contrary, be to identify effective procedures with characteristics which cannot be attributed to computational procedures in the strong sense. We will call such procedures open or common sense effective procedures.

3. Effective Procedures and Common Sense Computation

"Effective" and "procedure" are common words, whose everyday meanings are respectively "successful in producing a desired or intended result" (Stevenson, 2015) and "an established or official way of doing something" (Stevenson, 2015). An effective procedure, according to a first approximate description, is therefore an established way of doing something which is successful in producing its intended result.

A procedure, being an *established* way of doing something, is meant to be followed more than once; it must be finitely describable, to allow its intended followers to learn it, and it must prescribe the action types[3] to be performed by its intended followers. An effective procedure, according to a second approximate description, is therefore an established way of doing something, finitely describable, which prescribes to its intended followers the action types to be performed to produce the intended result.

The effective procedures as defined by Church and Turing (from now on "computational procedures") are a subset of effective procedures in the sense just defined. Experimental procedures and qualitative medical procedures are effective procedures too. It's worth elucidating the most significant differences.

3.1. *Computational procedures (closed procedures)*

First of all, it's worth mentioning Turing's analysis of human computation in his seminal 1936 paper (Turing, 1936).[4]

Turing starts by considering a person (from now on "computor"[5]) who observes, writes and erases symbols on paper divided into squares (Turing, 1936, p. 240).

---

[3]As opposed to action tokens, or simply actions.

[4]This very analysis is the main reason for the intuitive appeal of the aforementioned Church–Turing thesis (Gandy, 1988; Sieg, 1994).

[5]As opposed to "computer", which now stands for a computing *machine*. See Sieg (1994) for this usage.

He then shows that the computor's activity must satisfy the following constraint:

Finiteness

- The computor may use a finite number of symbol types (Turing, 1936, p. 249).[6]
- In every single moment, the computor may observe a finite number of symbols and squares (Turing, 1936, p. 250).
- The computor may have a finite number of states of mind (Turing, 1936, p. 250).
- In every single moment, the computor must have a single state of mind (Turing, 1936, p. 250).

Compositionality, Locality and Boundedness

In every single moment, the computor may perform a sequence of actions of two basic types (Turing, 1936, pp. 250–251):

- a possible change of symbol on one of the observed squares;
- a possible change of observed squares within a finite number of squares of one of the previously observed squares.

The performance of the actions may determine a change of mind.

Determinacy

- In every single moment, the computor's behavior is determined uniquely by his current state of mind and the current observed symbols (Turing, 1936, p. 250).
- In every single moment, the state of progress of the computation is determined uniquely by the computor's behavior (Turing, 1936, pp. 253–254).

3.1.1. *Time insensitivity*

The computor may pause and continue the computation at will, provided he writes a note of instructions describing how to continue the computation (Turing, 1936, p. 253).

[6]As opposed to symbol tokens, or simply symbols.

3.1.2. *Multiple realizability*

It is possible to replace the computor with a machine similar to the aforementioned Turing machine (Turing, 1936, pp. 251–252).

Turing's analysis makes clear the possible form of computational procedures.

By Finiteness, there is only a finite number of possible combinations of states of mind and observed symbols (Gandy, 1988; Sieg, 1994). Therefore, it's possible to describe the rules which govern the computor's behavior through a finite number of conditional instructions: each instruction is conditional on a certain state of mind and certain observed symbols; it prescribes the sequence of action types to be performed, and it indicates the state of mind determined by the performance of the actions. If the computation produces the intended result, then the conditional instructions are an effective procedure to produce that result both in Turing's sense and, as soon as the effectiveness is accepted, in the sense defined before.

At the same time, Turing's analysis discloses four significant properties of computational procedures.

3.1.3. *Predetermination of the intended result*

By Determinacy, in every single moment the state of progress of the computation must be uniquely determined by the previous state of progress and by the computor's behavior, which in turn must be uniquely determined by the procedure that he is following and by his initial state of mind. The intended result of the computation is therefore predetermined by the procedure that the computor will follow and by the initial symbols and state of mind.

3.1.4. *Context insensitivity*

The mere performance of the basic actions according to the procedure must be sufficient, as well as necessary, to produce the intended result. That is, the intended result must be produced every time the computer performs the basic actions according to the procedure. Intuitively, the computor's behavior directly brings forth all the effects required to produce the intended result: "write" ends when the computor stops writing, "erase" ends when the computor stops erasing, etc. If that is the case, the computor may produce the intended result in any possible world in which he knows how to write, erase, etc.: the computation is self-contained, ideally isolated from its context.

3.1.5. *Time insensitivity*

This is explicitly stated by Turing. Anyway, it's worth mentioning why the computation must be insensitive to time. If the state of progress of the computation, as required, is uniquely determined by the computor's behavior, it doesn't matter when he performs the actions prescribed by the procedure: if he is not acting, the state of progress doesn't undergo any change.

3.1.6. *Content insensitivity*

Turing assumes that any symbol used in the computation supports the same basic action types. Moreover, by Determinacy, the computor's behavior must depend only on the identity of the observed symbols: they just need to be somehow distinguishable when observed. Thus, the meaning or the exact physical realization of the symbols must not be significant: the computor should be able to systematically swap symbols belonging to two types at will, while still producing the intended result.

3.2. ***Experimental procedures (open quantitative procedures)***

It would be advisable to describe experimental procedures starting from a detailed analysis of what a person does when engaging in science. However, there is no such analysis that is not still under discussion. It's better, therefore, to just note that according to everyday usage,[7] doing quantitative experiments to test hypotheses is surely part of doing science (Stevenson, 2015).[8] When the descriptions of the experiments are published and the

[7]And also with the reality of research funding (O'Malley *et al.*, 2009).

[8]In 1623, Galileo distinguished between primary and secondary qualities: "I think that tastes, odors, colors, and so on are no more than mere names so far as the object in which we locate them are concerned, and that they reside in consciousness. Hence, if the living creature were removed, all these qualities would be wiped away and annihilated" (Galileo, 1957, p. 274). His distinction has been instrumental in the development of modern science and has contributed to the prevalence of quantitative experimentation, i.e., experimentation focused on measurable properties of physical objects.

experiments are meant to be reproducible, then these descriptions are effective procedures in the sense defined earlier.[9]

The form of experimental procedures is best exemplified by the work of Simon Stevin (1955, pp. 509–511):

> Aristotle ... and his followers hold that if two similar bodies fall through the air, as the gravity of the one is to that of the other, so is the time of passage of the latter to that of the former, i.e., as gravity is to gravity, so is impediment to impediment.... The experience against Aristotle is the following: Let us take (as the very learned Mr. Jan Cornets de Groot, most industrious investigator of the secrets of Nature, and myself have done) two spheres of lead, the one ten times larger and heavier than the other, and drop them together from a height of 30 feet on to a board or something on which they give a perceptible sound. Then it will be found that the lighter will not be ten times longer on its way than the heavier, but that they fall together on to the board so simultaneously that their two sounds seem to be one and the same rap.... The same is found also to happen in practice with two equally large bodies whose gravities are in the ratio of one to ten; therefore Aristotle's aforesaid proportion is incorrect.

This example discloses significant properties of experimental procedures, symmetrical to those of computational procedures.[10]

3.2.1. *Non-predetermination of the intended result*

The intended result of an experiment, unlike the results of computational procedures, is not predetermined by the experimental procedure being followed because the state of progress of the experiment doesn't depend uniquely on the scientist's behavior. In Stevin's example, the scientist drops the spheres to check what they do independently from his behavior. Although the usual result is compatible with the followed experimental procedure, a different result would be compatible with it too.

---

[9]See, Challenges in irreproducible research. URL http://www.nature.com/nature/focus/reproducibility (2014), on the actual reproducibility of experiments.

[10]The following analysis is somewhat inspired by Carol Cleland's analysis of mundane procedures: see Cleland (1993, 2002) for details. Seligman (2002) mentions the possible extension of Cleland's analysis, whose original subject was cooking recipes, to experimental procedures.

3.2.2. *Context sensitivity*

The result of the experiment is determined by the scientist's behavior *and* by the physical laws of the world where the experimental procedures are followed. By performing the actions prescribed by the experimental procedure, the scientist does not only bring forth their typical effects, he also starts physical processes which are then independently modulated by the physical laws of the world. In Stevin's example, "drop" has no typical effects which, when brought forth, somehow directly change the dropped spheres. By dropping the spheres, the scientist just lets them fall. Their fall is a physical process which is then independently modulated by gravity.

3.2.3. *Time sensitivity*

Given that the scientist starts physical processes which evolve independently from his behavior, it is not possible for him to pause and continue the experiment at will. Indeed, the started processes may well be modulated by the laws of physics on the basis of the passage of time. In Stevin's example, by accelerating the fall of the spheres, gravity makes them collide with the board below within a finite interval of time. When they eventually collide with the board, the scientist must still be present to hear the perceptible sound that the collision produces. Moreover, in Stevin's example, time is also an integral part of the hypothesis which is being tested: the scientist listens to the produced perceptible sound to check whether the spheres end their fall at the same time.

3.2.4. *Content sensitivity*

Finally, the physical laws may modulate the started physical process on the basis of the content of the objects involved in the experiment. In Stevin's example, the two spheres differ in their weight, which is an integral part of the hypothesis to be tested. The usual result of the experiment shows that the weight of the spheres is not a relevant factor: gravity accelerates the fall of the spheres in the same way irrespective of their weight. However, the experimental procedure is compatible with a different result: if Aristotle was right, weight would have been a relevant factor.

3.3. *Qualitative medical procedures (open qualitative procedures)*

Medicine, according to everyday usage, concerns itself with diagnosing, treating and preventing diseases (Stevenson, 2015) using both quantitative and qualitative methods. Among the qualitative methods are questionnaires, which, being an established and finitely described way of collecting data, are effective procedures in the sense defined earlier. A prime example of a qualitative questionnaire is the McGill short-form pain questionnaire (Melzack, 1987).

	None	Mild	Moderate	Severe
Throbbing	0) ----------	1) ----------	2) ----------	3) ----------
Shooting	0) ----------	1) ----------	2) ----------	3) ----------
Stabbing	0) ----------	1) ----------	2) ----------	3) ----------
⋮	⋮	⋮	⋮	⋮
Sickening	0) ----------	1) ----------	2) ----------	3) ----------
Fearful	0) ----------	1) ----------	2) ----------	3) ----------
Punishing-cruel	0) ----------	1) ----------	2) ----------	3) ----------

Present pain intensity (PPI)   No pain |------------------------------| Worst possible pain

0	No pain	----------
1	Mild	----------
2	Discomforting	----------
3	Distressing	----------
4	Horrible	----------
5	Excruciating	----------

Each line of the first table contains a descriptor of a dimension of pain experience and its intensity in the range 0–3. The intermediate line, on which the patient puts a mark, indicates the general intensity of pain. The last table indicates the general intensity in the range 0–5 (Melzack, 1987, p. 193). The physician reads each descriptor aloud and asks the patient whether the descriptor describes his pain and, if that is the case, what is its relative intensity in the range 0–3 (Melzack, 1987). He also asks the patient to put his mark on the intermediate line and to indicate his general pain intensity in the range 0–5 (Melzack, 1987).

By looking at this example, it is possible to disclose the properties of qualitative medical procedures.

3.3.1. *Non-predetermination of the intended result*

As in experimental procedures, the intended result of a qualitative medical procedure is not predetermined by the procedure itself. The state of progress of the McGill questionnaire depends on the collected values of pain intensity, which are different from patient to patient and may be collected only by interacting with patients.

3.3.2. *Context sensitivity*

The result of a qualitative medical procedure is determined by interacting with a patient, who is asked by the physician to access that qualitative experience dimension which the other types of procedures ignore. In a broad sense, the physician starts a process which then evolves independently and which is modulated not by physical laws but by the intrinsic laws of the patient's qualitative experience, whatever they are. Indeed, the McGill questionnaire depends on the existence of such laws: the patient must be able to apply the "sickening" descriptor to his pain in a meaningful way and to indicate its intensity in a meaningful way, otherwise the questionnaire would be useless.

3.3.3. *Time sensitivity*

Given the dependence on the patient's qualitative experience, the physician is not able to pause and continue a qualitative medical procedure at will. The patient's qualitative experience may evolve independently, thus invalidating the data initially collected. In the case of the McGill questionnaire, a physician may not wait hours between completing the first part of the questionnaire and completing the other two parts: it is not far-fetched to imagine that the patient would take a pain-killer in the meantime, making the data already collected useless.

3.3.4. *Content sensitivity*

Finally, the content of qualitative experience is clearly relevant for qualitative medical procedures. When the physician reads aloud the descriptors of the McGill questionnaire, the patient has to repeatedly experience *what* the content of his pain *is like*: stabbing and mild, throbbing and severe. If the content of the pain were to change, the answers of the patient would change too.

The analysis shows clearly that the concept of effective procedures is broader, practically and not only in principle, than the concept of computational procedure. Experimental procedures and qualitative medical procedures are under less constraints than computational procedures. This makes it harder to describe them formally, but, at the same time, it makes them more powerful than computational procedures, e.g., by allowing them to take content and context into account. The analysis, however, also shows that all the analyzed procedures — but not necessarily all the effective procedures — share a key property: their result may be represented symbolically. Generalizing a little, it's worth labeling as "common sense procedures" all those effective procedures that share the aforementioned property. In light of the well-know connection of symbolic output with computation, "common sense computation" may stand for all the activities in which common sense procedures are used.

4. Conclusions and Future Works

The importance of qualitative experience makes it worthwhile to assume a qualitative stance, thus rejecting — in view at least of the benefits in terms of making modeling simpler — the subordination of the qualitative to the quantitative dimension.

Computational procedures as defined in the Church–Turing thesis consider only formal procedures. However, qualitative procedures are some of the most used in everyday life and share at least a key property with formal procedures: their results may be represented symbolically. It's possible to define an effective procedure as a way of doing something that is reliable in successfully producing the intended result. This definition is broad enough to include within the spectrum of effective procedures both the formal procedures, invariant on the content and context, and, at the opposite end of the spectrum, the qualitative procedures commonly used in everyday life. Therefore, this definition of effective procedures elucidates the seemingly computational character of some mental processes. Even if there is no computation going on, in a strict sense, there may be a common sense computation: an activity that involves procedures which share the aforementioned key property with computational procedures.

In future, perhaps, it will be possible to show that common sense procedures are informative and the sense in which they are so. For the time being, it suffices to say that the original meaning of "to inform" was "to in-form", "to give form": according to Plato, the form of the horse impresses itself in someone who observes it for the first time; that form then enables the observer to recognize other horses. Common sense procedures seem to be informative at least in this sense: they produce a result which grants new abilities to its recipients. A computor may use the result to prepare a budget, the scientist may use it to revise his theories, the physician may decide how to treat a patient on the basis of the result.

Finally, it is worth noting that algorithmic information theory quantifies the information contained in a string of symbols as the program of minimal complexity which can compute that string on a universal Turing machine. Algorithmic information theory is thus relevant only for computational procedures. However, it is worth asking if its key intuition could be used for all the common sense procedures: in other words, if it is possible to quantify the information contained in the symbolic solution of a problem as the complexity of the simplest common sense procedure able to produce that solution, when computed by a human being. To assess this possibility, it will be necessary to consider, along with the problem and the procedure used to solve it, the *need* the problem is formulated to satisfy. Therefore, the information, as defined above, and its quantification could be inherently bound to the acquisition of the ability to satisfy this need.

References

Ackermann, W. and Hilbert, D. (1928). *Grundzüge der Theoretischen Logik*, Springer-Verlag.

Block, N. (1980a). Are absent qualia impossible? *Philosophical Review* 89(2), 257–274.

Block, N. (1980b). Troubles with functionalism. In *Readings in Philosophy of Psychology*, Vol. 1, pp. 268–305. Harvard University Press.

Boden, M. A. (1988). *Computer Models of Mind: Computational Approaches in Theoretical Psychology.* Cambridge University Press.

Church, A. (1932). A set of postulates for the foundation of logic, *Annals of Mathematics* 33(2), 346–366.

Church, A. (1936a). A note on the entscheidungsproblem, *Journal of Symbolic Logic* 1(1), 40–41.

Church, A. (1936b). An unsolvable problem of elementary number theory, *American Journal of Mathematics* 58(2), 345–363.

Churchland, P. M. and Churchland, P. S. (1990). Could a machine think? *Scientific American* 262(1), 32–39.

Cleland, C. E. (1993). Is the church-turing thesis true? *Minds and Machines* 3(3), 283–312.

Cleland, C. E. (2002). On effective procedures, *Minds and Machines* 12(2), 159–179.

Copeland, J. B. (2000). Narrow versus wide mechanism: Including a re-examination of turing's views on the mind–machine issue, *Journal of Philosophy* 97(1), 5–32 .

Copeland, J. B. (2007). The church-turing thesis, *NeuroQuantology* 2(2), 101–115.

Dennett, D. (1991). *Consciousness Explained.* Little, Brown and Co.

Dennett, D. (2009). Intentional systems theory. In *The Oxford Handbook of Philosophy of Mind,* pp. 339–350. Oxford University Press.

Deutsch, D. (1985). Quantum theory, the church–turing principle and the universal quantum computer, *Proceedings Royal Society of London A* 400(1818), 97–117 (1985).

Dreyfus, H. L. (1992). *What Computers Still Can't Do: A Critique of Artificial Reason.* MIT Press.

Dreyfus, H. L., Dreyfus, S. E. and Zadeh, L. A. (1987). Mind over machine: The power of human intuition and expertise in the era of the computer, *IEEE Expert* 2(2), 110–111.

Fodor, J. A. (1981). The mind-body problem, *Scientific American* 244(1), 114–123.

Galileo (1957). The assayer. In Drake, S. (ed.) *Discoveries and Opinions of Galileo,* pp. 229–280. Doubleday.

Gandy, R. (1980). Church's thesis and principles for mechanisms, *Studies in Logic and the Foundations of Mathematics,* 101, 123–148.

Gandy, R. (1988). The confluence of ideas in 1936. In *A Half-century Survey on The Universal Turing Machine,* pp. 55–111, Oxford University Press.

Guttenplan, S. (1994). *A Companion to the Philosophy of Mind.* Blackwell Publishing.

Jackson, F. (1982). Epiphenomenal qualia, *Philosophical Quarterly* 32(127), 127–136.

Jackson, F. (1986). What mary didn't know, *Journal of Philosophy* 83(5), 291–295.

Kleene, S. C. (1936). λ-definability and recursiveness, *Duke Mathematics Journal* 2(2), 340–353,

Kleene, S. C. (1967). *Mathematical Logic.* John Wiley & Sons, Inc.

Melzack, R. (1987). The short-form mcgill pain questionnaire, *Pain* 30(2), 191–197.
Nagel, T. (1974). What is it like to be a bat? *The Philosophical Review* 83(4), 435–450.
Newell, A. (1980). Physical symbol systems, *Cognitive Science* 4(2), 135–183.
O'Malley, M. A., Elliott, K. C., Haufe, C. and Burian, R. M. (2009). Philosophies of funding, *Cell* 138(4), 611–615.
Putnam, H. (1960). Minds and machines. In Hook, S. (ed.) *Dimensions of Mind: A Symposium*, pp. 148–179. New York University Press.
Putnam, H. (1991). *Representation and Reality*. MIT Press.
Searle, J. R. (1980). Minds, brains, and programs, *Behavioral and Brain Sciences* 3(3), 417–424.
Searle, J. R. (1992). *The Rediscovery of the Mind*. MIT Press.
Seligman, J. (2002). The scope of turing's analysis of effective procedures, *Minds and Machines*, 12(2), 203–220.
Sieg, W. (1994). Mechanical procedures and mathematical experience. In George, A. (ed.) *Mathematics and Mind*, pp. 71–117. Oxford University Press.
Smolensky, P. (1988). On the proper treatment of connectionism, *Behavioral and Brain Sciences* 11(1), 1–23.
Stevenson, A. (2015). ed., *Oxford Dictionary of English* (2010). URL https://en.oxforddictionaries.com. Online edition.
Stevin, S. (1955). *Principal Works: General Introduction. Mechanics.* C. V. Swets & Zeitlinger.
Turing, A. M. (1936). On computable numbers, with an application to the Entscheidungsproblem, *Proceedings of London Mathematical Society* 2(42), 230–265.
Turing, A. M. (1937). Computability and λ-definability, *Journal of Symbolic Logic* 2(4), 153–163.
Turing, A. M. (1948). *Intelligent Machinery*, p. 395, Edinburgh University Press.
Wittgenstein, L. (1980). *Remarks on the Philosophy of Psychology*, Vol. 1, Blackwell Publishing.

Chapter 5

A Theory of Semantic Information in the Context of its Ecology

Yixin Zhong[*,‡] and Gordana Dodig-Crnković[†,§]

[*] *University of Posts & Telecommunications, Beijing, China*

[†] *Chalmers University of Technology, Sweden*

[‡] *zyx@bupt.edu.cn*

[§] *dodig@chalmers.se*

In the theory of human communication, information is described as the trinity of aspects, or dimensions: syntactic, semantic, and pragmatic, corresponding, respectively, to *the form*, *the meaning*, and *the utility* of information. The theory of information proposed by Shannon in 1948, often considered as "the" theory of information, is a purely statistical theory of communication of syntactic information. Both the semantic and pragmatic theories of information, which are in the first place significant for human understanding, are still lacking. In this chapter, an ecological theory of semantic information is presented in the context of human communication and in the case of symbolic artificial intelligence. The novelty of the approach is that information is studied in the context of its ecology and not reduced to only one of its aspects. It is shown that the semantic information can be used as the unique representative for the trinity of information aspects. The same ecological approach can be applied in the case of embodied intelligence without representation, where the world is considered the best model of itself, which remains for future research. The same goes for the generalization to the system of communicating cognizing/intelligent agents. Another domain of high interest for further exploration is the temporal aspect of ecological information, including its evolution. Related to this, the development in the direction of a new, more appropriate logical basis is necessary.

1. Introduction

Information has been recognized as one of the major resources that humans could effectively utilize for raising their standards of living in our age of information. The establishment of the theory of information has thus become one of the most urgent demands of society.

From the theory of human communication, based on the insights of linguistics and semiotics, it is known that the really useful information for humans is the trinity of syntactic, semantic, and pragmatic aspects, which enables the users of information to know the form, the meaning, and the utility of the information. For a detailed account, see Burgin (2010).

Claude E. Shannon's seminal paper "A Mathematical Theory of Communication" (Shannon, 1948) successfully deals with such issues as the calculation of information amount produced by the source of the communication system, the calculation of the information capacity of the channel, the coding and decoding for matching the source and the channel, etc., making great contributions to communication theory. Considering that all these issues are related to information to some extent, the theory later changed its name to the Mathematical Theory of Information.

In the case of communication engineering, however, only the form factor of signals (the *carrier* of syntactic information) needs to be considered and it has nothing to do with either the semantic or pragmatic information. Hence, the Shannon theory of information is indeed a statistical theory of *syntactic information communication* only, and not the complete general theory of information. This implies that the theories of semantic and pragmatic information still need to be developed.

In addition to Shannon's theory of information are the Combinatory Information Theory and Algorithmic Information Theory, known in academic literature (Kolmogorov, 1968a, 1968b; Chaitin, 1987). Both of these are varieties of theory of *syntactic* information, as they do not address semantic and pragmatic aspects of information.

2. A Brief Review of the Classical Theory of Semantic Information

The earliest effort in developing the theory of semantic information was made by Carnap and Bar-Hillel. The framework of the theory that they proposed (Carnap and Bar-Hillel, 1953) can very briefly be described as follows, which can also be found in Brillouin (1956) and Bar-Hillel (1964).

First, they define an ideal language model that contains n (a finite positive number) nouns and k (another finite positive number) adjectives as well as 1 verb "have, or be". If "a" is a noun and "P" is an adjective, then "Pa" reads as "a has *the property* P" or "a is P". The model has five logical connectives:

Connective		Use	Meaning
$\sim$ (Not)	negation	"$\sim Pa$"	a is not P
$\vee$ (Or)	disjunction	"$Pa \vee Qb$"	a is P or b is Q
$\wedge$ (and)	conjunction	"$Pa \wedge Qb$"	a is P and b is Q
$\rightarrow$ (if ... then)	implication		
$\equiv$ (if and only if)	equivalence		

Based on the model above, one can produce a number of sentences that may be true ($Pa \vee \sim Pa$, for example), or false ($Pa \wedge \sim Pa$, for instance), or infinitive (Pa, for example) in logic. The amount of semantic information contained in a well-formed expression i is thus defined as *a function of the number of sentences in the ideal model of language that the expression i can imply.* The higher the number of sentences that an expression can imply in the model of language, the larger the amount of semantic information it has.

More specifically, a concept called "state descriptor Z" is defined as the conjunction of one noun and one adjective (positive, or negative, but not both). As a result, there are kn possible such conjunctions and 2^{kn} possible state descriptions.

Another concept called "range" is set up as follows: The range for a sentence i is defined as a set of the state descriptors in which the sentence i is valid and is denoted by $R(i)$. Further, a measurement function $m(Z)$ is introduced with the following constraints:

(1) for each Z, there is $0 \leq m(Z) \leq 1$;
(2) for all kn state descriptors, there is $\sum m(Z) = 1$;
(3) for any non-false sentence i, $m(i)$ is the sum of all $m(Z)$ within $R(i)$.

The amount of semantic information that sentence i contains is defined as

$$I(i) = -\log_2 m(i).$$

It can be seen from the brief review of the classical theory of semantic information that the concept and measurement here is very similar to that of the Shannon theory of syntactic information. In fact, various kinds of classical theories of semantic information have made much efforts in *quantitative measurement of semantic information*. Moreover, the basic idea of semantic information measurement of any sentence in classical theory is dependent on the number of sentences in the language model that can be excluded by the sentence.

There are, at least, two shortcomings in the classical theory of semantic information. One is the ideal model of language that is way too far from the natural language and is a completely unrealistic model of language. The other shortcoming is that the theory *concerns only the quantitative measure of the semantic information* and yet has no interest in the core of semantic information, its *meaning*. In Schroeder (2012), another detailed analysis of the failure of Carnap and Bar-Hillel to formulate an adequate theory of semantic information is provided.

2.1. *Syntactic vs. semantic information*

It is very important to point out the major difference between the theories of syntactic and semantic information. The theory of syntactic information needs to take accurate account of the numeric measurement of syntactic information because the engineers in communication systems must be able to calculate the precise amount of communication resources consumed, such as the bandwidth of communication channel, the energy of transmission, etc. On the other hand, the most central concern in the theory of semantic information is the *meaning* of information. This is because the users of information should be able to not only efficiently *communicate* information but, even more importantly, also *understand* the meaning of information so as to be able to effectively use the information to solve problems.

The majority of researchers in classical theory of semantic information have not realized the radical difference between the two kinds of theories outlined above. Not even the researchers involved in the semantic information studies in recent time (Millikan, 2002; Smokler, 1966; Gottinger, 1973; Floridi, 2011) recognize the essential difference between *measuring the amount of (meaningful) information* and *understanding (and using) its meaning*. However, Floridi (2017) presents a more nuanced view of semantic information.

2.2. Fundamental concepts related to semantic information

The basic concepts constituting semantic information are those of *information* and *semantics*. Therefore, to start with, it is necessary to make clear the concept of information and that of semantics, as well as the relationship between them.

2.3. Classical concepts of semantics and information

The classical concept of semantics can be found in the studies on semiotics, with Saussure, Peirce, and Morris, among many others, as its major contributors.

Saussure proposed a dualistic notion of signs, relating the signifier as the form of the word or phrase uttered, to the signified as the mental concept (Saussure, 2001). In the approach proposed by Peirce, a sign is something that stands to somebody for something in some respect or capacity. He considered further that the science of semiotics has three branches: pure grammar, logic proper, and pure rhetoric (Peirce, 1934). More clearly, Morris defined semiotics as establishing the triad syntax, semantics, and pragmatics, where syntax studies the interrelation of the signs, without regard to meaning, semantics studies the relation between the signs and the objects to which they apply, and pragmatics studies the relation between the sign system and its user (Morris, 1971).

In semiotics, the terms syntactic, semantic, and pragmatic have been used for expressing, respectively, the formal features, the meaning, and the utility of the signs with regard to the user of the signs. Interestingly, the three terms form a comprehension of a sign. But semiotics does not give an analysis on the mutual relations among the three, as we will do later.

Apart from the study of communication and sign processes, there has been a lot of work concerning the concept of information in history. The most representative views on the nature and concept of information include the following:

(1) Norbert Wiener's assertion that "information is information, neither matter, nor energy" (Wiener, 1948). More generally, Cybernetics focuses on the role of information in control systems with goals (anticipation).

(2) Shannon's quantitative information described by Weaver as being "measured by the logarithm of the number of available choices" (Shannon and Weaver, 1949). This approach to information found its applications in the processes of information communication in different media, especially in telecommunications.
(3) Gregory Bateson's definition of information as "the difference that makes a difference" (for an agent) (Bateson, 1972). This is an agent-based approach that does not analyze the mechanisms of identification and use of differences by an agent.
(4) Ludwig von Bertalanffy's approach, where information is a measure of a system's complexity (Bertalanffy, 1968). This is a very useful approach, focused more on the nature of complex systems than on the nature of information.

Apart from an encyclopaedic volume of Burgin (2010), more works in this area can be found in Zhong (1988), Adriaans and van Benthem (2008), Schroeder (2011), and Dodig-Crnković (2006). As can be seen, the concept of information is still open to debate. We will have more discussions on this concept in the next subsection.

3. Concepts Related to Semantic Information From the Point of View of its Ecosystem

It is easy to see that the concept of semantic information not only has its roots but also its ecological system. Therefore, the understanding of the concept of semantic information must strictly rely on the understanding of its root concepts and its ecological system. Through the investigation of its roots and ecological system, all the constraints that the concept of semantic information should observe will be clear.

What is the ecological system that semantic information belongs to? As the root concept of semantic information is that of information, the ecological system of semantic information is the full process of information–knowledge–intelligence conversion (Zhong, 2013b), which is shown in Figure 1.

In the model of Figure 1, the part at the bottom stands for the object in the environment that presents the object information, while the part on the top stands for the subject interacting with the object through the following processes: the object information is converted to perceived information (through perception), and the latter is converted to knowledge (through

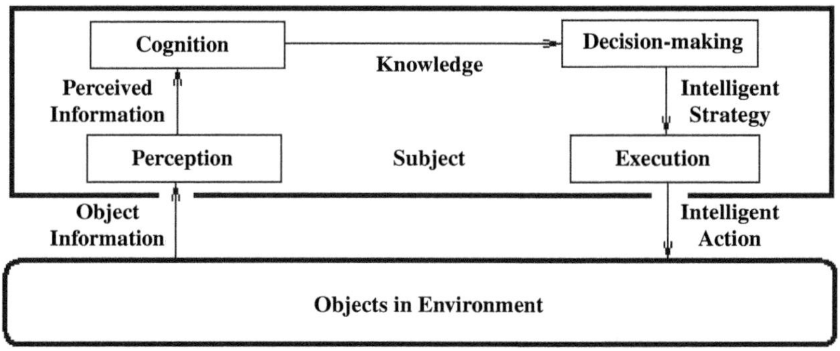

Figure 1. Model of information ecosystem defining the relationship between the intelligent subject and the environment.

cognition), and further converted into intelligent strategy (through decision-making), finally being converted into intelligent action (through execution of decisions) that is applied on the object, forming the fundamental process of an information ecosystem.

The most basic expectation from the theory of semantic information is that it must meet the constraints, or the requirements, imposed by the ecological process of information as described above.

It is worth pointing out that, among all kinds of object information, only those participating in the process of subject–object interaction (see Figure 1) will be regarded as meaningful and will therefore be carefully studied. Others having not entered into the subject–object interaction will, naturally, be neglected by the subject. In other words, the model of information ecosystem shown in Figure 1 is representative, necessary, and sufficient.

It is indicated from the model in Figure 1 that the most significant value that information could provide to humans (and human society) is not merely information itself, but even more, its ecological products, the knowledge and intelligence. This is because information itself is *about* something and can tell us *what it is*, while knowledge can tell us *why it is so* and intelligence is the strategy that can tell us *how to do it.*

So, it is not wise to only consider information itself in isolation, neglecting its strong connections to knowledge and intelligence. In other words, one who studies information should not stop at the level of information, but should continue to study with the view of better understanding the ecological process of information. This is an important insight for information studies.

Now let us start a specific investigation on all the concepts related to the one of semantic information along the lines of the ecological system of information shown in Figure 1.

Definition 1 (Object Information/Ontological Information). The object information concerning an object is defined as "the set of states at which the object may stay and the pattern with which the states vary", all presented by the object itself.

The term "ontological information" (ontology is concerned with what exists) is synonymous with "object information" because this information is completely determined by the object and in its description has no reference to the subject. The two names, object information and ontological information, are mutually equivalent, but the term "object information" will be more frequently used in this chapter.

Referring back to Definition 1, the argument of object information can briefly be expressed as the "state-pattern of variation". For example, given an object whose possible set of states is X and the pattern with which the states vary is P, the corresponding object information can briefly be represented as $\{X, P\}$. Object information is referred by Dodig-Crnković (2006) as *potential information*, as it denotes the informational content of the world before the subject starts to interact with it.

Definition 2 (Perceived Information/Epistemological Information). The perceived information a subject possesses about an object, which is also termed epistemological information, is defined as the trinity of the form (the syntactic information), the meaning (the semantic information), and the utility (the pragmatic information), all of which are perceived by the subject from the object information.

The term "epistemological information" is adopted because this information is determined not only by the factors of the object (the "states-pattern") but also by the factors of the subject (the subject's knowledge and goals).

It is clear by comparing Definitions 1 and 2 that the object information is the real source coming from the real world, while the perceived information is the product of the object information via the subject's perception. As a result, the perceived information should contain more abundant intensions than the object information does.

Definition 3 (Comprehensive Information). The trinity of syntactic, semantic, and pragmatic information is named comprehensive information.

So, the concept of "perceived information"/"epistemological information" has three components, syntactic, semantic, and pragmatic, whereas the trinity itself is named "comprehensive information".

Note that although the concepts of syntactics, semantics, and pragmatics have been defined in semiotics, the mutual relationship among them has never been studied. As we can see later, the mutual relationship among them is the most important issue in the study of information. Without an understanding of the mutual relationships, the concepts of syntactic, semantic, and pragmatic information remain ambiguous. This is why there still remain many controversies in semiotics, linguistics, and in the study of information in particular.

Definition 4 (Mutual Relation among Syntactic, Semantic, and Pragmatic Information). Syntactic information is specific in nature and can directly be produced through a subject's sensing function, while pragmatic information is also specific in nature and can directly be produced through a subject's experiencing. On the other hand, semantic information is abstract in nature and thus cannot be produced via the subject's sensing organs or direct experience. Semantic information can only be produced based on both syntactic and pragmatic information just produced already, that is, by mapping the joint of syntactic and pragmatic information into the semantic information space, and then naming it.

Further, the mutual relationship of the semantic, syntactic, and pragmatic information can be expressed as in Figure 2. This will be explained in more detail in Figure 5, in Section 4.1.

It can be concluded from the discussion above that the semantic information can serve as a proper representative of the perceived information. This is why semantic information possesses the highest importance compared to syntactic and pragmatic information.

From all definitions and discussions given above, we conclude the following:

(1) The object/ontological information of Definition 1, the "states-pattern presented by the object", is neither matter nor energy and is thus in agreement with Wiener's statement. However, Definition 1 is more specific than the statement.

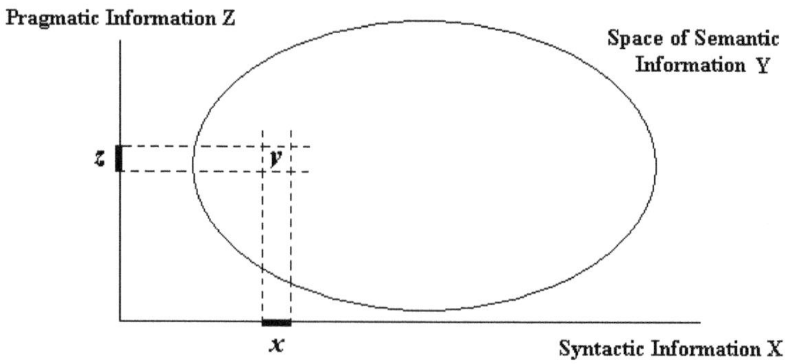

Figure 2. Relation of semantic, syntactic, and pragmatic information.

(2) The concept of perceived/epistemological information of Definition 2 is just what can be used to remove the uncertainties concerning the "states-pattern". It is easy to see that the concept of information in the Shannon theory is the statistically syntactic information, which is only one component of the perceived information. So, Definition 2 is more inclusive and more complete than Shannon's definition, which is expected given the purpose of Shannon's original theory of communication.
(3) As for Bateson's statement, it is pointed out that the most fundamental "difference" among objects is their own "states-pattern" presented by the objects. Hence, Definition 1 and Bateson's statement are equivalent to each other. But Definition 1 is more general.
(4) Bertalanffy regarded information as the measure of the "complexity of a system". In fact, the complexity of a system is just the complexity of its "states-pattern" presented.

All in all, Definitions 1 and 2 enjoy the advantages of more regular, more standardized, more universal, and more scientific definitions compared with the statements given by Wiener, Shannon, Bateson, Bertalanffy, and others. So, we will use Definitions 1 and 2 for defining object/ontological information and perceived/epistemological information. Up to now, the concept of semantic information has been well defined. According to the view of information ecology, however, the discussion on fundamental concepts cannot stop here. It is necessary to investigate how the later processes of information ecology (see Figure 1) may exert influence and impose constraints on semantic information.

Definition 5 (Knowledge). Knowledge that subjects possess in their minds concerning a certain class of events is defined as the "set of states at which the class of events may stay and the common rule with which the states vary" that have been summed up from a sufficiently large set of samples of perceived information related to the class of events.

For brevity, knowledge can be described by using the phrase of "state-rule" just as perceived information is described by using the phrase of "state-pattern". Since knowledge is the product of abstraction from perceived information, knowledge also has its own three components: form, content, and value of knowledge, respectively, corresponding to syntactic information, semantic information, and pragmatic information. Similar to the case of perceived information, the content of knowledge can serve as the rightful representative of the entire knowledge. It is obvious that semantic information plays a fundamental role in the studies of knowledge.

It is necessary to note that knowledge itself forms an ecological process: from the perceived information to empirical knowledge via the process of induction, and then to regular knowledge via the process of validation, and further to commonsense knowledge via the process of precipitation, all based on instinct knowledge. This is shown in Figure 3 and is called the internal chain of knowledge ecology.

It is easy to understand that empirical knowledge is the knowledge with the under-maturity status, regular knowledge is the one with the normal-matured status, and commonsense knowledge is the one in the over-matured status. As is well known, the development of the status from the under-matured to the normal-matured and further to the over-matured is a typical ecological process.

On the other hand, there is an external chain of knowledge ecology that is from (perceived) information to knowledge via inductive operations and then from (perceived) information and knowledge to intelligence via

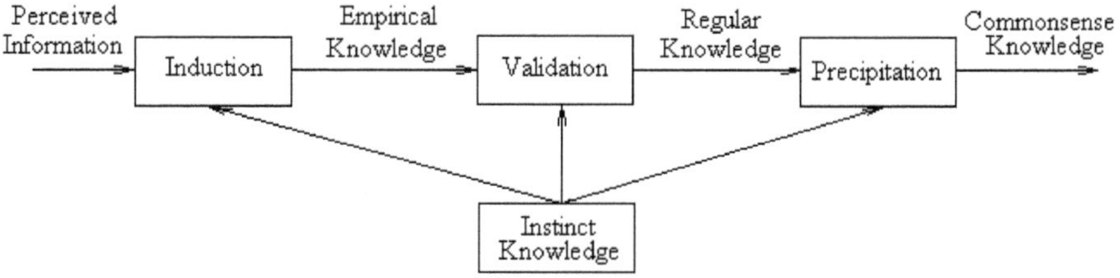

Figure 3. Internal chain of knowledge ecology.

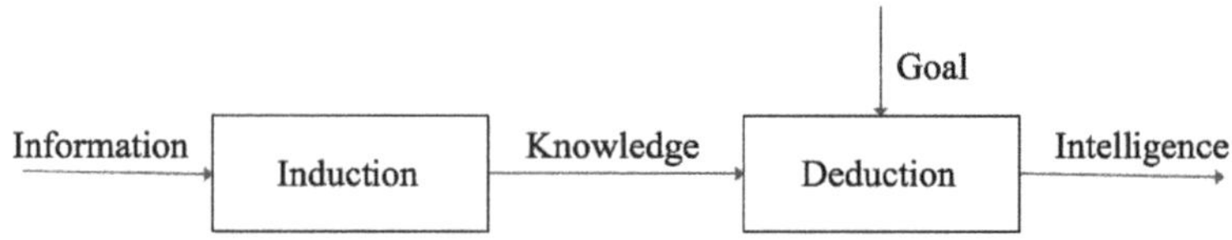

Figure 4. External chain of knowledge ecology.

deductive operations under the guidance of the system's goal, as is shown in Figure 4.

The internal chain and external chain of knowledge ecology, as well as the definition of semantic information (which is the representative of perceived information), are of great significance in the studies of artificial intelligence that will be seen later on in the chapter.

Definition 6 (Human Wisdom). Human wisdom is the special capacity of humans. It is the ability of humans to discover, to define, and to solve the problems faced in the environment by using knowledge and goals of better life. When the old problem is solved, the new problem will be discovered, defined, and solved. By virtue of this ability, the living standard of humankind and the human ability itself are continuously improved and strengthened.

It is clear from Definition 6 that human wisdom consists of two interacting parts: (1) the ability to *discover and to define* the problem that must be solved to improve life, which is termed "implicit wisdom", and (2) the ability to *solve* that problem, which is named "explicit wisdom". The ability of implicit wisdom, that is the ability to identify and frame the relevant problem, is supported by such factors as the subject's goal for better living, knowledge of problem solving, intuition, imagination, inspiration, aesthetics, and the like, whereas the ability of explicit wisdom is supported by operational abilities such as information acquiring, information processing, understanding, reasoning, and execution. It is possible to understand and simulate explicit wisdom (well-defined problem solving) on machines, while it is currently impossible to understand and simulate implicit wisdom on machines as they lack autonomous agency and can identify only very simple problems (like orientation of a robot in space) as they have no general ideas about the world and no developed value systems, or motivations.

Definition 7 (Human Intelligence). The exclusively human capacity named human intelligence is the subset of human wisdom.

Earlier we mentioned that information answers the question "what", knowledge answers the question "why", and intelligence answers the question "how". There are several different concepts of human intelligence, some broader than the others. The most specific view is that intelligence is a purely mental ability of problem solving. According to Encyclopedia Britannica, it consists of learning and adaptation, understanding of abstract concepts, and use of knowledge for manipulation of the environment. However, contemporary studies show the inseparable connection between intelligence and embodiment. For example, Gardner (2006) proposes multiple human intelligences that are closely related to biological functions: rhythmic-musical, visual-spatial, bodily-kinesthetic, verbal-linguistic, logical-mathematical, interpersonal, intrapersonal, and naturalistic, with later additions of existential and moral intelligence, covering different fields of human activities.

Definition 8 (Degrees of Intelligence in Other Living Beings). New insights into behaviors of different living beings, from the very simple unicellular organisms up to mammals, show that problem-solving capacities exist in all living beings in different degrees. Even the simplest bacteria learn and adapt, use "molecular language" for communication in swarms, and use previous experience for improving the chances of survival. So clearly, they possess elements of intelligent behavior (Ben-Jacob, 2014). Maturana and Varela were the first scientists to argue that life=cognition. (Stewart, 1996). Taking into account the different levels of intelligent behavior in living organisms, from single cell to the human level, it becomes easier to understand intelligent behavior in human-made artifacts, also called Artificial Intelligence (AI). Interestingly, the field of AI developed historically based on the goal to mimic human intelligence and only recently problem-solving behavior in other living organisms has been used in robotics in a biomimicry design of simple robots such as "Big dog", which actually is far from the level of intelligence of a living dog.

Definition 9 (Artificial Intelligence). The man-designed intelligence in machines inspired by human intelligence is named artificial intelligence. It relies on knowledge and therefore on semantic information.

It is easy to see that there are close links between artificial intelligence and human wisdom (see Definitions 6–8) on the one hand, and there is also a great difference between them on the other hand. As for the working performance like working speed, working precision, strength of working load, etc., artificial intelligence may be very much superior to human wisdom. However, where creative power is concerned, artificial intelligence cannot compete with human wisdom because of the fact that machines are not living beings, they do not have their own goals for living, and thus do not have the ability to discover and to define for themselves the problems relevant for solving.

Till now we have the following conclusions drawn from the aforementioned discussions:

(1) Semantic information is the result of mapping the joint of syntactic and pragmatic information into the space of semantic information and naming.
(2) Semantic information is the rightful representative of perceived/ comprehensive information.
(3) Semantic information plays a very important role in knowledge and intelligence research.

4. Nucleus of Semantic Information Theory

As the necessary foundation of semantic information theory, the concepts related to semantic information have been thoroughly analyzed in Section 3. In the following section of the chapter, the generative theory, the representation theory, and the applications of semantic information will be discussed successively, following Zhong (2014).

4.1. *Generative theory of semantic information*

Referring back to Definition 2, we realize that syntactic, semantic, and pragmatic information are three components of the perceived information, which are perceived by the subject based on the related object information. So, object information is the real source of semantic information. From Definition 4 it follows that, in contrast to the syntactic and pragmatic information, which are both specific in nature and thus can directly be produced via a subject's sensing and experiencing function, semantic information is abstract in nature and thus cannot directly be produced via a subject's sensing and experiencing. Semantic information is a concept produced by a subject's process of abstraction. Based on Definitions 2 and 4, the model for the semantic information generation process can then be implemented as follows (see Figure 4).

First, let the object (ontological) information be denoted by S, while we denote syntactic, semantic, and pragmatic information by X, Y, and Z, respectively. We assume that there is a comprehensive knowledge base (representing the subject's memory system) that has already stored a big set of knowledge pairs: $\{X_n, Z_n\}$, $n = 1, 2, \ldots, N$, where N is a large positive integer. This means that the nth syntactic information is associated with the nth pragmatic information. In other words, the subject has an abundance of *a priori* knowledge: what kind of syntactic information will have what kind of pragmatic information. X_n and $Z_n, n = 1, 2, \ldots, N$, are in a one-to-one correspondence. The specific process for generating Y from S can be described in the following steps as is shown in Figure 5.

Step 1: ***Generating X from S via Sensing Function***
Mathematically, this is the mapping function from S to X. This function can be realized via sensing organs for humans. It can also be implemented via sensor systems for machines.

Step 2-1: ***Generating Z from X via Retrieval Operation***
Using X, the just-produced syntactic information in step 1, as the key word to search the set $\{X_n, Z_n\}$ in the knowledge base, find the X_n that is matched with X. If such an X_n is found, the Z_n corresponding to X_n in $\{X_n, Z_n\}$ is accepted as the pragmatic information Z. This operation can be realized via the recalling/memorizing function for humans. It can also be implemented via information retrieval system for machines.

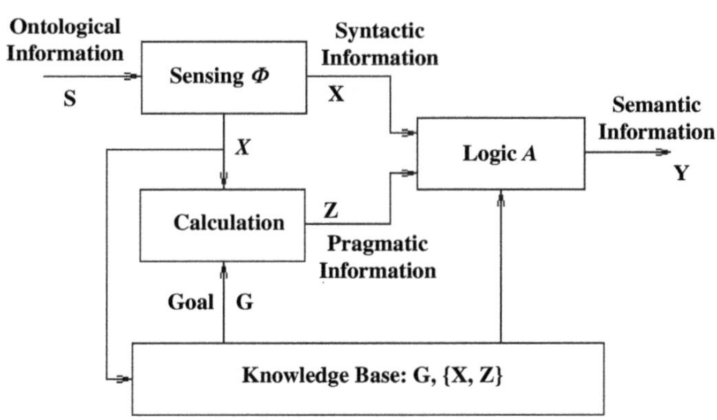

Figure 5. Model for generation of semantic information.

Step 2-2: *Generating Z from X via Calculation Function*
If the X_n matching with X cannot be found from $\{X_n, Z_n\}$, this means that the system is facing a new object never seen before. In this case, the pragmatic information can be obtained via calculating the correlation between X and G, the system's goal:

$$Z \sim \text{Cor. } (X, G). \tag{1}$$

This function can be performed via utility evaluation for humans. It can also be implemented via calculations for machines.

Step 3: *Generating Y from X and Z via Mapping and Naming*
Map (X, Z) into the space of semantic information and name the result:

$$Y = \lambda(X, Z). \tag{2}$$

This function can be performed via abstraction and naming for humans. It can also be implemented via mapping and naming for machines.

It is evident that both the model in Figure 5 and the steps 1–3 clearly describe the principle and the processes for semantic information (with the accompanying syntactic as well as pragmatic information) generation from the object (ontological) information via the system's functions of sensing, retrieving (recalling), calculation (evaluation), and mapping and naming. All of the functions are theoretically valid while being technically effective and feasible.

Equation (2) can also be expressed as

$$y = \lambda(x, z), \quad x \in X, \; y \in Y, \; z \in Z, \tag{3}$$

where X, Y, and Z stand, respectively, for the spaces of syntactic, semantic, and pragmatic information and λ, the operations of mapping and naming.

The correctness of Eq. (2), or Eq. (3), can also be illustrated by a number of practical examples, such as the following:

Suppose a piece of syntactic information (the letter sequence) is given, the corresponding semantic information (y = meaning of the letter sequence) will then depend only on z, the pragmatic information associated according to Eq. (3). For instance, given that x = apple, then y will be different, depending on different z:

If $\{x = \text{apple}, \; z = \text{nutritious}\}$, then $y = \text{fruit}$.

If $\{x = \text{apple}, \; z = \text{useful for broad range of information processing}\}$, then $y = \text{apple computer}$.

If $\{x = \text{apple}, \; z = \text{information processor in a pocket}\}$, then $y = \text{iPhone}$.

The model in Figure 5 and Eqs. (2) and (3) have identified that the concept of semantic information stated in Definition 2 and the mutual interrelationship among syntactic, semantic, and pragmatic information described in Definition 4 are appropriate and legitimate. It is also identified that semantic information so defined can practically be generated either by human subjects or by intelligent systems.

From now on, we will use the concept of semantic information to represent the comprehensive/epistemological/perceived information.

4.2. *The representation and measurement of semantic information*

As has been emphasized before, different from the Shannon theory of information (as a statistically syntactic information theory) where the measure of communication of information is the issue of most concern, here in the case of semantic information theory the most important issue is generally not its quantitative measurement but its qualitative assessment — its ability to provide the meaning of the information and support logical reasoning based on the meaning. The quantitative measurement is in the second place in semantic information theory.

Yet, no matter for quantitative measurement or for qualitative assessment, the representation of semantic information is a common need and an unavoidable foundation. Hence, the first issue before any others we need to discuss is the one of semantic information representation.

4.2.1. *Semantic information without representation — generated by physical computing/morphological computing*

There is an exception to this requirement of explicit representation, see Brooks (1991) "Intelligence without representation", which achieves meaningful behavior in a robotic agent through direct signal processing and morphological computation (such as done for Rumba robot). The idea is that "the best model of the world is the world itself", so the robot does not use internal representations, in contrast to classical representational AI. It uses direct signal transduction and physical transformation obtained automatically by its physical structures. Similar examples are Braitenberg vehicles (Braitenberg, 1984) or Passive dynamic walker robots (McGeer, 1990). Similar processes can be found in simple biological organisms. The theory behind the role of embodiment for behavior and cognition in animals and robots can be found in Pfeifer and Bongard (2006) and Hoffmann and

Pfeifer (2011). In this chapter, we focus on the further development of the classical approach and not on the broadening of the domains of cognition and intelligence, which is a different task.

4.2.2. *Representation of semantic information*

Suppose we have a piece of semantic information $y \in Y$, which may be a word, a phrase, or a sentence, even a paragraph of voice, or video; it can be represented by an N-dimensional vector, where N is a finite and positive integer:

$$y = \{y_n \,|\, n = 1, 2, \ldots, N\}. \tag{4}$$

If $N = 1$, y is the semantic information with a single element and is named atomic semantic information, the minimum unit of semantic information which can be judged as true or false; in other words, is the corresponding syntactic information x_n really associated with the pragmatic information z_n? If $N > 1$, y is named the composite semantic information.

For atomic semantic information y_n, it can be expressed by using the following parameters: its logical meaning y_n and its logical truth t_n, where y_n can be defined by Eq. (3) while t_n can be defined by

$$0 \leq t_n \leq 1, \quad n = 1, 2, \ldots, N. \tag{5}$$

It is clear that the parameter is a kind of fuzzy quantity (Zadeh, 1965):

$$i_n = \{u \in (0, 1), \text{ fuzzy}\}. \tag{6}$$

Thus, for atomic semantic information y_n, $n = 1, 2, \ldots, N$, the comprehensive expression is

$$(y_n; t_n), \quad n = 1, 2, \ldots, N, \tag{7}$$

where y_n is determined by Eq. (3) and t_n by Eqs. (5) and (6).

For N-dimensional semantic information vector y, its truth vector is expressed as

$$t = \{t_n \,|\, n = 1, 2, \ldots, N\}. \tag{8}$$

Correspondingly, its comprehensive expression is

$$\{y_n, t_n \,|\, n = 1, \ldots, N\}. \tag{9}$$

4.2.3. *Measurement of semantic information*

Generally speaking, the quantitative measurement of semantic information is not an important issue. So this subsection is ignored for the time being.

5. Major Applications of Semantic Information

As stated above, the most important issue related to semantic information theory is its ability for expressing the meaning of information and for logical operation and reasoning rather than for its quantitative measurement.

More specifically, the most fundamental applications of semantic information include (1) knowledge organization based on semantic information, (2) learning and cognition based on semantic information, and (3) strategy creation based on semantic information.

5.1. *Knowledge base: Knowledge organization based on semantic information*

The role that knowledge base plays in advanced information systems, like artificial intelligence systems, is almost the same as the role that the long-term memory system of the brain plays in human information system. They are really fundamental in knowledge and intelligence systems.

In the fields of computing and artificial intelligence research, the concept of knowledge base is not new. Hence, we will not discuss the concept of knowledge base itself. But note that the approaches to organizing knowledge in those fields have all been based on the formal data (syntactic information). Here we will show how the knowledge is organized via semantic information, which is the legal representative of comprehensive information, instead of syntactic information alone. In other words, the knowledge base formed on semantic information will be more reasonable than that formed only on the syntactic one.

It has been pointed out in Section 3 of this chapter that the equation $y = \lambda(x, z)$, $x \in X$, $y \in Y$, $z \in Z$ can be regarded as the existential definition of semantic information and can also be seen as the constructive definition of semantic information. This indicates that what determines the semantic information, y, is exactly its accompanied 2-tuple array of syntactic and pragmatic information, (x, z).

For example, if given that the 2-tuple array is $x =$ something with round shape like a small ball, red color, more or less 0.2 kilo weight, $z =$ nutritious, good taste, and "good for health", the 2-tuple array (x, z) can then be mapped into the space of semantic information and named as "$y =$ fruit" (see Figures 2 and 5).

Once again, the meaning of a piece of semantic information, y, can well be expressed by its associated 2-tuple array of syntactic and pragmatic information, (x, z). Therefore, no matter whether it is an object or an action, its semantic information (meaning) can well be expressed by its associated 2-tuple array of syntactic and pragmatic information:

$$\text{Name of an object} = (\text{feature set of its form, feature set of its utility}). \tag{10}$$

$$\text{Name of an action} = (\text{feature set of its form, feature set of its function}). \tag{11}$$

The more precise the description of the feature sets for the 2-tuple array, the more precise the description of the semantic information. There must, of course, be a compromise between the requirement of preciseness and that of briefness.

The method of knowledge representation based on semantic information is especially suitable for the description of "knowledge refining and abstraction", which is the real nucleus for human thinking and for organizing the large amount of knowledge into a large-scale organic system of knowledge — the knowledge base.

Figure 6 shows the principle of knowledge organization based on semantic information. There are n pieces of knowledge, name-1, name-2, ..., name-n, each of which has its form feature and utility feature. The form feature in common among all form features and the utility feature in common among all utility features are abstracted to the new name — the class name in a higher level. Evidently, the class name of the organization is more abstract than that in a lower level.

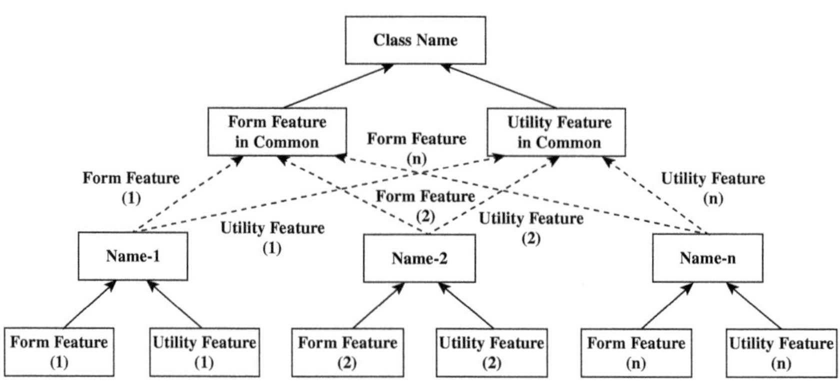

Figure 6. Principle of knowledge organization based on semantic information.

It is easy to find numerous practical examples of implementing the principle of knowledge organization based on semantic information, as shown in Figure 6, in academy and everyday life.

For example, ball-point-pen, brush-pen, and pencil, respectively, have their own form features and utility features, namely form feature (1) and utility feature (1); form feature (2) and utility feature (2); and form feature (3) and utility feature (3). Since they have certain form features in common and certain utility features in common, they have been named "pen" as their class name. The name of pen is more abstract than that of ball-point-pen, brush-pen, and pencil.

By utilizing the same principle of knowledge organization based on semantic information, it is possible to organize great amounts of knowledge into a large-scale and multi-level knowledge base. At the lower level of the base, the names are more specific, while at the higher level of the base the names are more abstract; all the entities under their class name within the base will have certain form and utility features in common. This is the principle of building up the knowledge base and is an important function that semantic information plays. Compared with the one based on syntactic information, this principle is apparently much more reasonable and effective.

5.2. *Cognition: Learning and understanding based on semantic information*

Learning, understanding, cognition, and evolution are a group of fundamental concepts that are mutually interlinked and are the most important abilities for humans as well as for machines. Without these abilities, humans would not be able to adapt to the environment and would not be able to make such great progress as they have already made.

There have been many, and yet different, explanations of the concept of learning in history. In the earlier days, learning was explained as based on memorizing and application of rules, while in the modern time, learning is explained as a statistical processing based on large volume of data and high-speed computing (Michalski *et al.*, 1983, 1986).

What we would like to propose here is the model that defines cognition as knowledge acquiring via learning and understanding based on semantic information. This is a new definition of cognition because all previous definitions of cognition are based only on syntactic information.

If referring back to the history of the development of human learning, the definition of cognition proposed here will turn out as a good choice against other definitions that existed before.

There are three major stages of the development of cognition during the lifetime of a human being (Zhong, 2013c).

(1) *Accepting and remembering: The infants' mode of learning and cognition*

In the infant stage, a child lives in a small world created by its parents and has no ability of independent thinking. The parents in the small world play the role of absolute, and also reliable, authority. As the result, what the parents teach is naturally accepted and remembered with no doubt. This is the most elementary mode of learning in the earliest stage of one's life during which a certain number of the very basic terms and rules are learned and stored in an infant's long-term memory as the base for further learning in later stages. This is also the mode of rule-based learning in a machine.

(2) *Following the public: The teenagers' mode of learning and cognition*

As youngsters, they begin to step into society and live both inside and outside of the family. Parents are no longer the only authorities for them. Instead, teachers, classmates, and various kinds of public leaders could tell them many things that they would never have heard before within their family. Consequently, teenagers could learn much more from the outside world than from their parents. On the other hand, however, they are still not mature enough to understand the complex problems of society. Therefore, a natural mode for them to adopt for learning is to follow the general trend of the public in society. This is just the mode of statistical learning in machines.

(3) *Autonomous understanding: The adults' mode of learning and cognition*

As adults, humans are mature enough and therefore able to analyze things, learning from normal education and various kinds of practical experiences. They establish the ability and habit of autonomous thinking and analysis and hence would accept something only based on their own understanding. It is by this mode of thinking and learning that one is capable of innovating technology and creating new theories. This is the most powerful and useful mode of thinking and learning for human cognition. Unfortunately, there has not been such a kind of mode of learning implemented in a machine so far.

It is worth pointing out that the three modes of learning discussed above are not independent of each other. In fact, through the mode of *accepting and remembering*, one learns much of the commonsense knowledge and lays the very foundation for further learning; through the mode of *following the public*, one learns much of the empirical and regular knowledge that is necessary for living life successfully in society; and finally, through the mode of *autonomously understanding*, one could be able to make contributions to social development. So, accepting and remembering, following the public, and autonomously understanding do form a kind of evolutionary ecological process of human learning following the ability of growth. In other words, autonomously understanding is the most advanced mode of learning.

Technologically, the mode of human learning based on *accepting and remembering* has been implemented as *mechanical learning* in the theory of machine learning, whereas the mode of human learning based on *following the public* has been implemented as *statistical learning* in the theory of machine learning. However, due to a lack of studies in semantic information theory in the past, there has never been, till the present, any theoretical study or technical implementation for the most advanced and the most useful mode of learning based on the *autonomous understanding*.

Owing to the development of computing and database technology, currently statistical learning has become the dominant approach to machine learning. However, the performance that statistical learning provides is far from satisfactory. This is because statistical learning is based only on syntactic information and is unable to use semantic information.

As has been emphasized many times earlier, semantic information is the legitimate representative of comprehensive information. Therefore, having semantic information implies knowing its form, meaning, and utility with respect to the subject and this is what we mean by a complete understanding of a piece of information.

The technological model of learning and cognition with autonomous understanding based on semantic information is shown in Figure 7.

The input of the model in Figure 7 is not syntactic information as usual, but semantic information; the central algorithm performing the function of learning from the semantic information is the inductive type, and the output is the knowledge, the result of cognition. The comprehensive knowledge base is the foundation of the model. It provides the system with various kinds of knowledge such as the algorithm for induction, the goal for learning, the set of {form feature; utility feature}, and other kinds of knowledge that

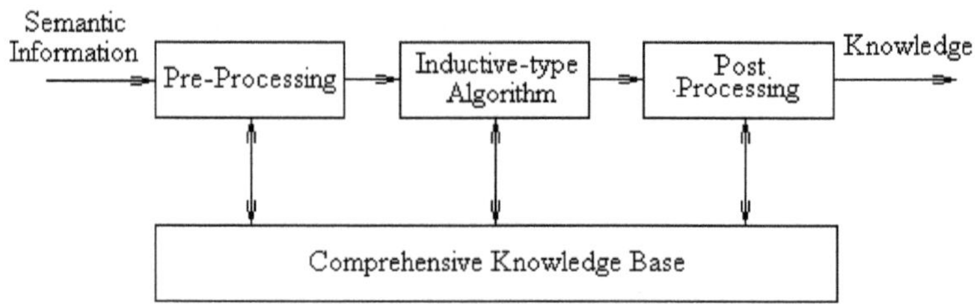

Figure 7. Cognition with autonomous understanding based on semantic information.

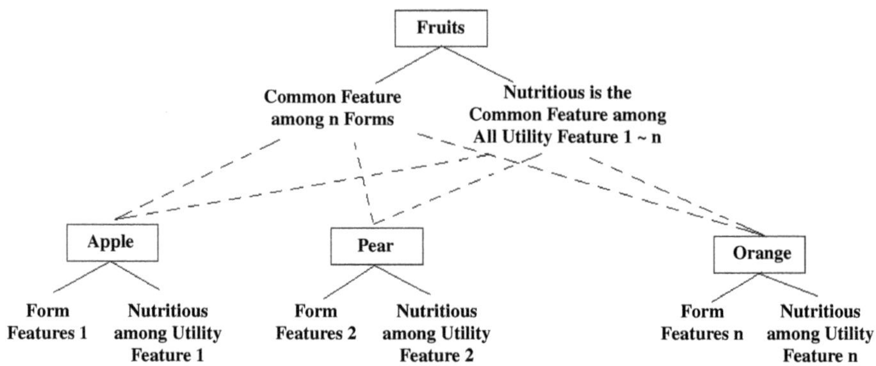

Figure 8. An example of inductive cognition based on semantic information.

are needed for inductive learning. The pre-processing and post-processing are units performing the auxiliary functions like information expression, knowledge expression, and the like.

The choice of induction as central algorithm of the model is natural because all kinds of information are facts and specific phenomena while all kinds of knowledge are rules and essences. Any rules and essences can be established only through the operation of induction and abstraction. Figure 8 is an example of cognition based on semantic information.

The inductive cognition based on semantic information expressed in Figure 8 consists of the following necessary steps:

(1) Given a number of samples of semantic information (names): apple, pear, ... and orange, each of them is described by the corresponding form features (the description of syntactic information) and utility

features (the description of pragmatic information) as is seen in the lower part in Figure 8.

(2) The common features for the forms and utilities among all the names are found through the operation of induction. The joints of the common features for form and utility are then mapped into the knowledge space and named a "fruit". It is found that all types of fruits have the same utility feature, which is being "nutritious". This can be seen in the higher part in Figure 8.

(3) Consequently, the piece of the knowledge that can be derived from step (2), operation of inductive cognition, is the sentence of "Fruits are nutritious". This is because there is a common feature of being "nutritious" in the utility features of "fruit".

It is not difficult to have many more examples showing the same principle of inductive cognition based on semantic information.

Comparing the three modes of cognition/learning, it is evident that the inductive cognition with "autonomous understanding" based on semantic information is the most advanced approach to learning and cognition among the three. This is true not only for human learning and cognition but also for machine learning and cognition. On the other hand, however, there is an organic relation among the three models of cognition/learning because the knowledge learned via the "accepting and remembering" and "following the public" has been regarded as the fundamental knowledge stored in the comprehensive knowledge base in Figure 7. In other words, the three models of cognition/learning are very well coordinated.

5.3. *Strategy creation based on semantic information*

Referring back to the model of information ecology shown in Figure 1 we can clearly see that intelligence (with intelligent strategy as its major expression) can only be created through the process of converting the semantic information to a strategy supported by the related knowledge and guided by the subject's goal. This process can be described by the following conversion steps: (semantic) information → knowledge → intelligence.

This conversion has already been discussed in the last subsection and will become the thematic topic of this subsection. The principle of the model for implementing this conversion, or the relationship between intelligent strategy and semantic information, is illustrated in Figure 9.

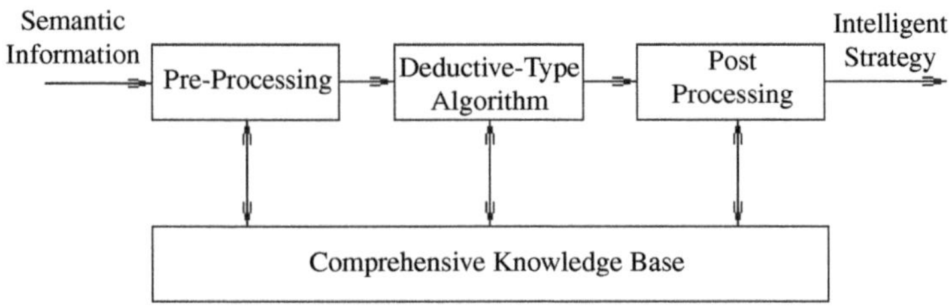

Figure 9. Strategy production (generation) based on semantic information.

As can be seen from the comparison between Figures 7 and 9, the models for the two kinds of information conversion (to knowledge vs. intelligent strategy) look very similar to each other. The major differences between them lie in the following: (1) The final product (output) is different; knowledge is the final product in a cognitive process (Figure 7) while intelligent strategy is the final product in the intelligence process (Figure 9). (2) The inductive-type algorithm is the central operation in the cognitive process (Figure 7), while the deductive-type algorithm is the central operation in the intelligence process (Figure 9). (3) The contents of the comprehensive knowledge base as well as the pre-processing and post-processing are also accordingly different for the two models.

The function of intelligent strategy is to solve the problem the subject is facing. As is well known, inductive-type algorithms are suitable for refining knowledge from semantic information, whereas deductive-type algorithms are suitable for creating intelligent strategy from knowledge, semantic information, and the subject's goal for problem solving.

Recalling the problem solving in Artificial Intelligence, there is, for example, a well-known algorithm called Production System, which can be described as follows (Nilsson, 1982):

1. DATA← initial database;
2. Until DATA satisfies the termination condition, do:
    i) Begin;
    ii) Select some rules, R, in the set of rules that can be applied to DATA;
    iii) DATA← result of applying R to DATA.

In our case, the initial database, DATA, the termination condition, and the set of rules (or rule base) in Production System, respectively, correspond to the following:

(1) The semantic information stands for the subject's understanding about the problem — the object information, which is the original state of the problem that the subject wants to solve and is the starting point for the intelligent strategy to run. This is just the "initial database" in Production System in AI.
(2) The subject's goal for problem solving stands for the final state of the problem to be solved and is therefore the final point for intelligent strategy. This is equivalent to the termination condition in Production System in AI. Note that the expression for the subject's goal should also have the format of {form feature of the goal, utility feature of the goal}. This means that the subject's goal is regarded as a kind of semantic information.
(3) The knowledge supplied by the comprehensive knowledge base presents the constraints to the finding of the intelligent strategy: what it can do and what it cannot do within the space of strategy finding. This is equivalent to the "Rule Base" in product System in AI. Again, the expressions for the rule should also follow the format: If (form features, utility features), Then (form feature, utility feature).

It is seen that both the starting state and the final state are given from the problem description. The task for the strategy creation is to find a path going from the starting state to the final state, subject to the constraints imposed by the knowledge stored in the knowledge base. The strategy can be implemented as a series of rules' selection and application where the rules are written as "If ..., Then ..." defined by the knowledge. This is a typical operation of a deductive process.

Obviously, the strategy created for intelligent solving of a given problem will be the result of some complex function of semantic information, knowledge, and the subject's goal. Using the symbols I_{sem}, K, and G to denote semantic information, knowledge, and the goal, respectively, we can express the strategy as

$$\text{St} = f(I_{\text{sem}}, \{K\}, G), \tag{12}$$

where St stands for the strategy and f for the complex function that needs the expression from operations of mathematical logic (Li, 2007) and universal logic (He, 2001).

Furthermore, as the strategy is created and applied to the problem, an error between the practical result of strategy exertion and the subject's goal assigned beforehand may occur. This error should be considered as the new information for problem solving and should be fed back to the input of the system for learning and optimizing the strategy. This kind of feedback loop of learning and optimizing may need to be performed many times to gradually decrease the error and finally reach a satisfactorily low level that can be accepted. This can be seen again from the model of information ecology in Figure 1. Naturally, the expression of error information here should have the format of {form feature, utility feature}, similar to the format of semantic information expression. This will be more convenient, and more helpful, for the learning and optimization processes.

It can be summarized based on the discussions carried out that the theory of semantic information does play the greatest role in the entire discipline of information science studies and cannot be replaced by any componential information theory, such as the Shannon (syntactic) theory of information communication, or classical theory of semantic information. The theory of information science is incomplete without a proper theory of semantic information. Proper theory of semantic information is only possible if information is seen as a part of the ecology in which it is produced, consumed, transformed, processed, and combined, in the circle from the data (object information/potential information), to syntactic, semantic, and pragmatic information, knowledge production, and generation of intelligent strategies in a cognitive/intelligent agent/subject.

6. Concluding Remarks

The most fundamental purpose for humans to acquire information is to utilize information for solving the problems they are facing. To this end, understanding the meaning of the information (semantic information) and evaluating the use value of the information (pragmatic information) are absolutely necessary. Without the semantic information and pragmatic information related to the problem, any decision for solving the problem would be blind. However, the unique theory established so far is the mathematical theory of communication by Shannon (1948), which is a statistical theory of communication of syntactic information, and has nothing to do with the theories of semantic and pragmatic information. The semantic and pragmatic aspects of Shannon information are left implicit: they concern only the problem of the transmission of a signal via a noisy channel.

Believing that it should not be acceptable for humans to live in a world that has only form factors while having no meaning, and no value aspects, we have made efforts to analyze the difficulties and setbacks that former approaches have encountered. We made detailed investigations of the methodologies suitable for information studies and presented the new methodology of ecological evolution in the information discipline as the methodology of semantic information ecology.

Based on the methodology of information ecology, the concepts related to information and particularly to semantic information have been re-examined, and the definitions of object information and perceived information have been more strictly and precisely clarified. The mutual relationships among the syntactic, semantic, and pragmatic information and the principles of genesis of semantic information for an agent have been exposed. The representations of semantic information have been set up, and the important applications of semantic information in such fields as knowledge expression and organization (knowledge base architecture), learning, understanding, cognition, and intelligent strategy creation have been re-explored. The advantages of these applications are obvious compared with the existing approaches. Based on all the above, the framework of semantic information theory has been formulated.

For future work, there are several immediate applications of this approach. One important field that could be explored with the same ecological methodology is non-representational semantic information. Starting with the world as its own best representation, one can follow the equivalent process of meaning generation, this time expressed through underlying physical processes such as natural computation/morphological computation (Dodig-Crnković, 2008, 2017). The holistic ecological approach even here will offer valuable new insights which are impossible to reach when information processing is studied partially and in isolation from its ecology.

The dynamics (time aspects) of information are an issue worth further exploration with our approach, taking into account both the short-time as well as long-time perspectives (the dynamic view and the evolutionary approach) within ecological theory of information.

In the case of non-representational semantic information, taking living organisms as a model of information processors with cognitive and intelligent abilities that develop increasing complexity as evolution proceeds, we can apply the ecological approach even to a lower level, non-symbolic information processing, in which organisms use physical properties of the world to directly morphologically compute their own behaviors and increasingly complex structures.

The present approach is focused on the information processes and structures within a cognizing/intelligent agent. The straightforward generalization is due to the distributed system of cognizing/intelligent agents.

Furthermore, there are still some open issues remaining in the development of the ecological approach itself. One is the development of the quantitative measurement of semantic information, although it is not a central issue. The other, more essential development area is the logic theory needed for supporting the inference based on semantic information. The current theory of mathematical logic is not sufficient, and the new and more powerful theories of logic, such as universal logic and dialectic logic, for instance, are eagerly expected.

Acknowledgments

The studies on semantic information theory presented here have gained financial support, respectively, from the China National Natural Science Foundation and the Swedish Research Council. The authors would like to express their sincere thanks to the Foundation and the Council. We would also like to acknowledge the constructive criticism of Marcin Schroeder on the previous version of this manuscript.

References

Adriaans, P. and van Benthem, J. (2008). *Handbook of Philosophy of Information*, Elsevier Science Publishers, Amsterdam.

Bar-Hillel, Y. (1964). *Language and Information*, Addison-Wesley, Reading, MA.

Ben-Jacob, E. (2014). My encounters with bacteria — learning about communication, cooperation and choice. *Phys. Biol.* 11, 053009.

Bateson, G. (1972). *Steps Towards an Ecology of Mind*, Jason Aronson Inc.

Bertalanffy, L. V. (1968). *General System Theory*, George Braziller Inc.

Braitenberg, V. (1984). *Vehicles: Experiments in Synthetic Psychology*, MIT Press, Cambridge, MA.

Brillouin, A. (1956). *Science and Information Theory*, Academic Press, New York.

Brooks, R. (1991). Intelligence without representation, *Artif. Intell.* 47, 139–159.

Burgin, M. (2010). *Theory of Information: Fundamentality, Diversity and Unification*, World Scientific, New York.

Carnap, R. and Bar-Hillel, Y. (1953). Semantic information, *Brit. J. Phil. Sci.* 4, 147–157.

Chaitin, G. J. (1987). *Algorithmic Information Theory*, Cambridge University Press, Cambridge.

Dodig-Crnković, G. (2006). *Investigations into Information Semantics and Ethics of Computing*, Mälardalen University Press.

Dodig-Crnković, G. (2008). Semantics of information as interactive computation, in *Fifth International Workshop on Philosophy and Informatics*, Moeller, M., Neuser, W. and Roth-Berghofer, T. (eds.), Kaiserslautern. DFKI Technical Reports. Springer, Berlin.

Dodig-Crnković, G. (2017). Nature as a network of morphological infocomputational processes for cognitive agents, *Eur. Phys. J. Special Topics*, doi:10.1140/epjst/e2016-60362-9; *Eur. Phys. J.* 226, 181–195.

Floridi, L. (2011). *The Philosophy of Information*, Oxford University Press, Oxford.

Floridi, L. (2017). Semantic conceptions of information, *The Stanford Encyclopedia of Philosophy* (Spring, 2017).

Gardner, H. (2006). *Multiple Intelligences: New Horizons in Theory and Practice*, Basic Books.

Gottinger, H. W. (1973). Qualitative Information and Comparative Informativeness, *Kybernetik*, 13, 81.

Hoffmann, M. and Pfeifer, R. (2011). The implications of embodiment for behavior and cognition: Animal and robotic case studies, in *The Implications of Embodiment: Cognition and Communication*, Tschacher, W. and Bergomi, C. (eds.), Imprint Academic, Exeter, pp. 31–58.

He, H. C. (2001). *Principles of Universal Logic*, Science Press.

Kolmogorov, A. N. (1968a). Three approaches to the quantitative definition of information, *Int. J. Comput. Mach.* 2, 157–168.

Kolmogorov, A. N. (1968b). Logic basis for information theory and probability theory, *IEEE Trans. Inform. Theory* 14, 662–664.

Li, W. (2007). *Mathematical Logic: Basic Principles and Formal Calculus*, Science Press.

Maturana, H. and Varela, F. J. (1987). *The Tree of Knowledge*, Shambhala, Boston.

McGeer, T. (1990). Passive dynamic walking, *Int. J. Robot. Res.* 9(2), 62–82.

Michalski, R. S., Carbonell, J. G. and Mitchell, T. M. (eds.) (1983). *Machine Learning: An Artificial Intelligence Approach*, Vol. 1, Morgan Kaufmann, California.

Michalski, R. S., Carbonell, J. G. and Mitchell, T. M. (eds.) (1986). *Machine Learning: An Artificial Intelligence Approach*, Vol. 2, Morgan Kaufmann, California.

Millikan, R. G. (2002). *Varieties of Meaning*, MIT Press, Cambridge.

Morris, C. W. (1971). *Writings on the General Theory of Signs*, De Gruyter Mouton, The Hague.

Nilsson, N. J. (1982). *Principles of Artificial Intelligence*, Springer-Verlag, New York.

Peirce, C. S. (1934). *Collect Papers: Volume V. Pragmatism and Pragmaticism*, Harvard University Press, Cambridge, MA.

Pfeifer, R. and Bongard, J. C. (2006). *How the Body Shapes the Way We Think: A New View of Intelligence*, MIT Press, Bradford Books.

Saussure, F. de (2001). *Course in General Linguistics*, Ryan, M. and Rivkin, J. (eds.) Blackwell Publishers.

Schroeder, M. J. (2011). Semantics of information. Meaning and truth as relationships between information carriers, in *The Computational Turn: Past, Presents, Futures? Proceedings of the IACAP*. Ess, C. and Hagengruber, R. (eds.), 4–6 July, 2011, Aarhus University. Verlagshaus Monsenstein Und Vannerdat, Münster, pp. 120–123.

Schroeder, M. J. (2012). Search for syllogistic structure of semantic information, *J. Appl. Non-Class. Log.* 22(1–2), 83–103, doi:10.1080/11663081.2012.682442.

Shannon, C. E. (1948). A mathematical theory of communication, *Bell Syst. Tech. J.*, 47, (Part I), 379–423; Part II, 632–656.

Shannon, C. E. and Weaver, W. (1949). *The Mathematical Theory of Communication*, University of Illinois Press, Urbana, IL.

Stewart, J. (1996). Cognition = Life: Implications for higher-level cognition, *Behav. Process.* 35, 311–326.

Smokler, H. (1966). Informational content: A problem of definition, *J. Philos.* 63(8), 201–211.

Wiener, N. (1948). *Cybernetics*, John Wiley & Sons, New York.

Zadeh, L. A. (1965). Fuzzy sets theory, *Inform. Control* 8, 338–353.

Zhong, Y. X. (2013a). *Principles of Information Science*, Beijing University of Posts and Telecommunications Press (Earlier editions 1988, 1996, 2002, 2005).

Zhong, Y. X. (2013b). Information conversion: The integrative theory of information, knowledge, and intelligence, *Sci. Bul.* 85(14), 1300–1306.

Zhong, Y. X. (2013c). Methodology for NLU based on comprehensive information, in *Some Issues in Chinese Information Processing*, Bo, X. (ed.), Science Press.

Zhong, Y. X. (2014). *Principles of Advanced Artificial Intelligence*, Science Press.

PART II

Information as a Natural Phenomenon

Chapter 6

The "No Information at a Distance" Principle and Local Mathematics: Some Effects on Physics and Geometry

Paul Benioff

Physics Division, Argonne National Laboratory, Argonne, IL, USA

pbenioff@anl.gov

Local mathematics assumes the existence of number structures of different types, vector spaces, etc. localized at each space-time point. Relations between number structures at different locations are based on two aspects: distinction between two so far conflated concepts, number and number value, and the "No information at a distance" principle. This principle forbids the choice of the value of a number at one location to determine the value of the same number at another location. Value changing connections, related to a real valued field, g, move numbers between structures at different locations. The effect of the g field, or its exponential equivalent, $g(y) = e^{\alpha(y)}$, on numbers extends to other mathematical structures, vector spaces, etc. The presence of α affects theoretical descriptions of quantities in physics and geometry. Two examples are described, the effect on the Dirac Lagrangian in gauge theory and the effect on path lengths and distances in geometry. The gradient field of α, $\vec{A}$, appears in the Lagrangian as a spin 0, real scalar field that couples to the fermion field. Any value for the mass of $\vec{A}$ is possible. The lack of direct experimental evidence for the presence of the g or α field means that the field must be essentially constant within a local region of the cosmological universe. Outside the local region, there are no restrictions on the field. Possible physical candidates (inflaton, dark matter, dark energy) for α are noted.

1. Introduction

The relation between the foundations of mathematics and physics is a subject of much interest. This is well expressed by Wigner (1960) and others (Omnes, 2011; Plotnitsky, 2011; Hitchin, 2007; Dirac, 1938; von Neumann, 1947)[1] in papers on the unreasonable and reasonable effectiveness of mathematics in science. This effectiveness appears to be a problem, especially if one accepts the Platonic views of mathematical elements as having an ideal existence outside of space and time. From this perspective, it is not understandable why mathematics should have anything to do with the description of physical systems that move and interact inside a space-time arena. This leads to the wish for a coherent theory of physics and mathematics together as one coherent whole rather than as two separate disciplines (Benioff, 2002, 2005).

One approach to this problem is to replace global mathematics existing outside of space and time with local mathematics. In this approach, mathematical systems of different types are considered to be structures (Shapiro, 2008). Here, these structures are considered to be local with separate structures of each type existing at each point of space-time. Relations or maps between systems of different types correspond to maps or relations between structures of different types at the same or different locations in space-time. This puts mathematical structures and physical systems on a more equal footing, in that they both exist in space and time.

The idea of local mathematical structures has a precedent, in that gauge theories are based on vector spaces localized at different points of space-time (Montvay and Münster, 1994). To each point, x, is associated a vector space, $\bar{V}_x$.

Relations between states in vector spaces at different locations are influenced by the "No information at a distance" principle (Montvay and Münster, 1994; Mack, 1981). This principle says that information used to choose a basis set of states in $\bar{V}_x$ cannot influence the choice of a basis set of states in another vector space, $\bar{V}_y$, at a different location, y. This condition leads to the existence of unitary gauge transformations between states at different locations. These transformations, first investigated by Yang and Mills (1954) for non-Abelian gauge theories, led to the description of the standard model for elementary particles in physics.

[1]Wikipedia: Relationship between mathematics and physics, see for these and additional references.

An aspect of this that seems strange is that, as mathematical structures, vector spaces do not exist by themselves. They are closely associated with scalar fields such as the real or complex numbers. The association of local vector spaces with a single global scalar field outside of space-time seems problematic (Benioff, 2013).

For this reason, it seemed worthwhile to replace a single global scalar field by separate local scalar fields. These fields, as structures for numbers of different types, become the scalars for the vector spaces, both at the same location (Benioff, 2011).

This leads directly to the idea that if number structures of different types are local, this locality should extend to mathematical structures for all types of systems that include numbers in their axiomatic descriptions. There are many of these types of structures in mathematics. Group representations, algebras, and matrices, all types of vector spaces, are examples of this type of structure.

An aspect of number structures that is relevant here is the observation that the concept of number is different from that of number value. For each type of number, there exist many structures in which the same number has different values. For real numbers, there is an infinity of real number structures, one for each real number. For complex number structures, there is an infinity of structures, one for each complex number.

These structures are referred to as scaled structures with associated real or complex scaling or value factors. These scaled structures have one base set of numbers in common. With one exception, the numbers in the base set have no intrinsic value. The value of a number is determined by the scaling factor and the position (for natural numbers and integers) of the number in the structure being considered. The exception is the number 0. Its value is 0 in number structures for all scaling factors.

The concept of scaling for number structures is easily extended to structures of all types that include numbers in their description. For example, one can have scaled vector spaces, scaled algebras, etc. Most of the work done so far has been limited to numbers of different types and to vector spaces.

This distinction between number and number value and the resulting existence of number structures with different scaling factors is an example of the distinction between Diophantine and non-diophantine arithmetic (Burgin, 1997, 2001). The number structure with a scaling factor of 1 would correspond to the usual or Diophantine arithmetic. The structures with scaling factors different from 1 correspond to non-diophantine arithmetic.

The description of natural number structures with different scaling factors given in this work is an example of projective arithmetic (Burgin, 2010).

As defined here, the scaling factors are linear, in that they preserve order relations between numbers. A more general functional type of scaling has been used by Czachor (2016) to suggest that the relativity of arithmetic may be a symmetry of physics. It was also suggested that this relativity may be a source of dark energy (Czachor, 2017). Details of the use of this more general type of scaling for physics and geometry remain to be investigated.

The description used in gauge theories to describe relations between vector spaces at different locations by unitary gauge transformations is extended here to number structures. The "No information at a distance" principle (Montvay and Münster, 1994; Mack, 1981) forbids the use of the informational choice of the value of a number at one location to determine the value of the same number at a different location. In other words, the choice of the value of a number at one location does not determine the value of the same number at a different location.

This leads to a location-dependent valuation of number structures. The valuation is implemented by the introduction of a scalar scaling or value field, g. This field associates a scaling value $g(y)$ to the local number structures at y. The relation between numbers and number structures at different locations is determined by a parallel transform or connection that maps numbers and structures at one location to those of another. The connection depends on the values of g at the different locations. The connection is the number equivalent of gauge transformations in gauge theory.

As might be expected, the presence of the g field affects theoretical descriptions of many properties of systems in physics and geometry. This is especially the case for properties described by integrals or derivatives over space or space-time. Reconciliation of these theory predictions with experiment results places restrictions on the g field.

This work has been described in detail elsewhere (Benioff, 2012, 2013, 2015, 2016). Here, two examples, one from physics and the other from geometry, will be given to illustrate the effects of local mathematics and the presence of the g field. The next section is a brief outline of the effect of scaling on numbers and on other mathematical structures. The following two sections describe the effects of g field on gauge theories and geometry. It will be seen that the presence of the g field introduces a new scalar field into the Dirac Lagrangian. The field also affects distances between points. The final section describes restrictions on the g field imposed by experiment.

2. Number Scaling

2.1. *Natural numbers*

The description of number scaling begins with the idea that the concept of numbers is separate from that of number value. The best way to see this is to consider the natural numbers,

$$0\ 1\ 2\ 3\ 4 \cdots = N_1. \tag{1}$$

Here, as symbols, $0, 1, 2, 3, 4$ are names for numbers.[2] From now on, the distinction between name and number will be dropped.

The numbers in Eq. (1) are well ordered with 0 the first, 1 the second, 2 the third, and so on in the well ordering. The value of a number is determined by its position in the well ordering. Thus 0 has value 0, 1 has value 1, 2 has value 3, and so on.

Consider the subset

$$N_2 = 0, 2, 4, 6 \cdots$$

of even numbers. This set inherits the well ordering of the numbers in Eq. (1). 0 is first, 2 is second, 4 is third, 6 is fourth, and so on. The values of the numbers in this set are determined by their position in the well ordering. Thus, 0 has value 0, 2 has value 1, 4 has value 2, 6 has value 3, and so on.

This shows that, with the exception of the number 0, the numbers have no intrinsic value. Their values are determined by the set containing them. The number 2 has value two in the set of Eq. (1). It has value 1 in the subset, $0, 2, 4, 6$. The number 4 has value 4 in Eq. (1). It has value 2 in the subset.

This description can be extended to subsets N_n that contain every *nth* number in N_1. For $n = 4$, the number 4 can have three possible values, 4 in N_1, 2 in N_2, and 1 in N_4. The number of number values a number can have is determined by the prime factors of the number. If the number is a prime number, it has just two values. If it is not a prime number, it can have more values. For example, the possible values of the number 30 can be expressed by the equation

$$30_1 = 15_2 = 10_3 = 6_5 = 5_6 = 3_{10} = 2_{15} = 1_{30}. \tag{2}$$

---

[2]Each of the numbers along with its name is a named set (Burgin, 2003).

Expressions such as 10_3 mean that the value of the number is 10 in N_3. This representation of numbers and their values will be used much from now on.

This simple example of the effects of separation of number from number value extends to the other number types, integers, rational, real, and complex numbers. For rational, real and, complex numbers, the base sets B_{Ra}, B_R, and B_C remain unchanged under scaling. Subsets are not needed. This is a result of the observation that, unlike the natural numbers, these number types are closed under division.

2.2. *Real and complex numbers*

In what follows, it is useful to consider mathematical systems of different types as structures (Shapiro, 2008) satisfying a set of relevant axioms. Relations between structures are included in the axiomatic descriptions. The closure of vector spaces under scalar multiplication is an example. Here, the description is limited to real and complex numbers as these are most relevant to physics and geometry. The structures for real and complex numbers are given by

$$\bar{R} = \{B_R, \pm, \times, \div, <, 0, 1\} \quad \text{and} \quad \bar{C} = \{B_C, \pm, \times, \div, ^*, 0, 1\}. \tag{3}$$

In these definitions, B_R and B_C are base sets, $\pm, \times, \div$ are the basic field operations, $<$ is an order relation, * denotes complex conjugation, and 0 and 1 are constants. The structures satisfy axioms for real and complex numbers.

Let r and c denote real and complex number values. Real and complex number structures scaled by r and c are given by

$$\bar{R}_r = \{B_R, \pm_r, \times_r, \div_r, <_r, 0_r, 1_r\} \text{ and } \bar{C}_c = \{B_C, \pm_c, \times_c, \div_c, (^*)_c, 0_c, 1_c\}. \tag{4}$$

In these structures, 1_r and 1_c are numbers in $\bar{R}^r$ and $\bar{C}^c$ with value 1. Also, $0_r = 0 = 0_c$. Let Op denote any one of the four binary field operations, $+, -, \times, \div$. Then Op denotes an operation value and Op_r and Op_c denote the operations in $\bar{R}^r$ and $\bar{C}^c$ with value Op. Note that for multiplication and division, if $r \neq s$, then Op_r differs from Op_s by a scaling factor.[3]

The fact that r and c can be any real and complex number values leads to collections of scaled structures, one for each scaling factor. The base

[3]Since rational numbers, integers, and the natural numbers are substructures of the real numbers, structures for these number types scaled by real numbers can be defined.

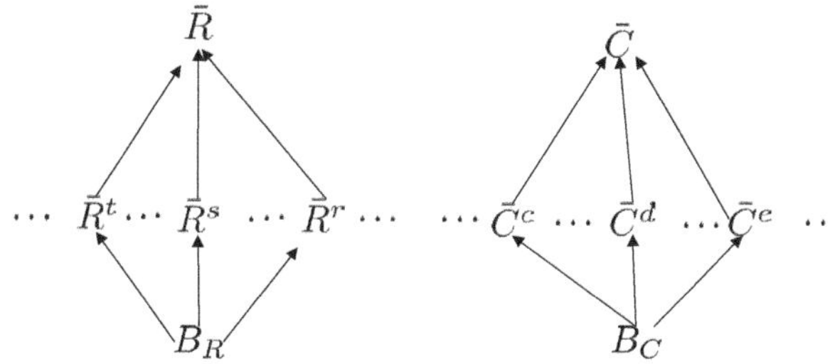

Figure 1. Representations of real and complex number structures for all non-zero real and complex scaling factors, respectively. Three out of an infinite number of structures are shown for both real and complex numbers. The base sets, B_R and B_C, and the number value structures, $\bar{R}$ and $\bar{C}$, are the same for all scaled real and complex number structures.

sets remain the same for all scaling factors. These collections and the real and complex number value structures for real and complex numbers are illustrated in Figure 1 as collections over the two common base sets. Real and complex number value structures, $\bar{R}$ and $\bar{C}$, which are the same for all the scaled structures, are included in the figure.

The collections can be defined by

$$F_R = \bigcup_t \bar{R}^t \quad \text{and} \quad F_C = \bigcup_c \bar{C}^c \tag{5}$$

for all real and complex scaling factors, t and c, different from 0. The base sets are included in the definitions.

There are two types of maps on F_R and F_C. One type is number-changing and value-preserving. The other is number-preserving and value-changing. For each real number s, let W_s be a number-changing value-preserving map on F_R. The action of W_s is defined by

$$W_s\bar{R}^t = \bar{R}^{ts}. \tag{6}$$

If s is negative, the direction of the order relation in $\bar{R}^{ts}$ is changed from that in $\bar{R}^t$. The map is number-changing, value-preserving, because 1_t is a different number from 1_{ts}. Both numbers have the same value in their respective structures. Note that the maps W_s for all $s \neq 0$ form a commutative group. The definition of W_c on F_C for $c \neq 0$ is defined in a similar fashion.

For example, the set of natural numbers scaled by π is given by

$$\bar{N}_\pi = \left\{B_N, +_\pi, \frac{1}{\pi}\times_\pi, <_\pi, 0_\pi, 1_\pi\right\}.$$

The definition of the number-preserving, value-changing map is more complex. Let s be a non-zero real number. Define Z_s to be the associated number-preserving, value-changing map. One has

$$Z_s\bar{R}^t = \bar{R}^t_u, \tag{7}$$

where $u = ts$. The components of $\bar{R}^t_u$ are given by

$$\begin{aligned} &Z_s(a_t) = \left(\frac{t}{u}a\right)_u, \quad Z_s(+_t) = +_u, \quad Z_s(\times_t) = \left(\frac{u}{t}\times\right)_u, \\ &Z_s(\div_t) = \left(\frac{t}{u}\div\right)_u, \quad Z_s(<_t) = <_u \text{ for } s > 0, \\ &Z_s(0_t) = 0_u, \quad Z_s(1_t) = \left(\frac{t}{u}1\right)_u. \end{aligned} \tag{8}$$

These definitions can be combined to write $\bar{R}^t_u$ as a structure,

$$\bar{R}^t_u = \left\{B_R, \pm_u, \left(\frac{u}{t}\times\right)_u, \left(\frac{t}{u}\div\right)_u, <_u, 0_u, \left(\frac{t}{u}1\right)_u\right\}. \tag{9}$$

$\bar{R}^t_u$ is a representation of the components of $\bar{R}^t$ in terms of those in $\bar{R}^u$. Scaling of the division and multiplication operations is necessary to preserve axiom validity under the action of Z_s.[4] Note that $((t/u)1)_u$ is the same base set number in $\bar{R}^u$ as is 1_t in $\bar{R}^t$. Its value differs from 1 by the factor t/u. As was the case for W_s, the maps, Z_s, for all non-zero s form a commutative group.

For complex numbers, the action of Z_c for c complex is given by

$$Z_c\bar{C}^d = \bar{C}^d_e, \tag{10}$$

where $e = cd$. The components of $\bar{C}^d_e$ are given by Eq. 8 with the replacement of $Z_s(<_t)$ with

$$Z_c((a_d)^{*_d}) = \left(\frac{d}{e}\right)_e (a_e)^{*_e} = \left(\frac{d}{e}(a^*)\right)_e.$$

[4]As an example of axiom preservation, $((t/u)1)_u$ satisfies the multiplicative identity axiom. From

$$\left(\frac{t}{u}a\right)_u \left(\frac{u}{t}\times\right)_u \left(\frac{t}{u}1\right)_u = \left(\frac{t}{u}a\right)_u,$$

one obtains

$$a_u \times_u 1_u = a_u,$$

by multiplying out the ratios of scaling factors.

These components can be combined to define a structure $\bar{C}^d_e$ by

$$\bar{C}^d_e = \left\{B_C, \pm_e, \left(\frac{e}{d}\times\right)_e, \left(\frac{d}{e}\div\right)_e, \left(\frac{d}{e}-\right)_e(-)^{*_e}, 0_e, \left(\frac{d}{e}1\right)_e\right\}. \tag{11}$$

The blank in the complex conjugation component denotes a number in $\bar{C}^e$.

It must seem strange that, with both d and e complex, $((d/e)1)_e$ is both the identity and is a real number.[5] This is an illustration of the fact that properties of numbers do not exist in the abstract, independent of structure membership. Their properties are determined by the structure containing them.

It should also be emphasized that operations such as multiplication and division are carried out on numbers before their transport from $\bar{C}^d$ to $\bar{C}^d_e$, not after. Implementing the operations in the target structures on transported numbers gives a different result. For example, for any complex number value, a,

$$\left(\frac{d}{e}a\right)_e \times_e \left(\frac{d}{e}1\right)_e \neq \left(\frac{d}{e}a\right)_e$$

and

$$Z_c((a_d)^{*_d}) = \left(\frac{d}{e}(a)^*\right)_e \neq \left(\left(\frac{d}{e}a\right)_e\right)^{*_e}.$$

2.3. *Vector spaces*

As noted before, number scaling extends to all mathematical structures whose description includes numbers. Vector spaces are an example. They can be based on either real or complex numbers as the scalar field. Here, the scalar field will be assumed to be complex. Spaces based on real scalar fields are a simplification, in that the real scalars are a subfield of the complex scalars. However, the scaling factor will be restricted to be a real number value.

For each non-zero real number value r, let $\bar{V}^r$ be a scaled normed vector space. It is defined by

$$\bar{V}^r = \{B_V, \pm_r, \cdot_r, |-|_r, \psi_r\}. \tag{12}$$

Here, ψ_r denotes an arbitrary vector with vector value ψ, $|-|_r$ denotes the norm of a vector value, $\cdot_r$ is scalar vector multiplication, and $\pm_r$ is linear

[5] Details are given in Benioff (2016).

superposition. The values of these three operations are represented by $|\underset{\cdot}{-}|$, $\cdot$, and $\pm$. The base set of vectors is represented by B_V. The associated scalar field is $\bar{C}^r$. Note that one can express the norm in the form $|\psi_r| = |\psi|_r$. This is a real number in $\bar{C}^r$.

The collection of vector spaces and associated complex number fields for all real non-zero scaling factors is represented by

$$F_{VC} = \bigcup_r (\bar{V}^r \times \bar{C}^r). \tag{13}$$

There is one base set pair, $B_V \times B_C$, and one pair of value structures, $\bar{V} \times \bar{C}$, for the collection of scaled spaces and fields. As was the case for numbers, there are two types of maps on F_{VC}. One is vector- and number-changing, vector and number value-preserving, and the other is vector- and number-preserving and vector value- and number value-changing. The first type is denoted by W_p where

$$W_p(\bar{V}^r \times \bar{C}^r) = \bar{V}^{rp} \times \bar{C}^{rp}. \tag{14}$$

As was the case for scalars, the collection of W_p for all non-zero p forms a commutative group.

The description of the map Z_p for number structures can be extended to vector spaces. Define the actions of Z_p on the components of $\bar{V}^r$ by

$$\begin{aligned} Z_p(\psi_r) &= \left(\frac{r}{q}\right)_q \psi_q = \left(\frac{r}{q}\psi\right)_q, \quad Z_p(\pm_r) = \pm_q, \\ Z_p(\cdot_r) &= \left(\frac{q}{r}\cdot\right)_q, \quad Z_p(|\psi|_r) = \left(\frac{r}{q}|\psi|\right)_q, \quad Z_p(\psi_r) = \left(\frac{r}{q}\psi\right)_q. \end{aligned} \tag{15}$$

Here, $q = pr$.

This definition can be combined with that for $\bar{C}^r_q$ to write

$$Z_p(\bar{C}^r \times \bar{V}^r) = \bar{C}^r_q \times \bar{V}^r_q, \tag{16}$$

where

$$\bar{V}^r_q = \left\{B_V, \pm_q, \left(\frac{q}{r}\cdot\right)_q, \left(\frac{r}{q}|\psi|\right)_q, \left(\frac{r}{q}\psi\right)_q\right\}. \tag{17}$$

Note that the norm is special in that

$$\left(\frac{r}{q}|\psi|\right)_q = \left|\frac{r}{q}\psi\right|_q. \tag{18}$$

As an example, this equality does not hold for scalar products in Hilbert spaces. For these spaces,

$$\left(\frac{r}{q}\langle\psi|\psi\rangle\right)_q \neq \left\langle\frac{r}{q}\psi|\frac{r}{q}\psi\right\rangle_q. \tag{19}$$

3. Local Mathematics and Number Scaling Fields

As was noted, gauge theories make use of local vector spaces in space-time. Relations between vectors in the different spaces are described by unitary gauge transformations between vector spaces at different locations. Here, the existence of local vector spaces is extended to include structures of numbers of different types. Locality can be further extended to mathematical systems of many different types, in particular those whose description includes numbers.

Here, the main emphasis is on vector spaces and real or complex numbers. For vector spaces and complex numbers, this leads to the existence of $\bar{C}_y \times \bar{V}_y$ as complex number structures and vector spaces at each location, y, of a manifold, M. Here, M will be taken to be the space-time of special relativity, although the results are applicable to more general manifolds.

In gauge theory, the "no information at a distance" principle (Montvay and Münster, 1994) has the consequence that the information used to choose a basis in a vector space at one location does not determine the choice of a basis in vector space at a different location. Here, this principle is extended to real and complex numbers. The result is that the information used to determine a value of a number at one location does not determine the value of the same number at a different location.

This is implemented here by first introducing a real scaling or number value field, g. For each y in M, $g(y)$ is the scaling or value factor for the structures at y. This field will be used to introduce a connection as a number-preserving, value-changing map between numbers in structures at different locations.

Before describing the connection, it is worthwhile to describe the collection of scaled complex number structures and vector spaces at each point of space-time. For each point y, these are given by

$$F_y^{g(y)} = \bar{C}_y^{g(y)} \times \bar{V}_y^{g(y)}. \tag{20}$$

The collection of the $F_y^{g(y)}$ for all points y in M with the scalars real or complex number structures is the mathematical arena for much of the rest of this work.

A good representation of this mathematical arena is by a fiber bundle, $\mathfrak{MF}^g$ where

$$\mathfrak{MF}^g = M \times F, \pi_g, M. \tag{21}$$

The fiber F is defined by

$$F = \bigcup_r F^r, \tag{22}$$

where

$$F^r = (\bar{C}^r \times \bar{V}^r). \tag{23}$$

The parameter r varies over all real numbers. The definition of the projection π_g differs from that in the usual description of fiber bundles (Husemöller, 1975). Here, π_g is a bijective map from M and a subset of the fiber, to M. It has the property such that for each y in M,

$$\pi_g^{-1}(y) = F_y^{g(y)}. \tag{24}$$

The fiber, $F_y^{g(y)}$, is given by Eq. (20). Also, π_g^{-1} is the inverse of π_g.

4. Effects of Local Mathematics and Value Fields on Physics and Geometry

The fiber bundle of local complex number and vector space structures with the value field g provides a good setting for the effect of local mathematical structures and the value field on many quantities in physics and geometry. The effect is seen for all quantities and system properties that are represented by integrals or derivatives over space or space-time. Examples include fields and wave functions in physics and paths, membranes, and solids in geometry. Here, two generic examples are given, one for integrals and the other for derivatives. Specific examples, gauge theory for physics and paths in geometry, will then be summarized.

Let ψ be a complex scalar field on M. For each location y in M, $\psi(y)$ is a complex number. This is the usual representation of ψ as a function with domain M and range in a single complex number structure. The distinction between $\psi(y)$ as a number and as a number value is not present.

This description is changed here to describe a field as a section on the fiber bundle $\mathfrak{MF}^g$. For each y in M, $\psi(y)$ is the value of a number in $\bar{C}_y^{g(y)}$. The number with value $\psi(y)$ in $\bar{C}_y^{g(y)}$ is denoted by $\psi(y)_{g(y)}$.

For the description of integrals, it suffices to let M be a three-dimensional Euclidean space. The integral of $\psi(y)_{g(y)}$ over M is not defined because the definition of an integral as the limit of sums of integrands for different y is not defined. Arithmetic operations are defined only within local structures, not between structures at different locations.

This is remedied by the use of a connection to parallel transform a number in $\bar{C}_y^{g(y)}$ to the same number in $\bar{C}_x^{g(x)}$ at an arbitrary reference location, x. The action of a connection, $C_g(x,y)$, on $\psi(y)_{g(y)}$ is given by

$$C_g(x,y)\psi(y)_{g(y)} = \left(\frac{g(y)}{g(x)}\right)_{g(x)} \psi(y)_{g(x)} = \left(\frac{g(y)}{g(x)}\psi(y)\right)_{g(x)}. \tag{25}$$

In this equation, $(g(y)/g(x))_{g(x)}$ is the number in $\bar{C}_x^{g(x)}$ with value $g(y)/g(x)$, and $\psi(y)_{g(x)}$ is a number in $\bar{C}_x^{g(x)}$ with value $\psi(y)$. A result of the presence of the g field is that $\psi(y)_{g(x)}$ is a different number than $\psi(y)_{g(y)}$.

The desired integral can now be defined as

$$I_x(\psi) = \left(\frac{1}{g(x)}\right)_{g(x)} \int (g(y)\psi(y)dy)_{g(x)}. \tag{26}$$

This integral is defined because the integrands $(g(y)\psi(y))_{g(x)}$ are all numbers in $\bar{C}_x^{g(x)}$. Also, $d(y)$ is an infinitesimal number value of a small number in $\bar{C}_x^{g(x)}$. The factor $g(x)$ is outside the integral as it is independent of y.

The description for scalar fields in the presence of g extends in a straightforward way to vector fields. If ψ is a field of vector values, then, for each y, $\psi(y)_{g(y)}$ is a vector in $\bar{V}_y^{g(y)}$. Parallel transport of these vectors to a reference location, x, gives a vector field with all values as vectors in $\bar{V}_x^{g(x)}$. The integral of the resulting localized field is also expressed by Eq. (26).

The description of derivatives also shows the effects of the g field. Here, M is assumed to be a space-time of special relativity. Let ψ be a complex number value field where for each y, $\psi(y)_{g(y)}$ is a number in $\bar{C}_y^{g(y)}$ with value $\psi(y)$. The partial derivative of $\psi(y)_{g(y)}$ in the direction $\mu = 0, 1, 2, 3$ is given by

$$\partial_{\mu,y}\psi(y)_{g(y)} = \lim_{d^\mu y \to 0} \frac{\psi(y + d^\mu y)_{g(y+d^\mu y)} - \psi(y)_{g(y)}}{d^\mu y}. \tag{27}$$

This expression is not defined. The reason is that $\psi(y + d^\mu y)_{g(y+d^\mu y)}$ is a number in $\bar{C}_{y+d^\mu y}^{g(y+d^\mu y)}$ and $\psi(y)_{g(y)}$ is a number in $\bar{C}_y^{g(y)}$. Subtraction is not defined between numbers in different structures.

As was the case for integrals, this problem is fixed by parallel transport of the number $\psi(y+d^\mu y)_{g(y+d^\mu y)}$ to the same number in $\bar{C}_y^{g(y)}$. The resulting partial derivative can be expressed by

$$\begin{aligned} D_{\mu,y}\psi(y)_{g(y)} &= \frac{C_g(y, y+d^\mu y)\psi(y+d^\mu y)_{g(y+d^\mu y)} - \psi(y)_{g(y)}}{d^\mu y} \\ &= \frac{\left(\frac{g(y+d^\mu y)}{g(y)}\psi(y+d^\mu y)\right)_{g(y)} - \psi(y)_{g(y)}}{d^\mu y}. \end{aligned} \tag{28}$$

The limit $d^\mu y \to 0$ is implied.

Taylor expansion of $g(y+d^\mu y)$ to first order in differentials gives

$$g(y+d^\mu y) \simeq g(y) + d_g^\mu y \partial_{\mu,y} g(y). \tag{29}$$

Use of this in Eq. (28) gives

$$D_{\mu,y}\psi(y)_{g(y)} = \left(\partial_{\mu,y} + \left(\frac{\partial_{\mu,y} g(y)}{g(y)}\right)_{g(y)}\right)\psi(y)_{g(y)}, \tag{30}$$

as the final result.

Equation (30) also holds for ψ a vector value field with $\psi(y)_{g(y)}$ a vector in $\bar{V}_y^{g(y)}$ with vector value $\psi(y)$. This will be used in the next section where an example from physics and an example from geometry are discussed. The examples will show how local mathematics and the presence of a value field affect theoretical descriptions of quantities in physics and geometry.

5. Examples

Two examples will be discussed. One will be the effect of the value field on Dirac Lagrangians in gauge theory. The other example will be the effect of the value field on distances in geometry. For both examples, the manifold M will be a four-dimensional space-time as Minkowski space. Local mathematical structures associated with each point of M are assumed.

It is also useful to replace the value field g by its exponential equivalent, as in

$$g(y) = e^{\alpha(y)}. \tag{31}$$

The real scalar field, α, will also be referred to as a scaling or value field.

In the following, the subscripts, $g(y)$ and $g(x)$, will be suppressed. This makes the mathematical expressions easier to read and more connected to the usual expressions of gauge theory and geometry. Subscripts will be used where needed to provide clarity.

5.1. *Gauge theory*

The Lagrangian density value for Dirac field values is given by

$$\mathcal{L}(\psi(y), \partial_{\mu,y}\psi(y)) = \bar{\psi}(y)i\gamma^{\mu}\partial_{\mu,y}\psi(y) - m\bar{\psi}(y)\psi(y). \tag{32}$$

Here, m is the mass of the field ψ and

$$\bar{\psi} = \gamma^5\psi^* \tag{33}$$

where γ^μ and γ^5 are the gamma matrices (Cheng and Li, 1984). The corresponding Lagrange density for Dirac fields is given by

$$(\mathcal{L}(\psi(y), \partial_{\mu,y}\psi(y)))_{g(y)} = (\bar{\psi}(y)i\gamma^{\mu})_{g(y)}\partial_{\mu,y}\psi(y)_{g(y)} - (m\bar{\psi}(y)\psi(y))_{g(y)}. \tag{34}$$

Here, the description will be given for Abelian gauge theory. The fermion fields, $\psi(y)_{g(y)}$ and $\bar{\psi}(y)_{g(y)}$, are taken to be sections on the fiber bundle, $\mathfrak{MF}^g$, with $\psi(y)$ and $\bar{\psi}(y)$ values of the vectors, $\psi(y)_{g(y)}$ and $\bar{\psi}(y)_{g(y)}$, in $\bar{V}_y^{g(y)}$.

So far, parallel transport of numbers has been limited to changes in scaling or valuation of scalars and vectors. The additional requirement that the Lagrangian density be invariant under local $U(1)$ gauge transformations (Montvay and Münster, 1994; Cheng and Li, 1984) results in the extension of parallel transport to apply directly to vectors.

This is taken care of by expanding the connection, $C_g(y, y + d^\mu y)$, in Equation (28) to include $U(y) = e^{i\phi(y)}$ as a local unitary gauge transformation. The connection in Eq. (28) is replaced by

$$C_g(y, y + d^\mu y) = \left(\frac{e^{\alpha(y+d^\mu y)}}{e^{\alpha(y)}}\frac{e^{i\phi(y+d^\mu y)}}{e^{i\phi(y)}}\right)_{g(y)}. \tag{35}$$

The exponential representation of g is used here. Taylor expansion of both exponents to first order in $d^\mu y$ gives

$$\begin{aligned} C_g(y, y + d^\mu y) &\simeq (1 + d^\mu y(A_\mu(y))(1 + d^\mu y i B_\mu(y)))_{g(y)} \\ &\simeq (1 + d^\mu y(A_\mu(y) + iB_\mu(y)))_{g(y)}. \end{aligned} \tag{36}$$

Use of this in Eq. (28) gives an expansion of Eq. (30) to include the effects of the B field. The result is the expansion of

Eq. (30) to

$$D_{\mu,y}\psi(y) = (\partial_{\mu,y} + A_\mu(y) + iB_\mu(y)))\psi(y), \tag{37}$$

(the subscript, $g(y)$, is suppressed).

Use of this in the Dirac Lagrangian density value gives

$$\begin{aligned}\mathcal{L}(\psi(y), \partial_{\mu,y}\psi(y)) &= \bar{\psi}(y)i\gamma^\mu D_{\mu,y}\psi(y) - m\bar{\psi}(y)\psi(y)\\ &= \bar{\psi}(y)i\gamma^\mu(\partial_{\mu,y} + aA_\mu(y) + ibB_\mu(y))\psi(y) - m\bar{\psi}(y)\psi(y).\end{aligned} \tag{38}$$

Coupling constants for the $\vec{A}$ and $i\vec{B}$ fields are denoted by a and b. Since $\vec{B}$ is the electromagnetic or photon field, b is the square root of the fine structure constant. The value of a is not known so far.

The requirement that the Lagrangian density value be invariant under local $U(1)$ transformations is expressed by

$$D'_{\mu,y}U\psi(y) = UD_{\mu,y}\psi(y). \tag{39}$$

If $U = e^{i\theta(y)}$, this equation gives the condition

$$A'_\mu(y) = A_\mu(y), \tag{40}$$

and

$$B'_\mu(y) = B_\mu(y) - \frac{1}{b}\partial_{\mu,y}\theta(y). \tag{41}$$

From these equations, one concludes the well-known fact that the photon mass is 0. The $\vec{A}$ field can have any mass, including 0.

The $\vec{B}$ field is non-integrable. This is a consequence of the Aharonov Bohm effect (Aharonov and Bohm, 1959). As defined here, the $\vec{A}$ field is integrable. The definition of the connection can be changed to allow for $\vec{A}$ to be nonintegrable.

One concludes from this that the $\vec{A}$ field is the gradient of a real scalar field. Any mass is possible. It is also assumed to be a spin 0 field. At present, it is not known which physical field, if any, corresponds to the $\vec{A}$ field. The Higgs field is the only spin 0 scalar field in the standard model. Are the two fields related?

The $\vec{A}$ field also appears as a real, vector field in non-Abelian gauge theories. The presence of the $\vec{A}$ field in these Lagrangians may be of help in deciding which physical field, if any, is represented by $\vec{A}$.

5.2. *Local scaled geometry*

The presence of the field α and local mathematics affects many geometric properties. For these properties, the fiber contents are different from those for gauge theories. Here, the fiber bundle has the same form given by $\mathfrak{MF}^g$. The contents of the fiber are different in that

$$F = \bigcup_r (\bar{R}^r \times \bar{T}^r). \tag{42}$$

The contents of the fiber at each location of M are given by

$$\pi_g^{-1}(y) = F_y^{g(y)} = \bar{R}_y^{g(y)} \times \bar{T}_y^{g(y)}. \tag{43}$$

Here, $\bar{R}_y^{g(y)}$ is a scaled real number structure and $\bar{T}_y^{g(y)}$ is a local scaled representation of M at y.

A good representation for $\bar{T}_y^{g(y)}$ is by the quadruple $\bar{R}_y^{4,g(y)}$ of scaled real numbers. The associated metric tensor is $(h_{\mu,\nu})_{g(y)} = ((1,-1,-1,-1)\delta_{\mu,\nu})_{g(y)}$. The space $\bar{T}_y^{g(y)}$ can be defined as a scaled chart representation of M. For each y,

$$\bar{T}_y^{g(y)} = \rho_y^{g(y)}(M). \tag{44}$$

Since M is flat, the chart, $\rho_y^{g(y)}$, is defined on all of M.

The effect of scaling of local real number structures and representations of M can be seen in the example of the length of a path. Let p be a path on M with beginning and end points, y and z. Let the path be parameterized by a real number s with $p(0) = y$ and $p(1) = z$. The length value of the path is defined by

$$L(p) = \int_0^1 [\partial_{\mu,s} p h_{\mu,\nu} \partial_{\nu,s} p]^{1/2} ds. \tag{45}$$

Sum over repeated indices is implied.

Under the assumption of global mathematics, this path integral is well defined. With the assumption of local mathematics and the presence of the value field, there is a vector space structure, $\bar{T}_{p(s)}^{g(p(s))}$, associated with each path point, $p(s)$, on M. The path gradients, $\nabla_s p$, form a field of vectors along the path. Treatment of this field as a section on the fiber bundle, defined by Eqs. 42 and 43, has the consequence that for each s, the vectors with components $\partial_{\mu,s} p$ are vectors in $\bar{T}_{p(s)}^{g(p(s))}$.

As a result, the integral of Eq. (45) is not defined. The reason is that for each s, the integrand is a number in $\bar{R}_{p(s)}^{(p(s))}$, and $\partial_{\mu,s} p$ is the μ component of a vector in $\bar{T}^{g(p(s))}_{p(s)}$. The integrand addition implied in the definition of an integral is not defined between numbers in different structures. It is defined only for numbers within a structure.

This problem is fixed by use of a connection to parallel transport the integrands to a real number and position space structure at a reference location x. The transport gives

$$\begin{aligned}[\partial_{\mu,s}ph_{\mu,\nu}\partial_{\nu,s}p]^{1/2}_{g(p(s))} &\to C_g(x,p(s))[\partial_{\mu,s}ph_{\mu,\nu}\partial_{\nu,s}p]^{1/2}_{g(p(s))} \\ &= (e^{\alpha(p(s))-\alpha(x)})_{g(x)}[\partial_{\mu,s}ph_{\mu,\nu}\partial_{\nu,s}p]^{1/2}_{g(x)}. \end{aligned} \tag{46}$$

Subscripts have been added to show that the factors with subscripts are numbers in $\bar{R}^{g(x)}_x$ with the indicated number values.

The integral of the transported path lengths is defined. It is given by

$$L(p)_x = (e^{-\alpha(x)}\int_0^1 e^{\alpha(p(s))}[\partial_{\mu,s}ph_{\mu,\nu}\partial_{\nu,s}p]^{1/2}ds)_{g(x)}. \tag{47}$$

This is the length of the path p from y to z in the presence of the scaling or value field, α. The path length is a real number quantity in $\bar{R}^{g(x)}_x$.

The distance or geodesic between y and z is the length of the shortest path between y and z. This is obtained by the use of the Euler Lagrange equations to find the minimum of $L(p)_x$ by variation of the path, p. The result is given by

$$\left[\frac{d}{d\tau} + \vec{A}(p(\tau))\cdot\nabla_\tau p\right]\frac{dp^\mu}{d\tau} - h^{\mu,\nu}A_\nu(p(\tau)) = 0. \tag{48}$$

This is the geodesic equation for the component p^μ of the minimal path. The term $\vec{A}(p(\tau))\cdot\nabla_\tau p$ is obtained as

$$\vec{A}(p(\tau))\cdot\nabla_\tau p = e^{-\alpha(p(\tau))}\frac{d}{d\tau}e^{\alpha(p(\tau))}.$$

Details of the derivation are given in Benioff (2015). The proper time is denoted by τ.

This equation shows that, if all components of $\vec{A}$ are 0, then Eq. (48) reduces to

$$\frac{d^2p^\mu}{d\tau^2} = 0. \tag{49}$$

This is the equation for a straight line in space-time. From Eq. (48) one sees that the presence of the $\vec{A}$ field affects the distance between locations in M. It would be interesting to see if this effect shows up in physics in some way — Dark energy?

6. Restrictions on the α Field

So far, there is no direct experimental evidence for the presence of the α field or its gradient, $\vec{A}$. The great accuracy of quantum electrodynamics shows that either the coupling of the $\vec{A}$ field to fermion fields must be extremely small compared to the fine structure constant, or $\vec{A}$ must be very close to zero. Other areas of physics also show no experimental evidence for the presence of $\vec{A}$.

This fact is to be combined with an important property of all experiments. This is the fact that all experiments carried out to date are implemented locally. They are done by us on or close to the surface of the earth. This includes experiments to determine properties of systems that exist locally as well as properties of cosmological systems such as galaxies and pulsars.

In the future, humans may inhabit other solar system planets and interplanetary space ships. It is assumed here that any experiment done by humans in these locations will find no direct experimental evidence for the presence of the $\vec{A}$ field.

This lack of experimental evidence for the presence of $\vec{A}$ should also be extended to potential inhabitants of planets around stars that are in effective communication with us. These observers can describe experiments they have carried out. They can also communicate to us the results and their interpretations of these results. It is assumed here that they also will find no direct evidence for the presence of $\vec{A}$.

These assumptions regarding the lack of direct experimental evidence for the lack of $\vec{A}$ can be combined into a restriction that

$$|\vec{A}(y)| < \epsilon \simeq 0 \tag{50}$$

for all space-time points, y, in a local region of the universe. Here, $|\vec{A}(y)|$ is the length of $\vec{A}(y)$. Within this local region, ϵ is smaller than the uncertainty of all experiments carried out to date and presumably in the future.

One may expect that in the future, experiments will be done with increasing accuracy in the results. However, uncertainties will never be exactly 0. At present, effective limits are presented by the Planck units of length, time, and mass. However, the possibility that even these limits will be surpassed in the future cannot be ruled out.

The important aspect of these restrictions is that they are limited to apply only to a local region of the universe. The size of the local region is determined by the maximum distance between us on the Earth and observers on other planets with whom we can effectively communicate.

The size of this local region is unknown. One literature estimate (Smith, 2011) is that it is restricted to those stars that are, at most, 1000 light years distant from us. Here, the size of the restricted region is not important. The only requirement is that it is a small fraction of the 14 billion light year size of the universe.

Outside the local region, there are no restrictions on the variation of $\vec{A}$. It can vary rapidly with location or slowly or not at all.

There are fields described in physics that share the restrictions of $\vec{A}$. These include the inflaton (Bezrukov and Shaposhnikov, 2008), quintessence (Zlatev *et al.*, 1999), and dark energy (Li *et al.*, 2013).[6] The effects of these fields are present globally. They are invisible locally. At present, it is an open question if $\vec{A}$, or α, is any of these fields, or none of these. This is a problem for future work.

7. Discussion

In this chapter, it was seen that the "no information at a distance" principle and the assumption of local number and vector structures led to the existence of a number scaling or value field, g. This field had two roles. It determined the scaling factor for number and vector structures at each point of space-time. It was also used to define the connection between structures at different locations. On number structures, the connection was a number-preserving, number value-changing operation. On vector spaces, the connection was a vector-preserving, vector value-changing operation. Its effect was to multiply a vector by a g-dependent scalar.

Theoretical quantities in physics that are defined by integrals or derivatives of some field or function are affected by the presence of g. This was seen from the need to parallel transport values of scalar or vector fields in different structures at different locations to a common location. Integrals and derivatives of sections of fields so transported were seen to be well defined. As a result, theoretical descriptions of many quantities in physics

[6]See Benioff (2016) for a discussion of the possible relation of A and α to the Higgs field.

are affected by the presence of the g field, or its equivalent, the α field, where $g(y) = e^{\alpha(y)}$.

Two examples of the effect of α on physical quantities were given. One example showed the effect on the Dirac Lagrangian in gauge theory. The other example showed the effect on path lengths and distances. It was seen that the presence of the α field resulted in an additional vector field in the Lagrangian that interacts with the fermion field. The same vector field appears in the geodesic equation for the distance between two locations in space-time. The vector field is the gradient of the scalar α field.

Other examples of the effect of the g or α field on physics and geometry are given in the cited references. These include the description of black and white holes in geometry, where $\alpha(y)$ approaches either $+$ or $-$ infinity as y approaches a point, x (Benioff, 2013), and the effect of α on quantum mechanical quantities (Benioff, 2016).

There is still much to do. The effect of number scaling or valuation needs to be extended to mathematical structures other than scalars and vector spaces. Examples are the operator algebras much used in quantum mechanics and the Lie algebras and matrices for unitary transformations used in gauge theories. In geometry, the description for paths needs to be extended to two-, three-, and four-dimensional systems. The effect also needs to be expanded to include general relativity. The relation of each fiber with greatly expanded mathematical content to the mathematics that is potentially available to a conscious observer at the location of a fiber is intriguing. This should also be investigated.

Acknowledgment

This material is based upon work supported by the U.S. Department of Energy, Office of Science, Office of Nuclear Physics, under contract number DE-AC02-06CH11357.

References

Aharonov, Y. and Bohm, D. (1959). *Phys. Rev.* 115, 485.

Benioff, P. (2002). Towards a coherent theory of physics and mathematics, *Found. Phys.* 32, 989–1029, arXiv:quant-ph/0201093.

Benioff, P. (2005). Towards a coherent theory of physics and mathematics: The theory-experiment connection, *Found. Phys.* 35, 1825–1856.

Benioff, P. (2011). New gauge field from extension of space time parallel transport of vector spaces to the underlying number systems, *Int. J. Theoret. Phys.* 50, 1887–1907, arXiv:1008.3134.

Benioff, P. (2012). Effects on quantum physics of the local availability of mathematics and space-time dependent scaling factors for number systems, in *Quantum Theory*, Cotaescu, I. (ed.), Intech open access publisher, Chapter 2, arXiv:1110.1388.

Benioff, P. (2013). Gauge theory extension to include number scaling by boson field: Effects on some aspects of physics and geometry, in *Recent Developments in Bosons Research*, Tremblay, I. (ed.), Nova Publishing Co., Chapter 3; arXiv:1211.3381.

Benioff, P. (2015). Fiber bundle description of number scaling in gauge theory and geometry, *Quant. Stud. Math. Found.* 2, 289–313, arXiv:1412.1493.

Benioff, P. (2016). Effects of a scalar scaling field on quantum mechanics, *Quant. Inform. Process.* 15(7), 3005–3034, arXiv:1512.05669.

Bezrukov, F. and Shaposhnikov, M. (2008). The standard model Higgs boson as the inflaton, *Phys. Lett. B* 659, 703–706.

Burgin, M. (2001). Diophantine and nondiophantine arithmetics: Operations with numbers in science and everyday life, arXiv:math/0108149.

Burgin, M. (1997). Nondiophantine arithmetics or is it possible that $2 + 2$ is not equal to 4, *Ukrainian Acad. Inform. Sci. Kiev* (in Russian).

Burgin, M. (2010). Introduction to projective arithmetics, arXiv:1010.3287.

Cheng, T. P. and Li, L. F. (1984). *Gauge Theory of Elementary Particle Physics*, Oxford University Press, Oxford, UK, Chapter 8.

Czachor, M. (2016). Relativity of arithmetics as a fundamental symmetry of physics, *Quantum Studies: Mathematics and Found.* 3, 123–133. arXiv:1412.8583.

Czachor, M. (2017). Dark energy as a manifestation of nontrivial arithmetic, *Int. J. Theor. Phys.* 56, 1364–1381, arXiv:1604.05738.

Dirac, P. A. (1938). The relation between mathematics and physics, *Proc. Roy. Soc. Edinburgh* 59(Part II), 122–129.

Hitchin, N. (2007). Interaction between mathematics and physics, *Arbor Ciencia, Pensamento y cultura, CLXXXIII* 725, 427–432.

Husemöller, D. (1975). *Fibre Bundles*, 2nd edn., Graduate texts in Mathematics, v. 20, Springer Verlag, New York.

Li, M., Li, X.-D., Wang, S. and Wang, Y. (2013). Dark energy, a brief review, *Front. Phys.* 8, 828–846.

Mack, G. (1981). Physical principles, geometrical aspects, and locality properties of gauge field theories, *Fortshritte der Physik* 29, 135.

Montvay, I. and Münster, G. (1994). *Quantum Fields on a Lattice*, Cambridge University Press, UK, Chapter 3.

Omnes, R. (2011). Wigner's unreasonable effectiveness of mathematics, *Revisited, Found. Phys.* 41, 1729–1739.

Plotnitsky, A. (2011). On the reasonable and unreasonable effectiveness of mathematics in classical and quantum physics, *Found. Phys.* 41, 466–491.

Shapiro, S. (2008). Mathematical objects. In Gold, B. and Simons, R. (eds), *Proof and Other Dilemmas, Mathematics and Philosophy*, Spectrum Series, Mathematical Association of America, Washington DC, Chapter III, pp. 157–178.

Smith, H. A. (2011). Alone in the universe, *Am. Sci.* 99(4), 320.

von Neumann, J. (1947). The Works of the Mind, 1, 180–196.

Wigner, E. (1960). The unreasonable effectiveness of mathematics in the natural sciences, *Commun. Pure Appl. Math.* 13(1) [Reprinted in E. Wigner, *Symmetries and Reflections*, Indiana Univ. Press, Bloomington, IN, 1966, pp. 222–237].

Yang, C. N. and Mills, R. L. (1954). Conservation of isotopic spin and isotopic gauge invariance, *Phys. Rev.* 96, 191–195.

Zlatev, I. Wang, L. and Steinhardt, L. P. (1999). Quintessence, cosmic coincidence, and the cosmological constant, *Phys. Rev. Lett.* 82(5), 896–899, arXiv:astro-ph/9807002.

Chapter 7

The Structure of the World, Unity of Nature and the Problem of Time

Rafael Capurro*,‡ and Mark Burgin†,§

*Stuttgart Media University, Stuttgart, Germany

†University of California, Los Angeles, CA, USA

‡rafael@capurro.de

§mburgin@math.ucla.edu

This chapter consists of two sections. In Section 1, Rafael Capurro analyzes Weizäcker's ideas about the unity of nature based on his long-standing dialog with Martin Heidegger about Heidegger's interpretation of being as three-dimensional time. This interpretation allows Weizäcker to face some of the challenges arising from Quantum Mechanics, questioning the traditional founding of physics on one-dimensional time. This, in turn, brings forth four questions, discussion of which has the form of self-interview, i.e., Capurro asks questions impersonating another person and then answers these questions. The *first question* about unity of nature is discussed from the perspective of the concept information and its ties to the Platonic and Aristotelian concepts of *eidos*, Ideas or Forms. In turn, the unity of nature is addressed from the methodological perspective of the unity of physics based on the search for a univocal meta-language. The *second question* is about connections between Weizsäcker's ideas and Wheeler's "It from bit". The third question concerns digital technology as today's driving force of intellectual innovation in thinking about the world. An answer to the *fourth question* about what would be the most important message from Weizsäcker to us is also obtained. Namely, it is the understanding of the unity of nature, as well as of language and of us in the world, as being embedded in three-dimensional time.

In Section 2, Mark Burgin demonstrates how new scientific discoveries support ideas of Weizäcker about relations between information and the Platonic and Aristotelian concepts of *eidos*, Ideas or Forms. Explication of these relations is based on the discovered structure of the world (Burgin, 2012). Many thinkers contributed to this discovery, which reflects unity of the world. Interestingly, this unity is triadic and not monadic. In addition, viable interpretations of Wheeler's metaphor "It from bit" are explained based on the general theory of information, which is constructed as an axiomatic theory on three levels: *conceptual*, *methodological* (also called *meta-theoretical*) and *theoretical* (Burgin, 2010, 2017a). The third topic of the Section 2 is a discussion and explanation of the idea of three-dimensional time based on the system theory of time in the context of the Existential Triad of the World. The system theory of time shows existence of a diversity of times extending relativity theory into the system domain and demonstrating the necessity of a multiplicity of time scales for adequate representation of time (Burgin, 2002).

1. On the Unity of Nature: A Question of Time

It was Weizsäcker who opened my eyes in the 1970s when he pointed to the concept of information "as something different from matter and consciousness", recalling "the Platonic *eidos* and the Aristotelian form, dressed in such a way that a human being in the 20th century can learn something about them" (Weizsäcker, 1974, p. 51). This gave rise to my research on the history of this concept (Capurro, 1978) that was the beginning of a long journey until today (Capurro, 2009). My colleague and friend, the Australian philosopher Michael Eldred, is a key personal interlocutor in this dialog with Weizsäcker (Eldred, 2014, 2015) dealing with the question of time as analyzed by Heidegger, particularly in his seminal work *Being and Time* (Heidegger, 1976) that influenced Weizsäcker over the years.

In the early 1980s, when I was working at FIZ Karlsruhe, a leading organization in the field of scientific information and documentation located on the premises of the Karlsruhe Nuclear Research Center, I suggested to a small group of engineers and physicists interested in Natural Philosophy that we read together Weizsäcker's works. After a while, we wanted to meet him personally. In 1982, he invited us to visit him in his home at Lake Starnberg near Munich. I took some notes of our conversation, which were later published in Spanish in the Uruguayan weekly newspaper *Opinar*, whose Founder and Director was Enrique Tarigo (1927–2002), Professor of Procedural Law at the Uruguay State University, Vice President of Uruguay under President Julio María Sanguinetti between 1985 and 1989, the first democratic government after the military intervention, 1973–1985 (Capurro, 1982).

I thank Gordana Dodig-Crnković and Marcin J. Schroeder for giving me the opportunity to answer some questions on Natural Philosophy in relation to a long-standing dialog with Carl Friedrich von Weizsäcker (1912–2007) (Capurro, 2018). In my responses, I will refer mainly to some of Weizsäcker's papers published in 1971 under the title *Die Einheit der Natur* (*The Unity of Nature*) (Weizsäcker, 1974) as well as to his late *magnum opus Zeit und Wissen* (*Time and Knowledge*) (Weizsäcker, 1992). All quotes from original German texts are my translations.

1.1. *On the Unity of Nature*

Question: *Professor Capurro, in the context of the contemporary Natural Philosophy, we would be interested in the project of unity of nature and your view of the ideas of Weizsäcker.*

Weizsäcker's book *The Unity of Nature* is structured in four parts: Part I. Science, Language and Method; Part II. The Unity of Physics; Part III. The Meaning of Cybernetics; and Part IV. On Classical Philosophy (Weizsäcker, 1974). Part I consists of preliminary thoughts on issues dealing with the unity of science that were discussed at that time mainly from a methodological perspective, leaving aside the question of the unity of nature itself (Weizsäcker, 1974, pp. 12–13). Weizsäcker's suggestion that the concept of information can be considered as a new way of understanding foundational concepts of Western metaphysics, such as *eidos* and form gave rise to my PhD inquiry into the etymology of the Latin term *informatio* (Capurro, 1978). The quote is embedded in a lecture dealing with "Language as information" from 1959. Weizsäcker distinguishes between the Platonic and Aristotelian concept(s) of form, on the one hand, and the modern concept of information in the context of human language and communication, on the other hand. Information *as* a structure of whatever kind, natural or artificial, is *prima facie* the opposite of information *as* language. But, as he points out, written language can be extended to artificial languages such as those used in the field of computer technology. There is a "circle" between language and information. Information in the context of scientific methodology concerns the search for certainty based on logic and calculation aiming at a univocal meta-language. But any meta-language remains dependent on natural language in order to be understood (Weizsäcker, 1974, pp. 59–60). Weizsäcker quotes Wittgenstein's famous *dictum* in the preface of the *Tractatus*: "Was sich überhaupt sagen läßt, läßt sich klar sagen, und wovon man nicht reden kann, darüber muß man schweigen" (Wittgenstein, 1984, p. 9; Weizsäcker, 1974, pp. 49–50). The standard English translation "What can be said at all can be said clearly; and what we cannot talk about we must pass over in silence" obliterates the difference between "speaking about" ("Sprechen über") and "speaking from" ("Sprechen von"), that is to say, between language as a tool and language as a source of meaning. Although Weizsäcker is aware of this difference, he does not reflect on the wording of Wittgenstein's *dictum*. Wittgenstein himself is not aware that he is already "speaking from" when he states that there is something "about which" we cannot speak without distorting the phenomenon at stake. This difference is analyzed by Heidegger in his essay from 1953–54 "Aus einem Gespräch von der Sprache" translated as "A dialogue on language" (Heidegger, 1975). Wittgenstein critically reviews his earlier position in the direction that our capacity of speaking "from language" allows a diversity of "language games" ("Sprachspiele") (Wittgenstein, 1984a). According to Weizsäcker, the unity of modern science is methodologically grounded on the quest for a universal univocal language that would correspond to the unity of its object, something Weizsäcker problematizes aporetically in Part II of this anthology.

In an unpublished text from 1969 with the title "Matter, Energy, Information", he comes back to the question of how the ontological and the epistemological meaning of information can be brought together in thought. Information is a bifurcated category: (1) information is only that which is understood; (2) information is only that which generates information. Definition (2) goes back to the Platonic and Aristotelian concept of form while, according to definition (1), information is a linguistically univocal concept. He underscores that information as "a measure for the amount of form" is something that can be potentially known (Definition (1)). An organism is conceived *as* the product of genetic information (Definition (2)). Weizsäcker calls such generated forms "objectivized semantics" (Weizsäcker, 1974, p. 351). He puts it concisely: "Matter *has* form, consciousness *knows* form" (Weizsäcker, 1985). This is a Platonic–Aristotelian thesis as well as a Kantian one. We cannot know things in themselves but only insofar as they can be grasped by a finite knower, which is the reason why we cannot have fully univocal concepts (Capurro and Hjørland, 2003, pp. 362–363). A corollary of this thesis is that the question concerning the unity of nature is epistemologically grounded on a finite human knower. This transcendental argument is cut across by the phenomenological perspective on time.

The essays in the book *The Unity of Nature* were written between 1959 and 1970 with one exception, namely, a short article published in 1939 in the *Annalen der Physik*, Vol. 36, p. 275, dealing with "The Second Law of Thermodynamics and the Difference between Past and Future" (Weizsäcker, 1974). The guiding thought is the question of irreversibility of natural processes that plays a key role in Quantum Mechanics. The difference between past and future is seen not just as a difference in "practical life", but as being foundational for physics as well. Weizsäcker writes that the statistical interpretation of the Second Law of Thermodynamics is the place where the (three-dimensionally open) structure of "real time" manifests itself in the "worldview of physics" (Weizsäcker, 1974, p. 172), (but only *as* one-dimensional, linear time). In the introduction to this article written 30 years later, Weizsäcker writes that when he wrote it, he had the feeling of saying something obvious, particularly for empiricists and positivists, describing temporal events as they are phenomenologically given. But, he adds, "I perceived with certain astonishment that most physicists regarded these thoughts as being rather strange" ("Mit einem gewissen Erstaunen habe ich dann bemerkt, daß die meisten Physiker diese Gedanken als eher fremdartig empfanden") (Weizsäcker, 1974, p. 172). Weizsäcker was influenced by Heidegger's phenomenology of time, more precisely, by the interpretation of being *as* three-dimensional time with its future, present and past dimensions, the opposite to the common understanding of time as a sequence of now-instants, taken for granted by "most physicists". Three-dimensional time becomes apparent and "obvious" for Weizsäcker in the Second Law of Thermodynamics if the observer dares to open his eyes to this obvious but hitherto "strange" perspective.

In 1977, one year after Heidegger's death, Weizsäcker wrote a contribution to a book *Remembering Martin Heidegger* with the title "Encounters over Four Decades" (Weizsäcker, 1977). He describes his first personal meeting with Heidegger (1889–1976) in his hut in Todtnauberg (Black Forest) in 1935 when he was 23 years old. Other participants of this meeting were the physiologist Johann Daniel Achelis (1898–1963), the physician and physiologist Viktor von Weizsäcker (1886–1956), an uncle to Carl Friedrich, the physicist Werner Heisenberg (1901–1976), and the art historian Kurt Bauch (1897–1975). None of them, as far as I know, were able to (fore-)see the imminent catastrophe, as announced in Adolf Hitler's *Mein Kampf* published ten years before. The discussion dealt mainly with the "inseparability of subject and object" in Quantum Theory as well as in medicine. After this meeting, Weizsäcker visited Heidegger usually every two years. Weizsäcker recounts that he read *Being and Time* in Winter 1933–34 in Copenhagen when he was working with Niels Bohr (1865–1962), while reading Kant at the same time. He invited Heidegger three times to be his guest in his seminar in Hamburg. Each time when he visited him, Heidegger, "who could not follow him in mathematics", paid particular attention when he spoke about the foundations of mathematics or about Weizsäcker's concept of "temporal logic". During his last visit in 1972, Heidegger asked him to report on his treatment of time in physics and logic, and pointed him to a passage in *Being and Time*. Weizsäcker reproduced this first encounter also in Weizsäcker (1992, p. 944).

It is no wonder that Weizsäcker's *magnum opus*, *Time and Knowledge*, starts with a quote from *Being and Time*, namely, the last sentence, "Does *time* itself manifest itself as the horizon of *being*?", preceded by a quote from Parmenides "The same is knowledge and being" (Weizsäcker, 1992, p. 27; Heidegger, 1976, p. 437). He acknowledges that Heidegger was for him more "a master" than "a partner" and that he would "run short of strength" in writing an essay on him for this book (Weizsäcker, 1992, p. 34). Nevertheless, Heidegger's thought is explicitly mentioned in many passages of the book. Heidegger's understanding of time as the horizon of being is a perspective "from above", or "from the human being", as distinguished from, but complementary to, his own perspective "from below", that is to say, from "the animal or from nature" (Weizsäcker, 1992, p. 186). Weizsäcker projects a kind of anthropocentrism in his interpretation of Heidegger's *Being and Time* that was intended to be only a first step toward a view of time as the horizon of being and not only of the human being. When reflecting about three-dimensional temporality as the foundation for understanding Quantum Mechanics (Weizsäcker, 1974, pp. 223–275), Weizsäcker follows the tradition that binds the question of time to the question of movement (*kinesis*) of beings, as stated prominently by Aristotle, to whom Weizsäcker refers (Weizsäcker, 1974, pp. 428–440, 1992, pp. 860–861). Heidegger, in fact, had already turned the question upside down in the last sentence of *Being and Time*. Being *as* (three-dimensional) time is what enables the presence or absence of beings and their particular movements in and as temporal. Human temporality is a prime example of the phenomenon of three-dimensional time, not an anthropological foundation for the question of being, as it may seem. Weizsäcker learns from Heidegger's criticism of Western metaphysics and particularly of Platonism that "our philosophy has what corresponds to the *eidos* in time, and to think the idea in history" and he quotes the last sentence of *Being and Time* again, and once more a few pages after (Weizsäcker, 1974, p. 504, 533). For Plato the ideas are beyond time, for Weizsäcker three-dimensional time is the very horizon of ideas. He conceives evolution in terms of "growth of information" ("Informationswachstum") being aware, once more, that not only is natural evolution subject to time but also human *logos*. Our understanding of nature is "logomorph", that is to say, subject to epochal changes and recastings. This is a critical stance with regard to Kantian transcendentalism. He acknowledges the contributions by Konrad Lorenz (1903–1989) and Karl Popper (1902–1994), while rejecting their "realism" that he considers "a kind of faith in an historical epoch of physics" (Weizsäcker, 1974, p. 877). He writes, "Reality ("Wirklichkeit") ("Being") means in Greek philosophy, as Heidegger has made it clear for us, first of all *presence* ("Gegenwart"). The fact that something is, can be strictly said if it is *now* the case. Presence is a mode of *time*" (Weizsäcker, 1974, p. 862). This being said, he immediately refers to the *ontic* issue of *beings* in movement. He writes, "What is, changes; what is, *is in movement*. Movement, that we expect, is *future-oriented*. It is *possible*, and the continuum is the field of possibilities. Insofar continuity is primarily continuity of future time, but at the same time, continuity of possible states, that can take place in time. Possibility is, as it is sometimes metaphorically said, the presence of the future. Instead of actuality, I say now *facticity*. It is the presence of the past, facts are irreversible, the past does not change any more. This is why when facts are described, it is possible to abstract from movement. This is why it is possible *to count* them in discrete form. In the *concept* we think that what is possible, but we think about it as what can be once described as a *fact*. Conceived in this way, the concept is the past form of the future" (Weizsäcker, 1974, p. 862). "Movement, that *we* expect" (my emphasis) — indeed! Expectation is not a quality that adheres to beings in movement, but a second-order category that comes into play when the observer is aware of her three-dimensional temporality that corresponds to but is not identical to the temporality of the phenomena unconcealed by Quantum Mechanics. Possibility *is* the presence of the future, not just metaphorically! He agrees with the philosopher Georg Picht (1913–1982) in his interpretation of *Being and Time* as excluding the analysis of the being of nature as well as of society (Weizsäcker, 1974, p. 1142). This criticism contradicts what is explicitly stated in the last sentence of the book quoted by Weizsäcker in the introduction of his book. Being as time is focused in *Being and time*, as a paramount example, on the mode of being of humans that he calls "Dasein", a being whose being is characterized by having to face the task of dealing with the temporality of her/his own being in its openness to a horizon of "unconcealment" for beings to be what they were, are and can be. The presentation of beings in their totality is traversed by three-dimensional temporality. Every form of temporal "*un*concealment" implies an "un*concealment*". What is unconcealed in the past is not of the same kind of what is unconcealed in the present or in the future. Likewise, what is concealed in the past is not of the same kind as what is concealed in the present or in the future. This three-dimensional structure of temporality does not allow us to have a full view of what was, is or will be, that is to say, to expect a total enlightenment on the totality of being. Heidegger follows *prima facie* the Kantian critique of pure reason while criticizing its transcendental foundation in subjectivity. We are, from the ground up, finite temporal beings sharing with others a world facing a unity of being(s) that unconceals and conceals itself in three-dimensional time. The critique of the traditional concept of time as presence-oriented, demonstrated with the empirical example of a being whose being is exposed to three-dimensional temporality, is, in Popperian terms, a falsification of the mainstream of the Western theory of one-dimensional linear time that underlies also modern science, where being as presence is implicitly a temporal interpretation in which past and future are homogenized into a linear series of now-instants. The degree of *temporal permanence* and *integrity of what is present*, the Platonic *eidos* and the Aristotelian form, is the temporal measure of being. The question about the unity of nature is a question of time, dealing critically with the scientific ambition that the *unity* of nature, understood as something permanent and present, can be objectively grasped by a meta-historical univocal language by an observer beyond three-dimensional time, knowing *at once* what was, is and will be — a delusory song of sirens.

Weizsäcker's paths of thinking show that what remains "excluded" in *Being and Time* (1927) — but was already addressed by Heidegger in the

1929–30 lectures, published only in 1983 (Heidegger, 1983), dedicated to his PhD student, the phenomenologist Eugen Fink (1905–1975) — could be included in a reflection on the foundations of physics, particularly when facing the phenomena unconcealed in Quantum Mechanics (Weizsäcker, 1992, pp. 896–903). This exclusion of the question of being as time with regard to nature is what Weizsäcker and Picht object to in the *provisional* foundation of the question of being in *Being and Time*, where the analysis deals with the way of being of a human as "Dasein", who is explicated *as* finding himself/herself facing an openness of past, present and future that he/she shares with others. Nothing is farther from the modern invention of a "world-less" isolated, encapsulated subjectivity facing objects in a so-called outside-world (Boss, 1979). The analysis of being as the three-dimensional time of human existence was chosen for the simple reason that we are the beings who, as far as we know, face three-dimensional time as a question concerning *who* and not just *what* we are in a world that we share with others from the ground up. The naturalistic fallacy consists in confusing *what* and *who* (Capurro *et al.*, 2013). The being of a who doing the observation lets other phenomena appear *in their own* three-dimensional temporal perspective, but *only as* interpreted by an observer of the kind we are. The question of being as the three-dimensional temporality of human "Dasein" does not mean some kind of human supremacy or hierarchy of beings as addressed by Western metaphysics in different forms, including Weizsäcker's bottom-up *ontic* perspective. Even the idea of some kind of complementarity between top-down and bottom-up approaches fails to see the underlying *ontological* question about being as three-dimensional time. On the basis of Weizsäcker's bifurcated concept of information, Holger Lyre has developed "a quantum theory of information" (*Ur-Theorie*) with "basic alternatives" (*Ur-Alternativen*) representing the information content of a yes/no decision or one bit of quantum-theoretic potential information (*Ur*). *Urs* are potential information (Lyre, 1998; Capurro and Hjørland, 2003, p. 363).

1.2. *Weizsäcker and Wheeler on Information*

Question: *How do you see the connection between Weizsäcker and Wheeler's "It from bit"?*

Weizsäcker deals with "The Meaning of Cybernetics" in Part II of the collection of articles on *The Unity of Nature* (Weizsäcker, 1974). He refers to Claude Shannon's seminal article "The Mathematical Theory of Communication" (Shannon, 1948). His focus is the cybernetic interpretation of living beings based on what he calls "objectivized semantics", that is to say, that "information measures the amount of form" and "according to ancient philosophy form (*eidos*) is exactly that which can be known" (Weizsäcker, 1974, p. 348). The form (*eidos*), as addressed by Weizsäcker in both the ontological and the epistemological sense, is a reversed Platonic perspective. For Plato, the ideas and their knowledge are beyond time. For Weizsäcker, time is the horizon for the development of forms as well as for our knowledge about them.

When John Archibald Wheeler (1911–2008) speaks about "it from bit", it is the opposite of Weizsäcker's forms embodied in natural evolution, being able to be potentially understood by a human knower. Forms for Wheeler are bits *underlying* any material evolution (Wheeler, 1989). Wheeler writes, "It from bit symbolizes the idea that every item of the Physical World has at bottom — at a very deep bottom, in most instances — an immaterial source and explanation; that what we call reality arises in the last analysis from the posing of yes–no questions and the registering of equipment-evoked responses; in short, that all things physical are information-theoretic in origin and this is a *participatory universe*" (Wheeler, 1989, p. 311).

Wheeler celebrates the fuzziness of his concept of bit as follows: "Finally: Deplore? No, celebrate the absence of a clean clear definition of the term "bit" as elementary unit in the establishment of meaning. We reject "that view of science which used to say, 'Define your terms before you proceed.' The truly creative nature of any forward step in human knowledge", we know, "is such that theory, concept, law and method of measurement — forever inseparable — are born into the world in union" (Wheeler, 1989). If and when we learn how to combine bits in fantastically large numbers to obtain what we call existence, we will know better what we mean both by bit and by existence. A single question animates this report: Can we ever expect to understand existence? Clues we have, and work to do, to make headway on that issue. Surely someday, we can believe, we will grasp the central idea of it all as so simple, so beautiful, so compelling that we will all say to each other, "Oh, how could it have been otherwise! How could we all have been so blind so long!" (Wheeler, 1989, p. 322). This is digital Platonism or Pythagoreanism. That "theory, concept, law and method of measurement — forever inseparable — are born into the world in union" is a step forward beyond the Cartesian split. A step back would take him to the "obvious" three-dimensional openness of time: "How could we all have been so blind so long!" Indeed!

1.3. *The Driving Force of Intellectual Innovation*

Question: *What is your view of the change of the role of physics in philosophical or intellectual discourse? Physics was at the time of the interview the driving force of intellectual innovation in thinking about the world. Now, physics shares this role with (or maybe has lost it to) biology, or rather to the study of life. Also, there is a strong interest in the contrast between artificial/synthetic and natural, and the possibility of their hybridization. It would be interesting to learn from you which of Weizsäcker's views remain intact in these changes of the roles of disciplines and themes. Which of the views lost either interest of philosophers or became challenged by scientific development?*

Neither physics nor biology but computer science is, I believe, "the driving force of intellectual innovation in thinking about the world". Digitization is today's challenge not only for society but for science and philosophy as well — the dissolution of "it", nature for instance, into "bit". This is different from Weizsäcker's time as well as from thinking "about" and "from" nature and language. Weizsäcker reflected on cybernetics against the background of natural science. We do the opposite, not only in physics but also in biology and other sciences, as well as in everyday life, in the organization and management of enterprises, in the economy and in shaping the *res publica* locally and globally. More fundamentally, we can say that underlying all these ontic areas of application of the stepwise, yes/no, algorithmic thinking of computer science, digitizability as such is the ontological horizon within which we — and it is of paramount importance to think about who is this "we" and who is not — understand not only beings, conceiving them *as* such-and-such, but also being itself. What can be *understood*, in its being, is what is digitizable. I call this foundational understanding of being *digital ontology* (Capurro, 2017). Ontology means, in this context, the way we, humans, *understand*, that is to say, *conceive*, the being of being through *logos*. It is, as traditionally spoken, an epistemological thesis that should not be misunderstood in the sense of digital Pythagoreanism or digital metaphysics as suggested by Wheeler, which aims at explaining "existence" based on what "bits" are supposed to be.

Weizsäcker's views on both levels — the ontic as related to nature, and the ontological dealing with being as three-dimensional time — remain intact as untimely meditations. This is particularly the case with regard to the present (one-dimensional linear) *time regime* on which computer technology is based within the framework of digital ontology. This time regime prolongs the one questioned by Weizsäcker in his 1939 article "The Second Law of Thermodynamics and the Difference between Past and Future" (Weizsäcker, 1974). One can object that digital technology is not present- but future-oriented and that it provides an almost infinite memory capacity. But in both cases what counts, in both senses of the verb, is the traditional understanding of time as a sequence of homogenous now-instants for which being as presence is foundational.

The issue of "hybridization" between the "artificial" and the "natural" provides evidence that the epochal changeover from the natural to the digital does not mean that this is a question of either/or, but of carefully reflecting what is being concealed or unconcealed in each case as well as in various combinations. This is particularly apparent when nature provides the model upon which the artificial can be built, as in the case of robotics, an issue that has been extensively analyzed by Negrotti (2002). The key question is not what *intelligence* is, but what is meant by *artificial* and how far the time regime of digital and/or hybrid artefacts of any kind affects, that is to say, distorts the "obvious" phenomenon of being itself as the openness of three-dimensional time.

1.4. *Digital Ontology*

Question: *What would be the most important message or messages from Weizsäcker to us?*

If we go back to the title of Weizsäcker's book *The Unity of Nature*, and if we understand this title as a question, then the answer is that the unity of nature is the question of time. Weizsäcker's short paper from 1939 was strange or even heretical at that time and it remains even more so today, in the digital age. It concerns also the very idea of the unity of nature no less than of the unity of our knowledge. Weizsäcker gave an answer to the last issue by considering the "circle" between language and information. But he did not address the kind of unity created by digital technology that we call globalization or simply the Internet. The philosophical challenge today is to give an answer to digital technology that retrieves the thesis of being *as* time. If we want to understand what being means in the digital age, we must go back to being as time, just as Weizsäcker did with regard to Quantum Mechanics. This ontological interpretation of the digital age is nothing technological or even digital but just the insight into what makes possible its presence-based temporality. What can we (who?) gain from this "obvious" insight? Nothing more and nothing less than the possibility of a free relationship to it that arises when life is lived as a three-dimensional temporal openness that we share with others in a common world (Boss, 1979). The question concerning the unity of nature is a question of time in the sense that nature conceals and unconceals itself in its own way(s). Its *unity* is in fact a temporal worldly *trinity*, not a metaphysical or even a theological one, but one to which the temporality of human existence corresponds in its own way. This kind of correspondence is what Weizsäcker was looking for all his life as a physicist and also as a *political being* taking responsibility for the consequences of his research within the limits of what was unconcealed and of his awareness thereof.

In his 1982 interview, Weizsäcker summarized his view on the unity of nature as well as of our knowledge of it as follows (Capurro, 1982):

Question: Heidegger says that science does not think.

Answer: Which must be understood in the sense that normal science — in Thomas Kuhn's terminology — does not think, that is to say, it does not question its own paradigm. In contrast to Edmund Husserl, I believe that the basic shape of a phenomenon is not something unchangeable but takes place in time and can change according to experience. In Plato's terminology, I would say that our challenge is to rethink the "idea" of time. We cannot escape three-dimensional time. I remember an anecdote about Einstein asking Carnap and Popper "what is time?" without receiving an acceptable answer. Einstein tried to eliminate time. Shortly before he died, when somebody told him that a friend had died a few weeks ago, he answered that "a few weeks ago" is an illusion. For a physicist, as a believer, the difference between past, present and future is an illusion. I think, quite the contrary, that the introduction of three-dimensional time, that was made explicit by Heidegger, can bring about a new revolution in physics that would also resonate in philosophy. Already in Plato and Aristotle we find the key concept of *kinesis*, that is to say, movement and change. *Chronos*, literally translated as time, is already for Plato a derivative representation of *aion*, a term that I would like to leave untranslated. It implies the temporal structure and refers, for instance, to a person's lifetime.

Question: Physics and philosophy meet here, in the question concerning time, ...

Answer: ... as if Magellan and Vasco da Gama, starting from different points and in opposite directions, would meet during circumnavigation.

Translating this conversation — originally in German, then published in Spanish — 35 years later is an amazing experience, not only because my thinking and, of course, my life have changed, but particularly because the world and our understanding of it have changed. Nobody could have predicted at the time the invention of the Internet, although it was not far away. As with other epochal changes, the digital age has the tendency to overestimate itself and even to believe that this is finally the truth bringing us (whom?) the foundation and the tools needed to understand the unity of nature and of ourselves.

The concept of information is multifaceted inside the "circle" of language mentioned by Weizsäcker. This is not something to deplore or to celebrate in a kind of postmodern attitude. It is just a question of contextualizing its different meanings in such a way that equivocity, analogy and univocity are analyzed as clearly as possible. This clarity depends also on the different languages in which information was and is being translated, starting with the concepts of *eidos* and form as Weizsäcker wrote in the late 1950s (Weizsäcker, 1974, pp. 39, 51). If information as *informatio* is a Latin translation of Greek concepts, it is also true that such concepts were translated in Late Antiquity into Arabic, Persian and Hebrew, which is a fascinating area of research (Capurro, 2014, 2015). Not only the unity of nature but also the unity of language is a question of time. Consequently, following Weizsäcker, Heidegger and Wittgenstein, language understood from the perspective of three-dimensional temporality, is something we can only speak "from" but not "about". This is the reason why translation is a core ethical issue going beyond the fact of language diversity and looking for ways of sharing understanding of what was concealed and unconcealed in what was said, what is being said and what can be said (Capurro, 2017). This is the most important message from Weizsäcker to us.

2. Unity of the World and its Temporal Dimensions

2.1. *Unity of the World*

To scientifically explain relations between information and the Platonic and Aristotelian concepts of *eidos*, Ideas or Forms, it is necessary to consider the structure of the world and the place in it for ideas, forms and information. This structure is represented by the *Existential Triad* composed of the top-level components of the world as a unified whole and reflecting unity of

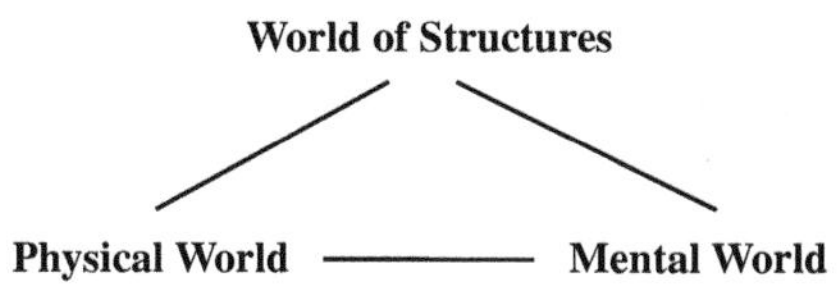

Figure 1. The Existential Triad of the world.

the world (Figure 1). This triadic structure is rooted in the long-standing tradition coming from Plato and Aristotle (Burgin, 2010, 2012).

In the Existential Triad, the Physical (or material) World represents the physical reality studied by natural and technological sciences, the Mental World encompasses different forms and levels of mentality, and the World of Structures consists of various kinds and types of structures.

According to physics and other natural sciences, people live in the *Physical (material) World* and this is the only reality that exists. In contrast to this opinion, idealistic and subjective philosophies as well as some religious teachings assert that only the Mental World is real, while the Physical World is just an appearance, a shadow without substance. For instance, Buddhism explains that the whole physical reality is a great illusion and the only reality is the spiritual world. In some forms of Buddhism (e.g., *Yogacara*), it is assumed that all things are created by Mind (Gethin, 1998). At the same time, achievements of physics brought scientists to the edge of physical reality where extremely small physical objects, such as quarks or electrons, and extremely large physical objects, such as our universe as a whole, do not allow direct comprehensive observation. As a result, both extremities become more structures than material things. In this context, the outstanding physicist Max Born admits that the notion of reality in the Physical World had become, during the last century, rather problematic (Born, 1953).

The Physical (material) World (PW) is the most familiar, accustomed and understandable of the three components comprising the Existential Triad because people comprehend material things mostly by their senses. Technical devices, such as microscopes and telescopes, extend cognitive abilities of people but are still tied to senses such as vision. That is why from the very beginning of physics as a science, physicists have explored this world trying to reduce everything to physical reality (cf., for example, Born, 1953). Nevertheless, many people encountered in their life various illusions and misapprehensions of physical things. This situation brought many thinkers to the conclusion about the illusory character of the Physical (material)

World or at least, about its imperfection and direct incomprehensibility. Such views are expressed in Indian Vedas (cf., for example, Balasubramanian, 2000) and the teaching of Buddhism (cf., for example, Gethin, 1998), as well as in works of Western philosophers such as Plato, John Locke, George Berkeley and David Hume. Besides, practically all religions assert that reality includes the Physical (material) World but cannot be reduced to it as it contains other levels of being.

In contrast to the radical views that exclude existence of the Physical (material) World, the Existential Triad of the World contains this World as one of its top-level components.

Coming to the Mental World, we see that for a long time people comprehended and studied it only at the level of individual mentality. Later this picture was extended to the first (lowest) three levels, which are studied by psychology and cognitive sciences and all of which are subcomponents of the Existential Triad (cf. Figure 2). In essence, as physics does not study the physical reality as a whole, psychology also separates and investigates only some components and facets of mental reality, such as intelligence, emotions or unconscious.

- The *first level* in the basic mental hierarchy encompasses mentality of individuals and is the subject of psychological and cognitive studies. As in the case of physical reality, now psychology knows much more about mentality/psyche of people than even the best minds knew at the time of Plato and Aristotle.
- The *second level* entails group mentality, including organization and institutional mentality in the domain of social psychology. This level includes

....

↑

Earth mentality

↑

Civilization mentality

↑

Social mentality

↑

Group mentality

↑

Individual mentality

Figure 2. The lower levels of the hierarchy of the Mental World (the *basic mental hierarchy*).

social conscience with its collective memory, collective wisdom, collective intelligence and computational collective intelligence.
- The *third level* encompasses mentality of society and is studied by social psychology.

Higher levels are discussed elsewhere because in addition, the Mental World from the Existential Triad comprises higher (than the third) levels of mentality, although they are not yet studied by science (Burgin, 2010). Upper levels of the Mental World are related to the spiritual worlds described in many religious and esoteric teachings.

The essential difference between the Mental and Physical Worlds is expressed by Putnam's *Thesis of Multiple Realization*, which suggests the total irrelevance of physico-material properties for distinguishing mental states (Putnam, 1960). Such a phenomenon as functional isomorphism when different material objects turn out to be equivalent of the mental perspective employing the same program gives supportive evidence for this thesis.

Note that although the most popular opinion is that only the Physical World exists, there are other opinions claiming that the Mental World is the overall reality of people (cf., for example, Kastrup, 2019).

The third component of the Existential Triad was foreseen by Plato as the World of Ideas/Forms, but was actually discovered only recently as the World of Structures (Burgin, 2011, 2012). In a similar way, atoms were foretold by ancient atomists but were discovered much later by physicists. By its essence, the World of Structures is a scientific counterpart of the World of Ideas/Forms, which are associated with structures in a natural way (Burgin, 2017). The exact scientific and mathematical concept of structure puts the notion of Idea/Form into the domain of science. Indeed, at the level of ideas and abstract entities, it is possible to link Ideas or Forms of Plato and Aristotle to structures in the same way as atoms of modern physics may be related to the atoms of Democritus. Only recently, modern science came to a new understanding of Plato's ideas, representing the global world structure as the Existential Triad of the world, in which the World of Structures is much more understandable, exact and explored in comparison with the World of Ideas/Forms of Plato and Aristotle. When Plato and other adherents of the World of Ideas/Forms were asked what an Idea or Form was, they did not have a satisfactory answer. In contrast to this, many researchers have been analyzing and developing the concept of a structure (Lautman, 1938; Bourbaki, 1957; Bastide, 1962; Corry, 1996; Landry, 1999; Cottrell and Pettifor, 2000; Burgin, 2011, 2012). It is possible to find the most comprehensive analysis of structures and the most forward-thinking concept of a structure in Burgin (2012) where elements of the general theory of structures are presented. As a result, in contrast to philosophy, science has been able to develop a sufficiently exact definition of a structure proving the existence of the World of Structures and demonstrating by means of observations and experiments that this World encompasses the structural level of the two other worlds. Indeed, each system, phenomenon or process either in nature, technology, society or mentality has some structure and usually not a single one. Forming the *structural level* of the Physical World, embodied material, e.g., physical or chemical, structures exist like material things, such as trees, rivers, planets, stars, mountains or oceans. When it is necessary to learn or to create a system or to start a process, it is done, as a rule, by means of knowledge of the corresponding structure. Structures determine the essence of things. This is what Plato asserted about Ideas and Aristotle about Forms.

Taking for granted that embodied structures exist in material things, we have to acknowledge that material things exist only in a structured form, i.e., matter (physical entities) cannot exist independently of structures. For instance, atomic structure influences how the atoms are bonded together, which in turn helps one to categorize materials as metals, ceramics and polymers permitting the drawing of general conclusions describing the mechanical properties and physical behavior of these three classes of materials. Even chaos has its structure and not a single one.

Besides, embodied mental structures exist in and like mental objects such as dreams, plans, emotions, instincts, needs or images because any mental object has and depends on its structure. The components of individual mentality have structures studied in psychology. For instance, psychoanalysis assumes that the human psyche has the structure of the fundamental triad presented in Figure 3.

In a similar way, the mind has the structure of the fundamental triad presented in Figure 4.

Unconscious ↔ Pre-conscious ↔ Conscious

Figure 3. The structure of the human psyche.

Id ↔ Ego ↔ Superego

Figure 4. The structure of the mind.

According to psychoanalysis, these parts have different characteristics: Id is the irrational and emotional part of the mind, the Ego is the rational part and the Superego is the moral part.

At the same time, it is necessary to understand that pure abstract structures from the World of Structures also exist but it is possible to reach them only through intelligence and contemplation. In other words, people can reach and discover structural "continents" using "vessels of intelligence".

Thus, it is pointless to search for the World of Structures in the surrounding people and physical reality in the same way as it is futile to look for Mount Chomolungma in France. Embodied structures can be treated as projections (shadows) of pure abstract structures from the World of Structures. It is possible to comprehend many embodied structures by the senses seeing and hearing some of them (Burgin, 2011, 2012).

After describing the three Worlds from the Existential Triad in themselves, it is necessary to explain that they are not separate and closed into themselves as systems: they interact, interrelate and interconnect, molding the unity of the world.

Indeed, we can see that individual mentality is based on the brain, which is a material thing, while according to the opinion of many physicists, mentality influences the Physical World (cf., for example, Herbert, 1987). At the same time, our knowledge of the Physical World mostly (if not completely) depends on the interaction between Mental and Material Worlds (cf., for example, von Bayer, 2004).

In addition, there is a projection of the Mental World into the Physical World in the form of material creations of human mentality (creativity), such as books, buildings, movies, vehicles, magazines, planes, newspapers, chairs, computers and cell phones. This projection determines the Extended Mental World, which consists of the Mental World and its projection.

Further, there is a projection of the Physical World into the Mental World in the form of mental images of physical objects and systems. These mental images can be simple reflections of physical objects or include (consist of) properties of these objects. Mental images can have an abstract character, being pure structures. For instance, physicists create images of physical systems in the form of differential equations. In a similar way, computer scientists create images of computing systems in the form of pure structures, such as computer architecture or algorithm.

Structural and material worlds are even more intertwined. Actually, no material thing exists without a structure. Even chaos has its chaotic structure. Structures do things what they are. For instance, it is possible to make a table from different materials, e.g., from wood, plastics, iron, aluminum, etc. What all these things have in common is not their material. Their common attribute is specific peculiarity of their structure. In a similar way, as Poincaré (1908) wrote, space is in reality amorphous, and it is only the things that are in it that give it a structure (form). As some physicists argue, physics studies not physical systems as they are but structures of these systems, or physical structures. This idea is the essence of the philosophical direction called structural realism, which maintains that people can learn (cognize) only structures of real things but not real things themselves (Maxwell, 1970; Zahar, 1996; Gower, 2000). In some sciences, such as chemistry, and areas of practical activity, such as engineering, structures play a leading role. For instance, the spatial structure of atoms, chemical elements and molecules determines many properties of these chemical systems.

Contemporary physics treats the Physical World as a net of interacting components (systems), where there is no physical meaning to the state of an isolated object. A physical system (or, more precisely, its contingent state) is represented by the net of relations it entertains with the surrounding objects. As a result, the physical structure of the world is identified with such a global net of system relationships.

Knowledge of the global structure of the world allows scientific explanation of the phenomenon of information and positioning it in the world structure. We will do this in the next section.

2.2. *The Place and Function of Information in the World*

The place and the role of information in the world are delineated by the ontological principles of the general theory of information, which is constructed as an axiomatic theory with three levels: *conceptual*, *methodological* (also called *meta-theoretical*) and *theoretical* (Burgin, 2010, 2017a).

At the conceptual level, the essence of information as a dynamic object playing a pivotal role in all walks of reality is explicated, clarifying a quantity of misconceptions, fallacies and illusions.

The methodological (meta-theoretical) level is based on two classes of principles and their interrelations. The first class contains ontological principles, which describe general properties and regularities of information and its functioning. Axiological principles form the second class explaining how to measure information.

At the theoretical level, axioms of theoretical (mostly mathematical) structures used and axioms reflecting features of information are introduced and utilized for building models of information and related phenomena, e.g., information flow or information processing (Burgin, 2010a, 2011a). These models are used in studies of information and various related systems and processes, e.g., information flow in society or information processing systems, such as computers and networks.

To position information in the structure of the world, we use the basic ontological principles. The first of them separates local and global perspectives on information, defining in what context information is considered.

2.2.1. *Ontological principle O1 (the locality principle)*

It is necessary to distinguish information in general from information (or a portion of information) for a system R.

It means that empirically it is possible to consider only information (or a portion of information) for a system. The system $\boldsymbol{R}$ with respect to which some information (a portion of information) I is considered is called the *receiver*, *receptor* or *recipient* of I. Such a receiver/recipient can be a person, organization, community, a group of people, animal, bird, fish, tree, computer, cell phone, network, database and so on.

The Locality Principle allows building mathematical models of information in the form of operators (Burgin, 2010a, 2011a).

We see that the Locality Principle describes an imperative property of information, but does not explain what information is. The essence of information is presented in the second ontological principle, which has several forms.

2.2.2. *Ontological principle O2 (the general transformation principle)*

In a broad sense, information for a system R is a capacity to cause changes in the system R.

It means that it is viable to understand information in a broad sense as a capacity (ability or potency) of things, either material or mental or abstract, to change other things. Information as a tangible essence exists in the form of portions of information. Informally, a *portion of information* is such information that can be treated as a coherent system.

The Ontological Principle O2 has several consequences. First, it demonstrates that information is closely connected to transformation (Burgin, 1998/1999). It means that information and transformation are functionally similar because they both point to changes. However, they are also different because information is potency for (or cause of) change, while transformation is the change itself. In other words, transformation is an operation, while information is what induces this operation.

Second, the Ontological Principle O2 explains *why* information influences society and individuals all the time, as well as why this influence grows with the development of society. Namely, reception of information by individuals and social groups induces transformation. In this sense, information is similar to energy. Moreover, according to the Ontological Principle O2, energy is a kind of information in a broad sense. This correlates well with von Weizsäcker's idea that *energy might in the end turn out to be information* (Weizsäcker, 1974).

Third, the Ontological Principle O2 makes it possible to separate different kinds of information. For instance, people, as well as any computer, have many kinds of memory. Moreover, it is conjectured that each part of the brain has several types of memory agencies that work differently to match particular purposes. Thus, it is possible to consider each of these memory agencies as a separate system and to study differences between information that change each type of memory. This might help to understand such properties as stability and flexibility of mind, in general, and memory, in particular. Thus, all kinds and types of information are encompassed by the Ontological Principle O2. In the most concise form, it is demonstrated in Burgin (2010).

However, the common usage of the word information does not imply such wide generalizations as the Ontological Principle O2 implies. Thus, we need a more restricted theoretical meaning because an adequate theory, whether of information or of anything else, must be in significant accord with our common ways of thinking and talking about what the theory is about, else there is the danger that the theory is not about what it purports to be.

Information in a proper sense is defined of structural infological systems. In essence, any subsystem of a system may be considered as its infological system. However, information in a proper sense acts on structural infological systems. An infological system is structural if all its elements are structures. For example, systems of knowledge are structures.

To achieve precision in the information definition, we follow two conceptual steps. At first, we make the concept of information relative to the chosen infological system IF(R) of the system R and then we select a specific class of infological systems to specify information in the strict sense. That is why it is impossible and counterproductive to give an exact and, thus, too rigid and restricted a definition of an infological system.

Infological system plays the role of a free parameter in the general theory of information, providing for representation of different kinds and types of information in this theory. That is why the concept of *infological system*, in general, should not be limited by boundaries of exact definitions. A free parameter must really be free. Identifying an infological system IF(R) of a system R, we can define information relative to this system. This definition is expressed in the following principle.

2.2.3. *Ontological principle O2g* (*the relativized transformation principle*)

Information for a system R *relative to the infological system* IF(R) *is a capacity to cause changes in the system* IF(R).

As a model example of an infological system IF(R) of an intelligent system R, we take the system of knowledge of R. It is called in cybernetics the *thesaurus* Th(R) of the system R. Another example of an infological system is the memory of a computer. Such a memory is a place in which data and programs are stored and is a complex system of diverse components and processes.

Elements from the infological system IF(R) are called *infological elements*. The nature of the infological elements depends on the nature of the infological system IF(R).

There is no exact definition of infological elements although there are various entities that are naturally considered as infological elements as they allow one to build theories of information that inherit conventional meanings of the word *information*. For instance, knowledge, data, images, ideas, algorithms, procedures, scenarios, schemas, values, goals, ideals, fantasies, abstractions, beliefs and similar objects are standard examples of cognitive infological elements.

When we take a physical system D as the infological system and allow only for physical changes, information with respect to D coincides with energy.

Taking a mental system B as the infological system IF(R) and considering only mental changes, information with respect to B coincides with mental energy. The concept of mental energy can be traced back to Aristotle who wrote about energy of the soul, which was called *psyche* ($\psi\upsilon\chi\acute{\eta}$) in Greek (Aristotle, 1984). In the 19th century, the concept of psychic energy, also called *psychological energy*, was developed in the field of psychodynamics by Ernst Wilhelm von Brücke who proposed that all living organisms are energy systems governed by the principle of the conservation of energy. Later Sigmund Freud adopted this new idea about energy and suggested that it was possible to apply both the first law of thermodynamics and the second law of thermodynamics to mental processes, describing functioning of a mental or psychic energy (cf. Freud, 1949). In *The Ego and the Id*, Freud argued that the Id was the source of the personality's desires, and therefore of the psychic energy that powered the mind. The psychoanalytic approach assumes that the psyche of people needs some kind of energy to make it work. This energy is used in mental work, such as thinking, feeling and remembering. It is assumed that psychic energy comes from the two main drives: Eros (or libido, the life and sexual instincts) and Thanatos (death instinct). Later Carl Gustav Jung published a seminal essay entitled "On Psychic Energy" (cf. Jung, 1969) while the theory of psychodynamics and the concept of "psychic energy" were further developed by Alfred Adler and Melanie Klein. In the context of the general theory of information, it is demonstrated that psychic energy is a special case of mental energy (Burgin, 2019).

These ideas, which connect concepts of information and energy, are summarized in the following principle which defines information in the strict sense. As it was explained in Section 1, Weizäcker professed that the concept *information* has intrinsic ties to the Platonic and Aristotelian concepts of *eidos*, Ideas or Forms. This forecast is fully explained and substantiated in the general theory of information where it is demonstrated that information *per se* or information in the strict sense belongs to the World of Structures (Burgin, 2010). This conclusion is implied by the following principle.

2.2.4. *Ontological principle O2a (the special transformation principle)*

Information in the strict sense or *proper information* or simply *information for a system R is a capacity to change structural infological elements from an infological system* IF(R) *of the system R.*

Existence of different kinds of infological systems implies existence of different types of structural infological systems. To better understand how infological systems allow explication of the concept of information in the strict sense, we consider cognitive infological systems.

An infological system IF(R) of the system R is called *cognitive* if it contains (stores) elements or constituents of cognition, such as knowledge, data, ideas, fantasies, abstractions, beliefs, etc. A cognitive infological system of a system R is denoted by CIF(R) and is related to cognitive information.

In this case, it looks like it is possible to give an exact definition of a cognitive infological system. However, now cognitive sciences do not know all structural elements involved in cognition. A straightforward definition specifies cognition as an activity (process) that gives knowledge. At the same time, we know that knowledge, as a rule, comes through data and with data. So, data are also involved in cognition and, thus, have to be included in cognitive infological systems. Besides, cognitive processes utilize such structures as ideas, algorithms, procedures, scenarios, images, beliefs, values, measures, problems, tasks, etc. Thus, to comprehensively represent cognitive information, it is imperative to include all such objects in cognitive infological systems.

In this context, the concept of information is considered on three basic levels of generality:

(1) *Information in a broad sense* is considered when there are no restrictions on the infological system (cf. Ontological Principle O2).
(2) *Information in the strict sense* is considered when the infological system consists of structural elements (cf. Ontological Principle O2a).
(3) *Cognitive information* is considered when the infological system consists of cognitive structures, such as knowledge, beliefs, ideas, images, etc. (cf. Ontological Principle O2c).

As a result, we come to three levels of information understanding:

(1) *Information in a broad sense* for a system R is a capability (potential) to change (transform) this system in any way.
(2) *Information in the strict sense* for a system R is a capability (potential) to change (transform) structural components of this system, e.g., cognitive information changes knowledge of the system, affective information changes the state of the system, while effective information changes system orientation.
(3) *Cognitive information* for a system R is a capability (potential) to change (transform) the cognitive subsystem of this system.

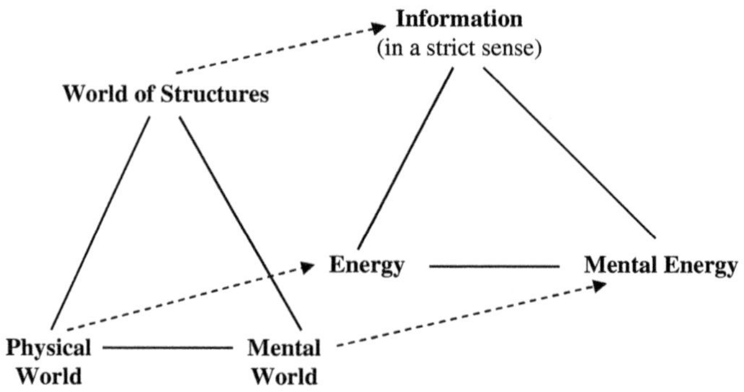

Figure 5. Information in the structure of the world.

This information stratification allows us to place information in a broad sense as a pivotal essence in the world as a whole and all its components. At the same time, information in a strict sense belongs to the World of Structures, also playing a pivotal role there and having its counterparts in two other Worlds. This situation is represented by Figure 5 (Burgin, 2011b). Cognitive information naturally belongs to the domain of cognitive systems.

It is interesting that defining information as "Information: that which determines form", MacKay also associated it with the World of Structures (MacKay, 1969) because *forms* form a special kind of *structures* (Burgin, 2010). It is possible to take the system of knowledge as an example of a cognitive infological system.

Cognitive infological systems are standard examples of infological systems, while their elements, such as knowledge, data, images, ideas, fantasies, abstractions and beliefs, are standard examples of infological elements. Cognitive infological systems are very important, especially for intelligent systems as the majority of researchers believe that information is intrinsically connected to knowledge.

2.2.5. *Ontological principle O2c (the cognitive transformation principle)*

Cognitive information for a system R is a capacity to cause changes in the cognitive infological system IFC(R) *of the system* R.

As the cognitive infological system contains knowledge of the system it belongs to, cognitive information is the source of knowledge changes.

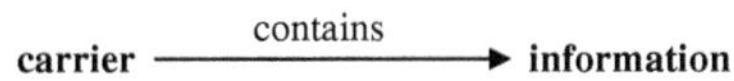

Figure 6. The information *carrier triad*.

Cognitive information is, as a rule, about something. What information is about is called the *object* of this information.

It is useful to understand that in the definition of cognitive information, as well as of other types of information in the strict sense, it is assumed that an infological system IF(R) of the system R is a part (subsystem) of the system R. However, people have always tried to extend their cognitive tools using different things from their environment. In ancient times, people made marks on stones and sticks. Then they used paper. Now they use computers and computer networks.

There are two ways to take this peculiarity into consideration. In one approach, it is suggested to consider *extended infological systems* that do not completely belong to the primary system R that receives information. For instance, taking an individual A, it is possible to include in the extended cognitive infological system IFC(A) of A not only the mind of A but also memory of the computer that A uses, books that A reads and cognitive objects used by A.

Another approach extends the primary system R as a cognitive object, including all objects used for cognitive purposes. In this case, when we regard an individual A as a cognitive system R, we have to include (in R) all cognitive tools used by A. The second approach does not demand to consider extended infological systems. In this case, all infological systems of R are parts (subsystems) of the primary system R.

Cognitive information belongs to the World of Cognitive Structures, such as knowledge, beliefs, ideas, concepts, images, hypotheses, etc.

Although information *per se* or pure information belongs to the World of Structures, it is present in two other worlds. In the Physical World, books, written and printed symbols, electric signals, computers, different memory devices, human brains and actually all physical objects are carriers of information (cf. Figure 6). There are also carriers of information at all levels of mentality in the Mental World.

2.2.6. *Ontological principle O3 (the embodiment principle)*

For any portion of information I, there is always a carrier C of this portion of information for a system R.

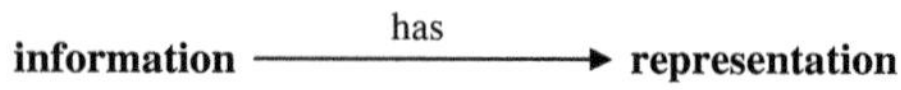

Figure 7. The information *representation triad.*

There are material carriers of information, such as memory, DNA or a book, and there are structural carriers of information, such as a text, symbols or ideas. A specific type of information carrier is formed by information representations (cf. Figure 7).

2.2.7. *Ontological principle O4 (the representability principle)*

For any portion of information I, there is always a representation C of this portion of information for a system R.

As any information representation is, in some sense, its carrier, the Ontological Principle O4 implies the Ontological Principle O3. Both representations and carriers have the structure of the fundamental triad (Burgin, 2011).

Thus, we come to the conclusion that information in the strict sense (information *per se*) is a part of the World of Structures having representations and carriers in two other worlds — the Physical World and the Mental World. This allows us to give scientific interpretation of Wheeler's far-reaching metaphor *It from Bit* (Wheeler). As we know, *It* here represents material (physical) reality, while *Bit* represents information.

Interpreting information (*Bit*) in the strict sense, we see that *It*, that is, the material world, as people comprehend it, comes from (is determined by) information (*Bit*) received by people.

Interpreting information (*Bit*) in the broad sense, we see that as Heisenberg wrote, all forms of matter (*It*) derive both from energy plus a simple yet profound mathematical structure (Heisenberg, 1971). Knowing that energy is information in the Physical World, we see that *It* derives from *Bit* (information).

The broad concept of information also allows us to give scientific interpretation of Heisenberg's in-depth assertion that all forms of matter derive both from energy and from a simple yet profound mathematical structure because "all the other relations, which we thus know in various fields of physics as laws of nature, should derive from this unique structure" (Heisenberg, 1971).

Indeed, it is proved (Burgin, 2011) that all mathematical and physical structures are named sets or are formed from named sets. Taking this into account, we can conclude that all forms of matter derive from both energy (information in the Physical World) and named sets.

2.3. *Time of Being*

As explained in Section 1, Heidegger suggested the idea of being embedded in three-dimensional time, while Weizäcker conceived it as the foundation for understanding Quantum Mechanics. To better understand this idea, let us analyze the meaning three-dimensional time using the system theory of time (Burgin, 2002).

The system theory of time has three roots:

- Aristotle's ideas about time;
- Poincare–Einstein's principle of time relativity;
- Bertalanffy's systems theory.

One of the principal additions of the system theory of time to Poincare–Einstein's principle of time relativity is relativity of the time scale. While Poincare, Einstein and the majority of other physicists treated time as a one-dimensional characteristic of the world, in the system theory of time, there is no only universal time but also no one universal time scale. In particular, this principle allows for three-dimensional time.

The system theory of time not only expanded the concept of time relativity from the spatial level to the system level but also discovered new structural constituents of time. For a system, in general there are the following categories of time:

- *Inner time* of a system R is determined by changes going on at the system level of R, e.g., changes in a cell of an organism are not on the system level of this organism.
- *Internal time* of a system R is inner time of some subsystem of the system R.
- *External time* of a system R is determined by interactions of the system R with other systems.
- *Outer time* of a system R is inner time of some supersystem Q of the system R, i.e., R is a subsystem of Q.

- *Upper time* of a system R is inner time of a system Q in which R is an element.
- *Lower time* of a system R is inner time of an element of R.

Note that inner time of a system depends on the decomposition of this system into subsystems. Thus, often inner time of a system is not unique. Namely, when a system has several concurrent subsystems, i.e., subsystems the functioning of which is not synchronized, internal time of this system is not unique (Burgin and Grathoff, 2018).

As a system can interact with different systems, it is possible to consider different external times of this system.

As a system can be a subsystem of different systems, outer time of this system is not unique.

As a system can be an element of different systems, upper time of this system is not unique. For instance, an individual can be an element of her family and of the company where she works.

When a system has several concurrent elements, i.e., elements the functioning of which is not synchronized, the lower time of this system is not unique.

To better understand the difference between inner and external times of a system, let us look at geological time on Earth.

As MacRae writes, few discussions in geology can occur without reference to geological time (MacRae, 1996–1997). Geological time is often discussed in two forms:

- *Relative* (*chronostratic*) *time* is subdivisions of the Earth's geology in a specific order based on relative age relationships (most commonly, vertical/stratigraphic position). These subdivisions are given names, most of which can be recognized globally, usually on the basis of fossils.
- *Absolute* (*chronometric*) *time* is numerical age in the scale of "millions of years" or some other measurement. These are most commonly obtained via radiometric dating methods performed on appropriate rock types. Absolute time measurements can be used to *calibrate* the relative time scale, producing an integrated geologic or *geochronologic* time scale.

We see that absolute (chronometric) geological time is inner time of the Earth or, more exactly, of the solid Earth, i.e., of the rock of which it is composed. In contrast to this, relative (chronostratic) geological time is external time of the Earth because it reflects interactions of the Earth with the biological species living on the Earth.

Other examples of external time of the Earth are the Solar time and Lunar time because they define time on the Earth according to its interaction with the Sun (Solar time) or with the Moon (Lunar time). As we know, some cultures have solar calendars while others have lunar calendars.

When in his philosophical writings Heidegger interpreted time as three-dimensional with its future, present and past dimensions, he considered unique absolute time. When this approach became apparent and palpable for Weizäcker, he also had in mind absolute time, which can have more than one dimension or, more exactly, can be adequately modeled by a three-dimensional mathematical space.

In contrast to this, the system theory of time explains that any system has its own time and as a rule, not a single one. That is why we analyze the hypothesis of three-dimensional time in each of the three top-level components of the world as a whole.

At first, we take the Physical World as the most explored constituent of people's reality. Analyzing physical time (of a system), we see that with respect to some temporal point or interval, the future and the past form two dimensions because the past (of a system), as a rule, does not uniquely determine the future (of this system), while the future (of a system) does not uniquely determine the past (of this system). For instance, Stephen Hawking writes, "One certainly cannot predict future events exactly if one cannot even measure the present state of the universe precisely!" (Hawking, 1988).

Note that these dimensions are not modeled by an infinite straight line or half-line as it is regularly done in physics. Actually, in any physical system, time is finite because physical systems do not exist forever and, at least in our universe, they emerged or were created at some finite moment of time, which cannot be further from us than the time of the emergence of the universe (creation).

While interpretation of the future and the past as separate physical dimensions is not very hard, what is really challenging is the puzzle of the present dimension of time because in physics, and in mundane comprehension, the present is treated as a point (one moment) in the infinite line of time. However, special relativity theory shows that this is not the case.

Indeed, two events simultaneously happening in two different locations in the reference frame of one inertial observer may occur non-simultaneously in the reference frame of another inertial observer. Thus, it is possible to connect events simultaneously happening in a given reference frame with this reference frame and assume that these events when they are happening form the present dimension in this reference frame. As the result, we obtain three-dimensional physical time.

In addition, the system theory of time goes even further in comparison with the special relativity stating that each system has its own time and often not a single one. Besides, subsystems in a complex system function concurrently (Burgin and Grathoff, 2018). Consequently, to delineate the present in a multisystem domain, it is necessary to bring together temporal points from different systems. These points together form the present system dimension in the Physical World. Combined with the future and past system dimensions, the present system dimension forms a three-dimensional time system in the Physical World.

Besides, time is determined by its measurement, which takes place during some time interval and not at an infinitely small moment (point) of time. This interval also gives a dimension of time because dimensions can have finite measures. This dimension of the present is studied by physicists as *imaginary time*, as a mathematical representation of time, which is used in some approaches to special relativity and quantum mechanics (Hawking, 2001).

Coming to the Mental World, we analyze only its first level, i.e., individual mentality, describing three-dimensional time in individual mentality. Starting with the present and considering various mundane situations, we see that the present is not a single moment (point) of time for an individual. Actually, it is always a temporal interval, the length of which depends on the situation when the present is reckoned.

In English, present time is represented by the word *now*. For instance, we can hear "Now is the time for action". In this statement, *now* can mean several minutes, several hours, several days and even a year depending on the situation. This interval of time forms the mental dimension of the present and the size of this dimension depends on the comprehension of the present, which is essentially contextual. However, this contextuality results in multiplicity of the mental present, the scale of which subsumes several intervals.

In people's mentality, the past is not unique and not linear because at different periods of life an individual creates her or his mental past and these pictures change during the life of this individual. Although people have an illusion that facts are irreversible and the past does not change any more after facts are described, the past is reconstructed in multiple ways.

For instance, in Western countries, such as USA, France or the United Kingdom, it was assumed that the World War II ended on May 8 while in the USSR and its satellites, it was determined that the World War II ended on May 9. As a result, the scale of the mental past time has the form of the graph that is called a tree or a forest.

The same is true for the future when an individual imagines or plans the mental future and these images (plans) change during the life of this individual. As a result, the scale of the mental future time is also a tree or a forest.

The capacity to mentally reconstruct personal events from the past as well as to construct possible scenarios in the future is called mental time travel or *chronosthesia* in psychology (Tulving, 2002; Suddendorf, 2010). The ability to travel mentally in time has been highlighted as a potential prime mover in human evolution, enabling humans to prepare, plan and shape the future to their advantage based on their past experience and knowledge (Suddendorf and Corballis, 1997; 2007).

After explication of mental temporal dimensions, we analyze time in the World of Structures. This World is the scientific counterpart of Plato's World of Ideas and according to Plato, Ideas are eternally unchanging and thus are beyond time. This is also true for abstract pure structures (Burgin, 2017). This property of structures means that there is no present, past and future in this world. However, this is true only for the inner time in the World of Structures while it also has external time determined by interaction with people. Researchers discover various structures from this world, similar to travelers who discovered new lands, new nations and new continents. It is an illusion, for example, that mathematicians construct or invent their proofs because in essence, construction or invention of proofs is traveling in domain of mathematical structures with the goal of finding the necessary proof. Physicists and other scientists discover not only structures embedded in material systems and processes but also abstract, often mathematical, structures, which they use for exploration of nature. Interestingly some new structures, such as the structure of DNA or the periodic table of chemical elements, were discovered in dreams when the researchers were sleeping.

The external structural time is determined by the time in which people live. For instance, external time of the domain of mathematical structures is time in the history of mathematics. Usually, this time is projected on the scale of the major mathematical discoveries or on the scale determined by lives of outstanding mathematicians. As we found, either physical time in which people live or their mental time has three dimensions. These three dimensions induce three dimensions of the external time in the World of Structures.

The new understanding of time in the World of Structures/Ideas allows elimination of contradictions between some conceptions of Plato and views of Weizsäcker. As it is explained in Section 2.3, the form (*eidos*), as addressed by Weizsäcker in both the ontological and epistemological sense, is a reversed Platonic perspective. For Plato, the ideas and their knowledge are beyond time. For Weizsäcker, time is the horizon for the development of forms as well as for our knowledge about them.

If we accept that ideas and forms represent the same essence, then it is possible to imagine that these perspectives are inconsistent being opposite. However, in a similar way to physics where contradiction between the wave and particle models of light was eliminated by synthesizing both approaches, the system theory of time eliminates contradiction between the perspectives of Plato and Weizsäcker. Plato's Ideas, aka abstract structures of contemporary science and mathematics, are unchanging and, thus, beyond inner time of the World of Structures. At the same time, people and now computers discover new (for them) structures and develop structures they already know. These processes go on in the external time of the World of Structures. Thus, Plato was writing about inner time of the World of Structures while Weizsäcker was writing about external time of the World of Structures. Consequently, there are no contradictions between these opinions.

3. Conclusion

We presented a scientific analysis and a theoretical explanation of some prolific ideas of Heidegger, Heisenberg, Weizsäcker and Wheeler, providing a novel vision of reality and opening new directions for future research. For instance, it would be interesting to describe scales of mental time in more detail. Exploration of external time of the World of Structures is another interesting direction of research.

References

Aristotle (1984). *The Complete Works of Aristotle*, Princeton University Press, Princeton.

Balasubramanian, R. (2000). *Introduction*, in *History of Science, Philosophy and Culture in Indian Civilization*. Chattopadhyana (ed.), Vol. II, Part 2: *Advaita Vedanta* Centre for Studies in Civilizations, Delhi.

Bastide, R. (ed.) (1962). *Sens et usages du terme Structure dans les sciences humaines et sociales*, Mouton, La Haye.

Born, M. (1953). Physical Reality, *The Philosophical Quarterly* 3(11), 139–149.

Boss, M. (1979). *Existential Foundations of Medicine and Psychology*, Jason Aronson, New York, London.

Bourbaki, N. (1957). *Structures*, Hermann, Paris.

Brown, P. and Lauder, H. (2000). Collective intelligence. In *Social Capital: Critical Perspectives*, Oxford University Press, New York.

Bulley, A. (2018). The history and future of human prospection, *Evolutionary Studies in Imaginative Culture*, 2(75), doi:10.26613/esic.2.1.75.

Burgin, M. (1998/1999). Information and transformation, *Transformation*, 1, 48–53 (in Polish).

Burgin, M. (2002). *Elements of the System Theory of Time*, LANL, Preprint in Physics 0207055 (electronic edition: http://arXiv.org).

Burgin, M. (2004). Data, information, and knowledge, *Information* 7(1), 47–57.

Burgin, M. (2010). *Theory of Information: Fundamentality, Diversity and Unification*, World Scientific, New York/London/Singapore.

Burgin, M. (2010a). Information operators in categorical information spaces, *Information* 1(1), 119–152.

Burgin, M. (2011). *Theory of Named Sets*, Nova Science Publishers, New York.

Burgin, M. (2011a). Epistemic information in stratified M-spaces, *Information* 2(2), 697–726.

Burgin, M. (2011b). Information in the structure of the world, *Information: Theories & Applications* 18(1), 16–32.

Burgin, M. (2012). *Structural Reality*, Nova Science Publishers, New York.

Burgin, M. (2017). Ideas of Plato in the context of contemporary science and mathematics, *Athens Journal of Humanities and Arts* 4(3), 161–182.

Burgin, M. (2017a). The General theory of information as a unifying factor for information studies: The noble eight-fold path. In *Proceedings*, 1(3), 164, p. 6, doi:10.3390/IS4SI-2017-04044.

Burgin, M. (2019). Mental energy in the context of the general theory of information, *Open Access Biostatistics & Bioinformatics* 2(3). doi: 10.31031/OABB.2019.02.000540.

Burgin, M. and Grathoff, A. (2018). *Concurrent systems and time synchronization, International Journal of General Systems* 47(4), 313–328.

Capurro, R. (1978). *Information*, Saur, Munich, PhD Thesis, University of Düsseldorf.

Capurro, R. (1982). La deuda de la ciencia natural. *Opinar*, p. 20. Available online: http://www.capurro.de/deuda.htm. Translated and updated as: The Debt of Natural Science (2018) unpublished). Available online: http://www.capurro.de/weizsaecker.html (accessed on 13 August 2018).

Capurro, R. (2009). Past, present, and future of the concept of information, *tripleC* 2, 125–141.

Capurro, R. (2015). Translating Information. In *Proceedings of the FIS/ISIS Conference 2015*: Information Society at the Crossroads — Response and Responsibility of the Sciences of Information, Vienna University of

Technology, Vienna, June 3–6. Available online: https://sciforum.net/paper/view/conference/2972 (accessed on 20 August 2018).

Capurro, R. (2017). In Search of Ariadne's Thread in Digital Labyrinths. In *ICIL 2016, Proceedings of the 7th International Conference on Information Law and Ethics*, University of Pretoria, South Africa, February 22–23, 2016. The University of Macedonia Press, Thessaloniki, pp. 1–19.

Capurro, R. (2017). *Homo Digitalis*, Springer VS, Wiesbaden.

Capurro, R. (2014). *Apud Arabes. Notes on the Greek, Latin, Arabic, Persian, and Hebrew Roots of the Concept of Information.* Available online: http://www.capurro.de/iran.html (accessed on 20 August 2018).

Capurro, R. "Wirf den Helden in Deiner Seele nicht weg!" (Nietzsche). Menschen, denen ich begegnet bin und mich geprägt haben. Available online: http://www.capurro.de/meinehelden.html (accessed on 13 August 2018, unpublished).

Capurro, R., Eldred, M. and Nagel, D. (2013). *Digital Whoness: Identity, Privacy and Freedom in the Cyberworld*; de Gruyter, Berlin.

Capurro, R. and Hjørland, B. (2003). The Concept of Information. In *Annual Review of Information Science and Technology*, 37, Information Today: Medford, New Jersey, pp. 343–411. Available online: http://www.capurro.de/infoconcept.html (accessed on 17 August 2018).

Corry, L. (1996). *Modern Algebra and the Rise Mathematical Structures*, Birkhäuser, Basel/Boston/Berlin.

Cottrell, A. H. and Pettifor, D. G. (2000). Models of structure. In *Structure in Science and Art*, Cambridge University Press, Cambridge, pp. 33–47.

Eldred, M. (2014). *Being Time Space: Heidegger's Casting of World.* Available online: https://www.arte-fact.org/untpltcl/bngtmspc.html (accessed on 13 August 2018).

Eldred, M. (2015). *A Question of Time*, CreateSpace, North Charleston.

Freud, S. (1949). *The Ego and the Id*, The Hogarth Press Ltd., London.

Gethin, R. (1998). *Foundations of Buddhism*, Oxford University Press.

Gower, B. (2000). Cassirer, Schlick and 'structural' realism: The philosophy of the exact sciences in the background to early logical empiricism, *British Journal for the History of Philosophy* 8, 71–106.

Hawking, S. (1988). *A Brief History Of Time: From Big Bang To Black Holes*, Bantam Books, New York.

Hawking, S. (2001). *The Universe in a Nutshell*, Bantam Books, United States & Canada.

Heidegger, M. (1975). Aus einem Gespräch von der Sprache. In *Unterwegs zur Sprache*, 5th edn., Neske, Pfullingen, pp. 83–155.

Heidegger, M. (1976). *Sein und Zeit*, 13th edn., Tübingen, Niemeyer.

Heidegger, M. (1983). *Die Grundbegriffe der Metaphysik*, Frankfurt am Main, Klostermann.

Heisenberg, W. (1971). *Schritte über Grenzen. Gesammelte Reden und Aufsätze*, Piper, München.

Herbert, N. (1987). *Quantum Reality: Beyond the New Physics*, Anchor Books, New York.

Jung, C. G. (1969). On Psychic Energy. In *On the Nature of the Psyche*, Princeton University Press, Princeton.

Kastrup, B. (2019). *The Idea of the World: A Multi-Disciplinary Argument for the Mental Nature of Reality*, John Hunt Publishing, Ropley, United Kingdom.

Klein, S. B. (2013). The complex act of projecting oneself into the future, *Wiley Interdisciplinary Reviews: Cognitive Science* 4(1), 63–79.

Landry, E. M. (2007). Shared structure need not be set-structure, *Synthese* 58, 1–17.

Lautman, A. (1938). *Essai sur les notions de structure et d' existence en mathématique*, Hermann, Paris.

Lyre, H. (1998). *Quantentheorie der Information*, Vienna, New York, Springer.

MacKay, D. M. (1969). *Information, Mechanism, and Meaning*, MIT Press, Cambridge, MA.

MacRae, A. (1996–1997). *Geological Time Scale* (http://www.talkorigins.org/faqs/timescale.html).

Maxwell, G. (1970). Theories, Perception and Structural Realism. In *The Nature and Function of Scientific Theories*, University of Pittsburgh Series in the Philosophy of Science, Vol. 4, University of Pittsburgh Press, Pittsburgh, pp. 3–34.

Negrotti, M. (2002). *Naturoids. On the Nature of the Artificial*, World Scientific Publishing, Singapore.

Nguen, N. T. (2008). Inconsistency of knowledge and collective intelligence, *Cybernetics and Systems* 39(6), 542–562.

Plato, (1961). *The Collected Dialogues of Plato*, Princeton University Press, Princeton.

Poincaré, H. (1908). *Science et Méthode*, Flamarion, Paris.

Putnam, H. (1960). Minds and machines. In *Dimensions of Mind*, New York University Press, pp. 148–179.

Shannon, C. (1948). A mathematical theory of communication, *The Bell System Technical Journal* 27, 379–423, 623–656.

Suddendorf, T. (2010). Episodic memory versus episodic foresight: Similarities and differences, Wiley Interdisciplinary Reviews, *Cognitive Science* 1(1), 99–107.

Suddendorf, T. and Corballis, M. C. (1997). Mental time travel and the evolution of the human mind, *Genetic, Social, and General Psychology Monographs* 123(2), 133–167.

Suddendorf, T. and Corballis, M. C. (2007). The evolution of foresight: What is mental time travel, and is it unique to humans? *The Behavioral and Brain Sciences* 30(3), 299–313.

Tulving, E. (2002). Chronesthesia: Conscious awareness of subjective time. In *Principles of Frontal Lobe Function*, pp. 311–325.

von Bayer, H. C. (2004). *Information: The New Language of Science*, Harvard University Press, Harvard.

Weizsäcker, C. F. (1974). v. *Die Einheit der Natur*, DTV, Munich.

Weizsäcker, C. F. (1977). Begegnungen in vier Jahrzehnten, in *Erinnerung an Martin Heidegger*, Günther Neske (ed.), Pfullingen, Neske, pp. 239–247.

Weizsäcker, C. F. (1985). v. *Aufbau der Physik*, Hanser, Munich.

Weizsäcker, C. F. (1992). v. *Zeit und Wissen*, Hanser, Munich.

Wheeler, J. A. (1989). Information, physics, quantum: The search for links. In *Proceedings III Symposium on Foundations of Quantum Mechanics*, Tokyo, pp. 354–368.

Wittgenstein, L. (1984). *Tractatus logico-philosophicus*, Suhrkamp, Frankfurt am Main.

Wittgenstein, L. (1984a). *Philosophische Untersuchungen*, Suhrkamp, Frankfurt am Main.

Zahar, E. G. (1996). Poincaré's structural realism and his logic of discovery. In *Henri Poincaré: Science and Philosophy*, Akademie Verlag & A. Blanchard, Berlin, pp. 45–68.

Chapter 8

The Debt of Natural Science: An Interview with Carl Friedrich von Weizsäcker

Rafael Capurro
Stuttgart Media University, Stuttgart, Germany
rafael@capurro.de

This chapter presents an interview of one of the prominent thinkers of the 20th century Carl Friedrich von Weizsäcker by Rafael Capurro.

1. Introduction

In 1982, I had the privilege of meeting Carl Friedrich von Weizsäcker (1912–2007) at his private home near Munich. While working at the *Center for Nuclear Energy Documentation* located in the premises of the *Karlsruhe Nuclear Research Center* (1980–1985), I invited a few engineers and physicists interested in natural philosophy to discuss von Weizsäcker's works. After a while, we wanted to meet him personally. He invited us to visit and I took some notes of the conversation that were published in Spanish in the Uruguayan weekly newspaper *Opinar* whose Founder and Director was Enrique Tarigo (1927–2002), Professor of Procedural Law at Uruguay State University (Capurro, 1982). Tarigo became Vice President of Uruguay under President Julio María Sanguinetti between 1985 and 1989, which was the first democratic government after the military intervention of

1973–1985. In the introduction to this interview, Enrique Tarigo wrote as follows:

> From Germany: Rafael Capurro interviews von Weizsäcker.
> Carl Friedrich von Weizsäcker is one of the great names in 20th century thinking. His life was dedicated to this: thinking. A disciple and friend of Werner Heisenberg, his original interest was in physics. In the thirties, he developed Quantum Theory together with Niels Bohr. This was the start of a deep controversy about some presuppositions of Einstein's thinking that still persists. But physics was not the only passion of this man. His yearning for deeper thinking took him to philosophy and particularly to the study of two thinkers, namely Plato and Kant. Religious thinking was also among his fields of interest. But his life was not only dedicated to thinking. The danger arising from nuclear weapons led him some years ago to setting up the *Max Planck Institute for the Study of the Scientific-Technical World.*
> Today, aged 70, von Weizsäcker has retired from teaching, but continues working with a group of scientists devoted to what remains his main research interest, namely Quantum Theory. What follows are the responses of a multifaceted and profound man given to the critical questions of our correspondent, Rafael Capurro. Ideas and men come across a privileged mind (Tarigo, 1982, p. 20).

2. Physics and Philosophy

Question. What is the contribution of *philosophical* thinking to modern physics and to science in general? Isn't it superfluous? Is it not enough to simply *do science*?

Answer. The fundamental concepts of physics (matter, space, time, energy, etc.) originate in the philosophical tradition. I became aware very early on that physicists often do not know what they are talking about. I decided to do research on the origin of such concepts. This took me first to Kant. But in order to understand Kant, it is necessary to have read Descartes who takes us to the Platonic and Aristotelian medieval traditions. Plato and Aristotle can be considered as the originators of such concepts, that is to say, they gave them an original and crucial shape. My first philosophical steps in Kant were under the guidance of my friend Georg Picht with whom I read without understanding much of it, to say the truth, the first 20 pages of the *Critique of Pure Reason.*

Question. How would you see the relation between philosophical thinking and a physicist's everyday research?

Answer. For this point, I think that the distinction made by the American philosopher and historian of science, Thomas S. Kuhn, in his book *The Structure of Scientific Revolutions* (Kuhn, 1962) might be helpful. Kuhn distinguishes between "normal" and "revolutionary science". Progress in science is achieved, according to Kuhn, not by accumulating knowledge, but is based on leaps or "revolutions" when a traditional "paradigm" used so far to explain phenomena is questioned and replaced by another one. This is the case of, for instance, the revolution brought about by the Copernican paradigm as opposed to the Ptolemaic one. During a period of "normal science," scientific progress aims at looking for solutions to a "puzzle" within a given paradigm. During such periods, philosophy has no direct influence on scientific research, and can even be deleterious, in contrast to a "revolutionary" period when a scientist can get an opportunity to question given presuppositions thanks to radical philosophical thinking. This is what we see in the case of Einstein, influenced by Ernst Mach who himself criticized Newton. Einstein did not believe in the Newtonian paradigm, although it worked.

Question. According to what you say, philosophical thinking has the quality of incessantly questioning.

Answer. This quality originates in Greece. Philosophy is a Greek experiment. Only analogically can we speak of Hindu "philosophy" with regard to, for instance, the Vedas. Philosophy is *per se* so beautiful that, similar to other activities, it does not need a why.

3. The Path of Natural Science

Question. How do you see the questions raised by the so-called alternative movements such as the ecological movement?

Answer. Ecology, in my understanding, is a branch of modern biology and cannot be understood without it. Ecologists criticize some devastating applications of modern biology. But such applications, I believe, are not a necessary consequence. A non-destructive path can also be found within the frame of modern natural science. Our technology is not at its end. The path of natural science is consistent and true although we cannot predict the next steps of evolution based on it. It was not possible, for instance, to predict the existence of the main religions before they arose or to predict Homer before the rise of epic poetry.

Question. But nevertheless, during the Renaissance...

Answer. Yes, there was a start, for instance, with Leonardo da Vinci's first attempts to predict. But the limits of such attempts are obvious.

Question. How would you see the relation, for instance, between the path of natural science and other experiences of reality, such as poetry or painting?

Answer. Such paths, well understood, are not opposed to natural science, but they address different ways of relating ourselves to reality that do not have the structure of empirical–rational science. Our perception of beauty concerns the connections of a whole that natural science does not deal with and that, in a certain way, cannot observe. Natural science is of recent origin. Humans as hunters, farmers, poets, etc. have a long history. From a historical point of view, the description of reality provided by natural science was not and is not the only one.

4. The Path of Meditation

Question. How in this context do you see the concept of meditation as a path of knowledge?

Answer. Meditation is an experience that arises originally in different religious traditions. Today, it is fashionable to speak about it. This is why I only dare deal with your question with some shyness. On the other hand, it is something that I do not feel competent to teach. In order to come near this phenomenon, the question, "what is?", is not very helpful. It is better to ask "how do you do it?" But in this case, as well as in other fields, one only sees what one previously knows.

Question. Can you give a simple example of "how do you do it?"

Answer. You view a flower, for instance. You view it. You remain a little bit longer. You close your eyes and continue viewing the flower. You think that you and the flower are alone in the world. It is a process of retreat. Imagine what this experience can mean if you do not do it just once but every day and if, instead of a flower, you experience yourself. Learning to see the things themselves is what religions have talked about. They are older than natural science. Its relation with them is not clear so far.

Question. In your writings, you mention the relativity of religious traditions.

Answer. Relativity in the sense of their interconnection. Think about the Jesuit priest Enomiya Lassalle, for instance, whom I know personally. He is also a ZEN monk. Religious experience is sometimes the same, but interpretations vary.

Question. And in the case of Christianity?

Answer. We have a high level of rationality due particularly to the influence of Greek philosophy, in contrast to Buddhism, for instance. The question about "salvation as such" has always had a historical answer. It is different in each case. The encounter among the religions is, without doubt, the most important event of our age. If natural science could become a partner in this dialogue on a parity basis, perhaps, after some world catastrophes, a new phase of consciousness could arise.

5. The Unification of Modern Physics

Question. In your writings and particularly in *The Unity of Nature* (von Weizsäcker, 1971a), you point to different paths that could lead to the unity of modern physics.

Answer. The present physical theory, valid in all fields and tested millions of times, is the Quantum Theory. What it basically deals with can be summarized on a single page. The project is to formulate the conditions of possibility of the quantum phenomenon. Providing the foundations for this theory is something to which I have pointed in the book you mentioned and also the one on which I am working at present.

Question. How do you see Einstein's position with regard to the Quantum Theory?

Answer. For Einstein, it was difficult to accept the fundamental probabilistic character of this theory. He looked for hidden parameters that would explain in a deterministic way all physical phenomena. It was Heisenberg who, proceeding from the indeterminacy principle, radically questioned the world or, as we previously said, the paradigm of classical physics. In quantum physics, we think with the category of probability which is nothing other than the quantification of what is possible, and what is possible is one of the dimensions of the structure of time, the dimension of the future.

A thinking of the possible is given only in relation to a factical past from a concrete present. Modern physics is based on this three-dimensional structure of time without which, as Heidegger says, we could not even formulate or understand the second principle of thermodynamics.

Question. How do you see modern currents in the philosophy of science?

Answer. I already mentioned Thomas Kuhn. But if we go back a little more, we will remember Rudolf Carnap and his criticism of apriorism: science must be empirical. But, what does "empirical" mean? Carnap defines it as *a priori*. Karl Popper remarked that the path of induction is wrong. The truth of our theories and general propositions cannot be achieved through generalization of singular experiments. Science proceeds through propositions that *prima facie* do not look well founded. But, on the other hand, in my view, Popper's thinking is oriented towards classical physics. This is the reason why I think that it is insufficient. Kuhn, by comparison, is more empirical by relying on the history of scientific revolutions. Scientific revolutions are, accordingly, paradigms that have not yet been falsified.

Question. The introduction of the temporal scheme would be one of these revolutions.

Answer. From a philosophical perspective, it was Heidegger who asked profoundly, again and again, about the phenomenon of temporality.

6. Physics and Temporality

Question. How are quantum physics and the experience of thinking addressed by Heidegger related to the phenomenon of temporality?

Answer. The closeness of the issues is, without doubt, greater with regard to Heidegger than to the philosophy of science without implying a total equivalence with Heidegger's thinking. Philosophy of science moves further and further from the real problems. This reminds me of Ernst Mach — we talked about him before — who denied the existence of atoms and relativity.

Question. Heidegger says that science does not think.

Answer. This must be understood in the sense that normal science — in Thomas Kuhn's terminology — does not think, that is to say, it does not question its own paradigm. In contrast to Edmund Husserl, I believe that the basic shape of a phenomenon is not something unchangeable, but takes place in time and can change according to experience. In Plato's terminology, I would say that our challenge is about thinking the "idea" of time. We cannot escape time. We are not free to do it. I remember an anecdote about Einstein asking Carnap and Popper "what is time?" without receiving an acceptable answer. Einstein tried to eliminate time. Shortly before he died, when somebody told him that a friend died a few weeks ago, he answered that "a few weeks ago" is an illusion. For a physicist, as a believer, the difference between past, present and future is an illusion. I think, quite the contrary, that the introduction of three-dimensional time, that was made explicit by Heidegger, can bring about a new revolution in physics that would also resonate in philosophy. Already in Plato and Aristotle, we find the key concept of *kinesis*, that is to say, movement and change. *Chronos*, literally translated as time, is already for Plato a derivative representation of *aion*, a term that I would like to leave untranslated. It implies the temporal structure and refers, for instance, to a person's lifetime.

Answer. Physics and philosophy meet here in the question concerning time, ...

Question. As if Magellan and Vasco da Gama, starting from different points and in opposite directions, would meet during circumnavigation.

7. Conclusion

Translating this conversation — originally in German, then published in Spanish — 35 years later is an amazing experience not only because my thinking and, of course, my life have changed, but also particularly because the world has changed with regard to what we are used to calling the digital revolution, a paradigmatic change in the sense addressed by Kuhn and von Weizsäcker, having impact not only on science and philosophy but also on everyday life. Nobody could have predicted at the time the invention of the Internet, although it was not far away. As with other paradigm changes, it has the tendency to overestimate itself and even to believe that this is eventually the final true paradigm change bringing us — whom? — the foundation and the tools needed to better understand nature and ourselves.

As in the case of physics, both classical and quantum, it is important not to forget, as von Weizsäcker remarks at the beginning of this interview, the roots of digital technology are not only in Modernity but also in Greek philosophy. von Weizsäcker himself mentions these roots with respect to one key concept of our age, namely, information. He remarks that information, as distinct from matter and consciousness, is a third foundational category whose roots are the "Platonic *eidos*" and the "Aristotelian form" "dressed up in a way that a human being in the 20th century might learn about them" (von Weizsäcker, 1971b, p. 51).

This remark sent me on a long journey of research on the concept of information, the first result of which was my PhD on the etymology and history of ideas of the concept of information (Capurro, 1978). von Weizsäcker mentions philosophy as being a Greek invention. This is true. But it is not less true that this invention became widespread in the Hellenistic age as well as in the Arabic world, with the translation of Greek philosophers and, through them, later on into Latin in the Middle Ages. Philosophy is originally intercultural, that is to say, it disseminates and hybridizes through the ages. If we want to understand who we are in the digital age, that is to say, how we understand nature and ourselves from a digital perspective, and what are the theoretical and practical gains and losses from such an enterprise, we must think about how foundational philosophical categories such as time, matter, information, etc. are "dressed up" — "eingekleidet" writes von Weizsäcker — in our time also in different languages and contexts (Capurro *et al.*, 2013).

von Weizsäcker addresses particularly the issue of three-dimensional time — as distinct from the pervading concept of time as a succession of 'nows' — being one of the main discoveries made by Heidegger and essential not only for understanding Quantum Theory but also for understanding nature beyond the cast of Modernity (Eldred, 2015). He also points to the potential impact of this discovery in philosophy, an impact that remains still unthought, also having high relevance in the field of information ethics (Nakada and Capurro, 2013). One needs no further evidence to become aware of the impact of digital technology in the lives of most people with different kinds of dependency on the *time regime* instantiated by digital technology that regulate and even rule life, thus becoming what is being called *onlife*. The gains and losses of *onlife* concern not only surveillance issues by different private and public agents but also a loss of freedom that we could get back by going *offlife* not only in everyday life but also in experiences of meditation as addressed by von Weizsäcker in order to distance oneself from the obsessions of the digital age. What is particularly relevant for future research in natural philosophy is to think about how the present one-dimensional time regime grounded in digital technology affects the way we understand nature and ourselves, the gains and losses thereof, and the possibility of a revolutionary cast based on three-dimensional time as addressed by von Weizsäcker 35 years ago.

References

Capurro, R. (1978). *Information. Ein Beitrag zur etymologischen und ideengeschichtlichen Begründung des Informationsbegriffs.* Munich, Saur.

Capurro, R. (1982). La deuda de la ciencia natural. *Opinar*, Dec. 30, p. 20. Online: http://www.capurro.de/deuda.htm.

Capurro, R., Eldred, M. and Nagel, D. (2013). *Digital Whoness: Identity, Privacy and Freedom in the Cyberworld.* de Gruyter, Berlin.

Eldred, M. (2015). Being time space: Heidegger's casting of world, in *A Question of Time*, CreateSpace, North Charleston.

Kuhn, T. S. (1962). *The Structure of Scientific Revolutions*, The University of Chicago Press, Chicago.

Nakada, M. and Capurro, R. (2013). An intercultural dialogue on roboethics, in *The Quest for Information Ethics and Roboethics in East and West. Research report on trends in information ethics and roboethics in Japan and the West*, Makoto N. and Capurro R. (eds.), Research Group on the Information Society (ReGIS), Tsukuba (Japan) and International Center for Information Ethics (ICIE) (Karlsruhe, Germany) (eds.), March 31, 2013, pp. 13–22, ISSN 2187-6061, Available online: http://www.capurro.de/intercultural_roboethics.html.

Tarigo, E. (1982). Introducción. *Opinar*, Dec. 30, p. 20. Online: http://www.capurro.de/deuda.htm.

von Weizsäcker, C. F. (1971a). *Die Einheit der Natur*, Hanser, Munich (Engl. *The Unity of Nature*, New York, 1980).

von Weizsäcker, C. F. (1971b). Sprache als information, in *Die Einheit der Natur*, Hanser, Munich.

Chapter 9

The Molecular Linguistics of DNA: *Letters, Words, Sentences, Texts, and their Meanings*

Sungchul Ji

Department of Pharmacology and Toxicology, Ernest Mario School of Pharmacy, Rutgers University, Piscataway, NJ, USA

sji.conformon@gmail.com

Natural language has served as a useful metaphor for cell biology at least since 1971 when E. Chargaff discussed his parity rules in DNA sequences. The first relatively rigorous qualitative comparison between cell language (cellese) and human language (humanese) appeared in 1997. Beginning around 2008, the availability of the Planckian Distribution Equation (PDE) that fitted both the mRNA distributions in cells and word-length distributions in English texts made it possible to compare cell linguistic and human linguistic structures quantitatively, leading, for example, to the conclusion that, whereas there is only one English alphabet consisting of 26 letters, there may be n different genetic alphabets in DNA (referred to as the nth-order alphabet), each alphabet consisting of 4^n letters, where n is the number of nucleotides in a letter and can be greater than 1. The mRNA levels encoding metabolic pathways measured with microarrays in many cell types under varying physiological conditions produce long-tailed histograms that fit PDE. When the mRNA levels of a given metabolic pathway fit PDE, two numbers can be computed based on the PDE — (i) the *Planckian Information of the Second kind* (I_{PS}) and (ii) *Shannon entropy*, H. When these numbers are plotted in the so-called *Planck–Shannon plot or space*, a linear regression line is observed if and only if the metabolic pathways being analyzed are functionally related, thus forming a "hypermetabolic pathway (HMP)" (also called a "supermetabolic pathway") which may be closely related to the "hyperstructures" of Norris *et al.* (1999).

These observations strongly support the hypothesis that there exist hyperstructures (also called superstructures) inside the living cell, whose functions are predicted to be "reasoning" or "computing" on the cellular level, according to the cell language theory.

The main objective of the present paper is to further elaborate on the structure and function of cellese taking into account the recent developments in information theory (Burgin, 2010; Diamant, 2015; Ji, 2015a, 2015b), the mathematical structures of DNA sequences (Petoukhov, 2016; Petoukhov, 2017a, 2017b), and the results of the lively discussions that went on in January, 2018 in the FIS (Fundamental of Information Sciences) list.

1. Introduction

Almost two decades have passed since I published the paper titled "The Linguistics of DNA: Words, Sentences, Grammar, Phonetics, and Semantics" (Ji, 1999). Many significant advances have been made during this period in understanding the molecular mechanisms underlying the so-called cell language (or *cellese* more briefly) as reviewed in Ji (2018a, Chapter 4). The main objective of the present chapter is to further elaborate on the structure and function of cellese, taking into account the recent developments in (i) information theory (Burgin, 2010; Diamant, 2015; Ji, 2015a, 2015b; Brenner, 2012), (ii) the mathematical structures of DNA sequences revealed by Petoukhov (Petoukhov, 2016; Petoukhov, 2017a, 2017b), (iii) the results of the lively discussions that took place in January 2018 in the FIS (Fundamental of Information Sciences) list, and (iv) the uncovering of the isomorphism between the Minimalist program in linguistics and the quantum field theory (Piattelli-Palmarini and Vitiello, 2015; Chomsky, 1995; Piattelli-Palmarini and Vitiello, 2017; Ji, 2019).

The following three results may be of particular interest: (i) There may be more than one genetic alphabet (defined as "the basic elements in a system which combine to form complex entities; https:en.oxforddictionaries/definition/alphabet) in DNA (i.e., other than the traditional one consisting of four bases, A, C, G, and T) (see Tables 5 and 6), (ii) the Planck–Shannon plots (PSP) of the mRNA levels of metabolic pathways may provide a novel mathematical method for investigating what V. Norris referred to as "hyperstructures" in 1999 (Norris *et al.*, 1999) and what are here called "hypermetabolic pathways" (see Section 13), and (iii) the *Planck–Shannon plots* may provide a new method for identifying "hyperbiomarkers" consisting of three or more metabolic pathways that function as units of action in responding to either toxicants or drugs.

The key points of the papers (Ji, 1999; Ji, 1997) were to establish the *isomorphism* (or the similarity in the operating principles) between cell language (later called *cellese*) and human language (*humanese*) based on the finding that, out of the 13 design features of humanese recognized by linguists (Hockett, 1960), 10 were found to have molecular counterparts in cellese, as summarized in Table 1.

Of the 10 design features of humanese listed in Table 1, I have found "Double Articulation" in Row 3 most useful, because it enabled me to distinguish between *letters*, *words*, and *sentences* and identify their molecular counterparts in the cell language (see Row 3 in Table 1). In Ji (2012), I extended the concept of double articulation to include what I called the "third articulation" which is defined in the last row of Table 2 (adapted from Ji (2012, Table 6.3)).

Although the concepts of the *first* and the *second articulations* were imported into biology from linguistics in 1997 (Ji, 1997; Ji, 2012), that of the "third articulation" was first invoked in biology in 2012 in order to account for the role of what Norris called "hyperstructures (HS)" (Norris *et al.*, 1999) and what is here referred to as "hypermetabolic/supermetabolic pathways (HMP)", both of which are examples of Prigogine's *dissipative structures* operating inside the living cell (Ji, 1985). The main difference between HS and HMP is that, whereas HS is inferred based on dynamic experimental data on cell metabolism (Norris *et al.*, 1999), HMP is suggested by the postulated isomorphism between cell and human languages (Ji, 1997; Ji, 2018a; Chapter 4). Until now, the concept of HS was not widely discussed in cell biology, but, with the availability of the Planck–Shannon plots (Section 13), it may be possible to study experimentally the role of HS and HMP in cell biology in health and diseases.

2. Physical vs. Semantic Information

There are at least three ways of defining what information is — (i) non-parametric (e.g., see Statement (1) below), (ii) parametric (Burgin, 2010), and (iii) diagrammatic (see Ji, 2018a, Figure 2.3). The non-parametric definition of information, Statement (1), which is attributed to Kolomogorov by Diamant (2015) is highly informative:

"Information is a linguistic description of structures in a given dataset". (1)

Table 1. The isomorphism between human and cell languages.

Design features	Human language (*Humanese*)	Cell language (*Cellese*)
1. *Communication channel*	Sound waves Written words	Conformational waves Concentration waves Covalent structures of DNA, etc.
2. *Arbitrariness of signs*	Arbitrary relation between the name and the named (e.g., polysemy)	Arbitrary relation between the linear structures of biopolymers and their 3D conformations or shapes (and hence functions)
3. *Duality* (or *Double articulation*)	1st: Words $\longrightarrow$ sentences 2nd: Letters $\longrightarrow$ words	1st: 1D biopolymer $\longrightarrow$ 3D biopolymer 2nd: Monomers $\longrightarrow$ 1D biopolymer
4. *Productivity* (or *Rule-governed creativity*)	Small number of letters, words, and grammatical rules can generate an almost infinite number of meaningful sentences.	Small number of monomers (20 amino acids, 4 nucleotides) can generate an almost infinite number of functional biopolymers (proteins, DNA, RNA).
5. *Discreteness*	Two word-forms (e.g., bear & pear) are absolutely different.	Elements of cell languages (e.g., hormones) are absolutely different because they are molecules with distinct shapes.
6. *Semanticity*	Each word (e.g., salt) has a fixed meaning (i.e., is not sand).	Extracellular messengers trigger specific gene-directed processes.
7. *Displacement*	Talking about things that are remote in space and time.	Watson–Crick genes (equilibrium structures) transmit information in time and Prigoginian genes (dissipative structures) transmit information in space (Ji, 1988).
8. *Interchangeability*	Humans can be both senders and receivers of messages.	Cells can be senders and receivers of messages.
9. *Complete feedback*	Humans can hear and monitor their own utterances to check their comprehensibility and correctness of formation as they are produced.	Certain cells in the human body utilize hormones produced and recognized by themselves called "autohormones". This phenomenon is known as "autocrine".
10. *Specialization*	A signal is said to be "specialized" if its direct physical effect on the behavior of the receiver is not functionally interrelated.	Hormones or DNA sequences show the phenomenon of specialization because these molecules and their target functions are not materially interrelated but largely arbitrary.

Source: Adopted from Ji (1999, 1997); Hockett (1960).

Table 2. A *formal* comparison between human and cell languages.

	Human language (*Humanese*)	Cell language (*Cellese*)
1. Alphabet (L)	Letters	4 Nucleotides (or 20 amino acids)*
2. Lexicon (W)	Words	Genes (or polypeptides)
3. Sentences (S)	Strings of words	Sets of genes (or polypeptides) expressed (or synthesized) coordinately in space and time dictated by DNA folds (cell states)
4. Grammar (G)	Rules of sentence formation	The *physical laws* and *biological rules* mapping DNA sequences to folding patterns of DNA (polypeptides) under biological conditions
5. Phonetics (P)	Physiological structures and processes underlying phonation, audition, interpretation, etc.	Concentration and mechanical waves responsible for information and energy transfer and transduction driven by *conformons* and *intracellular dissipative structures* (IDSs)
6. Semantics (M)	Meaning of words and sentences	*Codes* mapping molecular signs to gene-directed cell processes
7. First Articulation	Formation of sentences from words	Organization of gene expression events in space and time through *non-covalent interactions* between DNA and proteins (or space- and time-dependent non-covalent interactions among proteins, DNA, and RNA molecules). Thus, macromolecular complexes can be viewed as molecular analogs of sentences
8. Second Articulation	Formation of words from letters	Organization of nucleotides (or amino acids) into genes (or polypeptides) through covalent interactions
9. Third Articulation	Formation of texts from sentences	Organization of chemical concentration gradients in space and time called *dissipative structures or dissipatons* (Babloyantz, 1986; Kondepudi and Prigogine, 1998) in order to "reason" and "compute" (see Ji, 2012, Table 6.3)

Note: *Expanded in Tables 4 and 6 to include other alphabets, each composed of 4^n of n-nucleotide letters.
Source: Adapted from Ji (1999, 1997).

Table 3. The *arbitrariness* of the signs referring to "information". It doesn't matter what you call it, as long as your chosen label/name refers to the right reality, thing, entity, process, mechanisms, etc.

1	Type I Information	Type II information
2	Physical Information	Semantic information
3	Shannon information	Kolmogorov information, or Kolmogorov–Bateson information
4	"Meaningless" information	"Meaningful" information
5	I(−) information, or simply I(−)	I(+) information, or simply I(+)
6	Quantitative information	Qualitative information
7	Mathematical information	Linguistic information (see Statement (1))
8	Formal information	Phenomenological information
9	Classical information	New information
10	Interpretant-less sign* (see Ji, 2018, Table 2.1)	Triadic sign* (see Ji, 2018, Table 2.1)
11	Deterministic information	Stochastic information, or deterministically chaotic information (?)

Note: *Sign is thought to be the carrier of information, and C. S. Peirce (1839–1914) defined a sign as an irreducible triad of *representamen*, *object*, and *interpretant* (the effect a sign has on the mind of the interpreter).
Source: Adopted from Ji (2018a, 2001, 2002).

Four key concepts are embedded in Statement (1), which we may view as the definition of what may be called the "Kolmogorov information" for convenience:

(i) *dataset* (e.g., ACAGTCAACGGTCCAA);
(ii) *linguistic description* (e.g., Threonine, Valine, Asparagine, Glycine);
(iii) *structure* (e.g., 16 mononucleotides, 8 dinucleotides, 5 trinucleotides plus 1);
(iv) *mathematical description* (e.g., tensor product of two 2×2 matrices of 4 nucleotides).

The first three elements are evident in Statement (1), but the 4th is not so obvious, being justifiably included here to reflect the results of the recent mathematical analysis of DNA sequences carried out by Petoukhov (Petoukhov, 2016; Petoukhov, 2017a; 2017b).

If the above analysis of Statement (1) is valid, we can recognize two distinct kinds of descriptions of information — (i) *linguistic* and (ii) *mathematical.* Since we can identify Shannon information as belonging to the category of mathematical information, and since Shannon information is entirely about the average amount of information and has nothing to do with the meaning of information, it may be logical to conclude that the two kinds of informations, i.e., *linguistic* and *mathematical*, that are thought to be embedded in Statement (1) can be alternatively referred to as "meaningful" (denoted as I(+)) and "meaningless" informations (denoted as I(−)), respectively. There are other opposite pairs of names that can be employed to indicate the *dual nature* of information under discussion (see Table 3). For the sake of convenience, we will employ the pair of terms, "physical" and "semantic", used by Diamant (2015) to indicate the duality of information defined here, although other pairs of terms listed in Table 3 may be equally effective in most cases.

3. The Need for an *Essential Information Theory* in Contemporary Biology

In a recent post in the FIS Discussion Group, a leading discussant stated the following:

> "... essential informational ideas are missing..., and this absence of the informational perspective in the ongoing evodevo discussions is not a good thing..." (2)

I concurred and wondered if this unsatisfactory situation might have arisen in biology because biologists have blindly applied to their problems the information theory as introduced by Shannon almost seven decades ago in the context of communication engineering, without due attention given to the fact that the Shannon-type information theory is not designed to handle the "meaning" or semantics of messages but only the average AMOUNT of the information they carry in agreement with Diamant (2015).

There are three essential aspects to information, i.e., *amount* (e.g., my USB stores 3 Megabytes of information), *meaning* (e.g., the nucleotide triplet, ACG, encodes threonine), and *value* (e.g., the same message, "yes", can have different monetary values, depending on the context). We can readily see that the kind of information most useful for biologists is not the Shannon-type but whichever type that can handle the *semantics* and *pragmatics* of information.

One way to avoid the potential confusions in applying information theory to biology may be to recognize two distinct types of information, to be referred to as the "meaningless information" or I(−) and "meaningful information" or I(+) in Table 3 (see Row 5). I suggest that what is regarded as the missing "essential informational ideas" in Statement (2) above may be identified with I(+).

Based on my research results obtained over the past two decades (Ji, 1999; Ji, 1997; Ji, 2012; Ji, 2018a), I am inclined to suggest that "linguistics" can be viewed as an example of the I(+) theory. The term "linguistics" was once fashionable in Western philosophy and humanities (https://en.wikipedia.org/wiki/Linguistic_turn) under the rubric of "linguistic turn", but apparently it is now considered outmoded, but I am one of the many (including Chargaff, the discoverer of the two parity rules of DNA sequences; https://en.wikipedia.org/wiki/Chargaff%27s_rules) who believe that *linguistics* provide a valuable tool for elucidating the workings of living structures and processes. In fact, we may refer to the emerging trend in biology (Ji, 1999; Ji, 1997; Ji, 2018a; Piattelli-Palmarini and Vitiello, 2017) in the early 21st century of exploring the basic molecular and mathematical relations between linguistics and biology as the "Linguistic Return", in analogy to the "Linguistic Turn" referring to the "major development in Western philosophy during the early 20th century, the most important characteristic of which is the focusing of philosophy and the other humanities primarily on the relationship between philosophy and language" (https://en.wikipedia.org/wiki/Linguistic_turn). Evidently, linguistics played an important role in philosophy in the early 20th century and I suggest that it may again play a similarly important role in biology in the coming decades of the 21st century. But what about physics? Does physics need linguistics (Piattelli-Palmarini and Vitiello, 2015, 2017) to solve their basic problems as well? If not linguistics, perhaps semiotics, the study of signs? The latter possibility was emphasized by Brian Josephson in his lecture titled

"Biological Organization as the True Foundation of Reality"

given at the 66th Lindau Nobel Laureate Meeting held in 2016 in Lindau, Germany, stating that

> "Semiotics will eventually overtake quantum mechanics in the same way as quantum mechanics overtook classical physics". (3)

I referred to this potentially farsighted statement as the "Josephson conjecture" in (Ji, 2018, p. 177). If the concepts of the "Linguistic Return" in biology and the Josephson conjecture in physics prove to be correct in the coming decades, it may be possible to conclude that

> "Philosophy, biology, and physics can be finally united/integrated on the basis of the framework of semiotics viewed as a generalized communication theory". (4)

4. The Practical Consequences of the *Physical–Semantic Dual Theory or Physical–Semantic Complementary Theory of Information*

One practical application of the physical–semantic *dual theory of information* (PSDTI) as formulated by Diamant (2015) is in deducing the structure of cell language, or the structure of the linguistics of DNA, in a much more rigorous and detailed manner than was possible previously (Ji, 1999; Ji, 1997; Ji, 2012; Ji, 2018a; Ji, 2001; Ji, 2002). PSDTI seems also consonant with the information-energy complementarity principle (Ji, 2012, Section 4.13) and the *matter–form duality/complementarity* or the *hylomorphism* of Aristotle.[1]

It is the common practice in biology to use the terms "letters", "words", "sentences", and "texts" without any rigorous definitions. The general rule behind the use of these concepts is the principle of *concatenations* employed in linguistics leading to the following assumption:

(i) Just as 26 letters in the English alphabet are combined to form words (the process being called the second articulation (Ji, 1997)), so the four letters of the genetic alphabets, A, C, G, and T/U, combine in triplets to form genetic codons. Similarly, just as words form sentences and sentences form texts by the same concatenation procedure (or tensor product, mathematically speaking (Petoukhov, 2016)), i.e., linearly arranging words and sentences, respectively (see the second column in Table 4), so the 64 nucleotide triplets combine to form proteins and proteins combine to form metabolic pathways by continuing the concatenation process, or the tensor product of matrices of larger and larger sizes (see the fourth column in Table 4, which is based on the physical theory of information, i.e., without any involvement of *semantics* or the concept of the first articulation).

(ii) In contrast to the fourth column, we can justify an alternative structural assignment based on the *semantic theory of information* as shown in the fifth column of Table 4. Here the letters of the cell language alphabet are not always mononucleotides (as currently widely assumed), but thought to be *n*-nucleotides, where

[1]Form vs. Matter. Stanford Encyclopedia of Philosophy. https://plato.stanford.edu/entries/form-matter/.

Table 4. Two versions of the linguistics of DNA based on (i) the physical information theory (PIT) and (ii) the semantic information theory (SIT) (Diamant, 2015).

Organization level	*Human language* (*Cell language*)		*Cell language*	
	Structure	**Function/ Semantics**	**Structure Based on the Physical Information Theory (PIT) (Diamant, 2015)**	**Structure Based on the Semantic Information Theory (SIT) (Diamant, 2015)**
I	Letters (*Molecular letters*)	Basic building blocks or basic physical signals	4 Nucleotides (A, C, G, T/U); M = [C A; T G]	Mono-, di-, trinucleotides, 4-plets, 5-plets, . . ., n-plets of nucleotides.
II	Words| (*Molecular words*)	To denote	16 dinucleotides; M(×)M or M^2	Any combinations of the n-plets/amino acids to form genes/proteins with biological functions
III	Sentences (*Molecular sentences*)	To decide	64 trinucleotides/amino acids; M(×) M(×)M or M^3	Assembly of genes/proteins to form metabolic pathways (MPs)
IV	Texts (*Molecular texts*)	To argue/compute/ reason (e.g., syllogism)	254 tetranucleotides; Metabolic pathways (?); M(×)M(×) M(×)M or M^4	Networks of MP's forming superstructures (Ji, 2018)

Notes: M stands for a 2 × 2 matrix whose elements are the 4 genetic nucleotides, A, C, G, and T/U, i.e., M = [C A; T G] (see Figure 16 in (Petoukhov, 2017a, 2017b). The symbol, (×), indicates tensor multiplication (Petoukhov, 2017a, 2017b). The I to II transition is known in linguistics as the second articulation; the II to III transition as the first articulation (Ji, 1997); the III to IV transition was referred to as the third articulation (Ji, 2012, Table 6.3; Ji, 2018, Table 3.6, Section 4.2.1, Tables 4.5, 4.7, 4.17).

n can be 1 (i.e., mononucleotides), 2 (dinucleotides), 3 (trinucleotides), 4 (tetranucleotides), 5 (pentanucleotides), etc. That is, unlike in human language where the letters of an alphabet usually consist of one symbol, e.g., A, B, C, D, E, . . . , *I am claiming that in cell language, the letters can be mononucleotides* (i.e., *A, G, C, T/U*), *dinucleotides* (i.e., $AG, AC, \ldots$.), *trinucleotides* (i.e.,

$ACT, GTA, \ldots$), *tetranucleotides* (i.e., $ACTG, CCGT, \ldots$.), *pentanucleotides* (i.e., $ACCTG, TCGAT, \ldots$) *and, up to n-nucleotides (also called n-plets* (Petoukhov, 2016, 2017), *where n is a positive integer whose upper limit is not yet known.* If this conjecture turns out to be true, then the size of the cell language alphabet can be much larger (10^3–10^9?) than the size of a typical human linguistic alphabet which is usually less than 10^2, probably due to the limitation of the memory capacity of the human brain.

(iii) From linguistics, we learn that there are at least four levels of organization, each level being characterized by a unique function (see the second column, Table 4). Without presenting any detailed argument, it is suggested here that the linguistic structures deduced based on the *semantic information theory* (i.e., the fifth column, Table 4) agree with the human linguistic structures (i.e., the second column, Table 4) better than the linguistic structures based on the *physical/mathematical/quantitative information theory* (i.e., the fourth column, Table 4), when the functional hierarchy given in the third column is taken into account.

It is important to note that the process of concatenation is common to both of the cell linguistic structures postulated in columns 4 and 5 in Table 4, but the mechanisms underlying these concatenations are different. The concatenation mechanism in column 4 is *homogenous* across the four levels of organization from I to II to III to IV, i.e., the *tensor product of matrices* with increasing powers, from 1 to 2 to 3 to 4, etc. In contrast, the concatenation mechanisms implicated in column 5 are *heterogeneous*, involving (i) covalent interactions among nucleotides (or amino acids) to form dinucleotides (or dipeptides), trinucleotides (or tripeptides), 4-plets (or tetrapeptides), 5-plets (or pentapeptides), etc., during the I to II transition (corresponding to the *second articulation* in linguistics (Ji, 1997), and (ii) non-covalent interactions among n-plets including genes (or proteins) to form metabolic pathways during II to III transition (corresponding to the *first articulation* in linguistics (Ji, 1997)). I claim that the concatenation mechanisms underlying human languages are also heterogeneous, since the first articulation and second articulation are qualitatively distinct, the former being less constrained than the latter, i.e., it is easier to change the structure of a sentence than to change that of a word. Since non-covalently bonded structures (e.g., protein folds, protein complexes) are easier to change than covalently bonded ones (e.g., primary structure of a protein), the concatenation mechanism of biopolymer structures is heterogeneous just like the concatenation mechanism of human language, as postulated in the cell language theory (Ji, 1997).

Thus, it seems clear that the structure of cell language deduced based on the semantic information theory (see column 5, Table 4) agrees better, functionally, with that of the human language than the structure of cell language deduced based on the physical information theory (see column 4, Table 4).

5. Resonant Interactions Among Genetic Structures and the First Articulation in Linguistics

Based on combining the recent conclusions of Petoukhov (2016, 2017a, 2017b) that *genes are coupled via resonances* and those of mine indicating (i) that *cells use languages isomorphic with the human language* (Ji, 1997, 1999, 2001, 2002, 2012, 2018a) and (ii) that the *wave–particle duality first discovered in physics in the early decades of the 20th century is universal, applying not only to microphysics but also to molecular, cell, and human biology, and beyond* (Ji, 2015a, 2015b, 2018a), I was led to the following conclusion prior to constructing and hence independently of Table 4:

> "There are multiple genetic alphabets, each consisting of different number, 4^n, of molecular letters where n is the number of molecular components constituting a molecular letter and can range from 1 to 5 or more. Thus we may refer to the *multiple genetic alphabets* postulated here as the 1st-order, the 2nd-order, the 3rd-order, . . . , the nth-order genetic alphabets as summarized in Table 5". (5)

For the convenience of reference, we may refer to the proposed idea as the "multiple genetic alphabet (MGA) hypothesis". It is hoped that the validity of the MGA hypothesis can be tested through the computer-based analysis of currently available genetic data along the lines of research being carried out by Petoukhov (Petoukhov, 2016, 2017a, 2017b) and Trifonov (Trifonov, 1993).

Table 5. The structure and function of the cell language inferred on the basis of the postulated isomorphism between human and cell languages (Ji, 1997, 1999, 2012, 2018a) and the role of vibrational resonances in genetic structures (Petoukhov, 2016, 2017a, 2017b).

Human language	Cell language		
	Structure of molecular alphabets		Function
Alphabets	1st-order Alphabet	$4^1 = \mathbf{4}$ singlets (A, C, G, T)	Encoding 1-nucleotide frame shift?
	2nd-order Alphabet	$4^2 = \mathbf{16}$ doublets (AC, AG, AT, CA, CG, CT, GA, etc.)	Encoding 2-nucleotide frame shift? DNA shape code (Trifonov, 1993), chromatin code (Trifonov, 1993)
	3rd-order Alphabet	$4^3 = \mathbf{64}$ triplets (AAA, AAC, AAG, AAT, ACA, etc.)	Encoding amino acids, stop, and start codons
	4th-order Alphabet	$4^4 = \mathbf{256}$ tetrads (AAAA, AAAC, AAAG, AAAT, etc.)	Translation frame code? (Trifonov, 1993)
	5th-order Alphabet	$4^5 = \mathbf{1024}$ pentads (AAAAA, AAAAC, AAAAG, AAAAT, etc.)	Translation frame code? (Trifonov, 1993)
	nth-order Alphabet	$4^n \sim \mathbf{10^3}-\mathbf{10^6}$(?)-plets	Unit-functions yet to be discovered
Words	Genes		Encoding the primary structure of proteins (e.g., insulin)
Sentences	Gene systems		Encoding protein complexes (e.g., multisubunit glycolytic enzymes and metabolic pathways)
Texts	Systems of gene systems		Encoding systems of metabolic pathways (e.g., the signal transduction cascades underlying morphogenesis?)

6. The *Trifonov Mechanisms of DNA Information Storage and Transfer* (TM-DIST)

The hypothesis that there exist multiple alphabets in DNA sequences, each with a unique set of molecular letters numbering 4^n (see Table 5), is consistent with the conclusion reached by Trifonov (Trifonov, 1993) who stated in 1993 that DNA is a molecular language consisting of *letters*, *words*, and *texts*, carrying *overlapping messages* embodying multiple codes (as in our broadband communication channels of the Internet):

> "The classical triplet code is not the only code carried by the sequences. They contain, for example, the gene-splicing code, transcription codes and many other codes. By analyzing a large volume of the nucleotide sequences available, i.e., by performing various computer experiments with the sequences, one can decipher them and extract from them valuable biological information. At the DNA level, there are at least two more codes — the DNA shape code and the chromatin code. The overall DNA shape is sequence-dependent and can be described by a set of angles characteristic for various dinucleotide elements — codons of the DNA shape code. The chromatin code provides instructions for histone octamers where along the DNA to form the nucleosomes. This code is expressed as positional periodicity of primarily, AA and TT dinucleotides. A new RNA code has been described — the translation framing code. The frame seems to be maintained by a synchronizing pattern GCUGCUGCU... hidden in mRNA. Most enigmatic of all is, perhaps, the gene-splicing code. An interesting recent development indicates that the gene-splicing pattern in the sequences and the nucleosomal pattern have some common features. This has to do with superposition of the patterns that is characteristic of the sequence language in general, which carries simultaneously many codes in one and the same text. This results in an increased complexity of the sequences. Analysis of the protein-coding sequence complexity in eukaryotes and in prokaryotes revealed that the former are simpler. This is interpreted as the result of a spatial separation of the triplet code (carried by exons) and the chromatin code (carried by introns). Perhaps, the necessity of the separation of otherwise conflicting codes is one of the reasons why the intervening sequences had been introduced at all. The nucleotide sequences are written in an unbroken manner. One way to detect "words" in such a continuous text is to evaluate the degree of internal correlation by calculating contrast values for the words. This technique allows one to derive vocabularies, which are species- and function-specific. The nucleotide sequences, thus, carry numerous superimposed messages. We do understand only a few of these messages while many more are waiting for their turn to be deciphered".

Again, for the convenience of discussion, I am suggesting here that we refer to the ability of DNA sequences to encode and transmit multiple *overlapping* messages simultaneously as the "Trifonov mechanism of DNA information storage and transfer (TM-DIST)" or more simply the "Trifonov structure of DNA language (TSDL)".

Independently of Trifonov, I postulated in 1997 that cells use a molecular language that shares 10 out of the 13 design features of the human language (Ji, 1997, 1999, 2012, 2018a) as summarized in Tables 1 and 2.

7. The Lego Model of Genotype–Phenotype Coupling

Raanan *et al.* (2018) at Rutgers found that proteins that carry out oxidation-reduction (redox) reactions in living cells are made up of basic building blocks (BBBs, also called "legos"). They applied the so-called "smash-and search" method to about 10,000 proteins catalyzing redox reactions and found that there are only four BBBs, each consisting of 60–100 amino acids, that were common to all the oxidoreductases they examined. Since these BBBs are obviously heritable, they must be encoded in DNA sequences, each sequence at least 180–300 nucleotides long. They predicted that there are more BBBs to be discovered that catalyze chemical reactions other than redox reactions.

Inspired by the concept of the "legos" employed by Raanan *et al.* (2018) (see the upper panel in Figure 1) as applied to protein structure (to be denoted as $L_{proteins}$), I invoked the concept to the BBBs of genes (denoted as $L_{nucleotides}$), of metabolic pathways (denoted as $L_{\_MP}$), and of the cell function-specific networks of metabolic pathways that I elected to refer to as the "hyper-metabolic pathways" (denoted as $L_{HyperMP}$). The last is synonymous with the notion of "hyperstructures" proposed by Norris *et al.* (1999). These various BBBs are thought to be organized hierarchically within cellese and humanese in a similar manner (see the lower two panels in Figure 1).

The BBBs of oxidoreductases characterized by Raanan *et al.* (2018) consisting of 60 to 100 amino acids (or 180 to 300 nucleotides in DNA) may be examples of the higher-order genetic alphabets predicted here, namely the 3.74th-order alphabet (since 4^3.74 = 180) to the 4.11th-order (since 4^4.11 = 300) alphabet, encoding not only the individual structures of amino acids but their space-time behaviors within the four-dimensional structures of oxidoreductases in action. The non-coding sequences controlling the space- and time-dependent gene expressions were named the "spatiotemporal genes" in 1991 (Ji, 1991, pp. 176–178).

If this proposal turns out to be correct, the nucleotide triplets coding for the 20 amino acids are not genetic words, as currently widely thought (including by me at some point in time), but genetic letters of

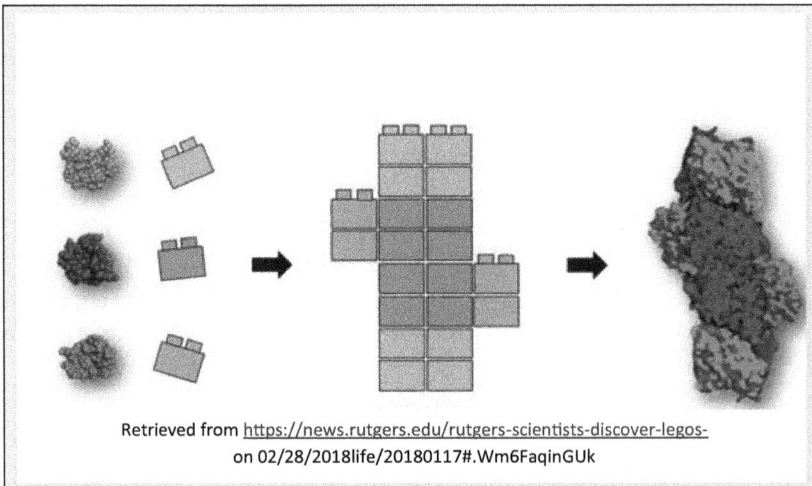

Cellese	$L_{Nucleotides} \xrightarrow{2^{nd}A} L_{Genes} \xrightarrow{1^{st}A} L_{MPs} \xrightarrow{3^{rd}A} L_{HyperMPs}$
Humanese	$L_{Letters} \xrightarrow{2^{nd}A} L_{Words} \xrightarrow{1^{st}A} L_{Sentences} \xrightarrow{3^{rd}A} L_{Texts}$

Figure 1. The lego (L) model of genotype–phenotype coupling.
Notes: MP = Metabolic pathways;
HyperMP = Hyper-metabolic pathways.
$1^{st}A$ = First articulation; words forming sentences (in order to decide).
$2^{nd}A$ = Second articulation; letters forming words (in order to build).
$3^{rd}A$ = Third articulation; sentences forming texts (in order to argue/reason/compute).

the 3rd order. This way of looking at the genetic alphabet indicates that a series of nucleotide triplets (viewed as letters) can form genes (or proteins) viewed as words via covalent interactions, consistent with the cell language theory summarized in Table 1 (Ji, 1997, 1999).

The hierarchically organized "legos" or cell *linguistic building blocks* as proposed in Figure 1 agree well with the isomorphism postulate between *cellese* and *humanese* first proposed in 1997 (Ji, 1997, 1999) and further developed in 2018 (Ji, 2018a), as summarized in Table 6 (which is a recapitulation of Table 4).

Probably, the most noteworthy novel feature of Table 6 is the concept of the "third articulation" (imported from linguistics (Ji, 1997)) and "hypermetabolic pathways" (described here for the first time). Due to the lack of

Table 6. The structure of the cell language (*cellese*) inferred based on the postulated functional isomorphism between *cellese* and *humanese* (or human language).

Organization level	Humanese	Function	Cellese
I	Letters	Basic building blocks	Nucleotides (A, C, G, T, AC, AG, TC, TTC, AGT, etc.)
II	Words	To denote	Genes (e.g., sets of nucleotide triplets)
III	Sentences	To decide	Metabolic pathways (MP)
IV	Texts	To argue/reason/compute	Hyper-metabolic pathways (HMP) or Hyperstructures of Norris *et al.* (2018)

Notes: Hyper-metabolic pathways are defined as sets of metabolic pathways that are functionally connected to carry out cell functions and are closely related to Norris's "hyperstructures" (Norris *et al.*, 1999).
1st Articulation = II → III (needed to decide);
2nd Articulation = I → II (needed to denote);
3rd Articulation = III → IV (needed to reason/argue/compute) (Ji, 2018a).

quantitative experimental methods, it has been very difficult, if not impossible, to study the "third articulation" and "hyper-metabolic pathways" in cell biology until now. However, the situation may change rapidly with the recent emergence of the method of "the Planck–Shannon plots (PSP)" described in Ji (2018b) and Section 13, if this method is adopted by cell biologists in analyzing their transcriptomes (i.e., the whole-cell mRNA levels) measured with microarrays or their equivalent. PSP can be applied to analyzing other "hyperstructural data" such as fMRI (functional Magnetic Resonance Imaging) data measured from different regions of the brain (see Ji, 2018a, Figure 8.6 (l)) that transiently cooperate to carry out brain functions and then disassemble.

Another potentially controversial proposal is that DNA may contain more than one genetic alphabet other than the traditional one with four nucleotides, A, C, G, and T. The predicted alphabets may be called the 2nd-order genetic alphabet consisting of $4^2 = 16$ "letters" of dinucleotides such as AT, CG, CT, etc., and the 3rd-order alphabet consisting of $4^3 = 64$ letters of trinucleotides such as ACT encoding threonine, CGT encoding arginine, GCG encoding alanine, etc.

The BBBs of oxidoreductases characterized by Raanan *et al.* (2018) consisting of 60–100 amino acids (or 180–300 nucleotides in DNA) may be examples of the higher-order genetic alphabets predicted here, namely the 180th-order alphabet to the 300th-order alphabet encoding not only the individual structures of amino acids but their space-time behaviors within in the four-dimensional structures of oxidoreductases in action.

If this proposal turns out to be correct, the nucleotide triplets coding for the 20 amino acids are not genetic words, as currently widely thought, but genetic letters of the 3rd order. This way of looking at the genetic alphabet indicates that a series of nucleotide triplets (viewed as letters) can form genes (or proteins) viewed as words via covalent interactions, consistent with the cell language theory summarized in Table 1 (Hockett, 1960; Ji, 1988, 1997).

Please note that, in some of my publications, I divided the structure of *cellese* and *humanese* into five levels. One way to reconcile this apparent discrepancy may be to split I in Table 1 into IA and IB, identifying the former as letters and the latter as syllables (i.e., a unit of pronunciation having one vowel sound, with or without surrounding consonants, forming the whole or a part of a word) and referring to the combination of IA and IB as the alphabet (defined as "the basic elements in a system which combine to form complex entities"; *English Oxford Living Dictionaries*) of *cellese* in agreement with Sergey Petoukhov (2017).

8. Molecular Mechanisms Underlying Information Processing Mediated by DNA

We now know that we are different from mice because our DNA sequences are different from those of mice (Ramsey and Lee, 2016). That is, we are different from mice because our DNA carries different kinds (both with respect to quality and quantity) of information than those of the mouse DNA:

> "When it comes to protein-encoding genes, mice are 85% similar to humans. For non-coding genes, it's only about 50%. The National Human Genome Research Institute attributes this similarity to a shared ancestor about 80 million years ago" (Ramsey and Lee, 2016).

We also know that our properties or behaviors are at least in part determined by both DNA sequences (i.e., *genetics*) and the way they are turned on or off by environmental conditions of our body (i.e., *epigenetics*). That is, we are the products of both our *genes* and our *environment.*

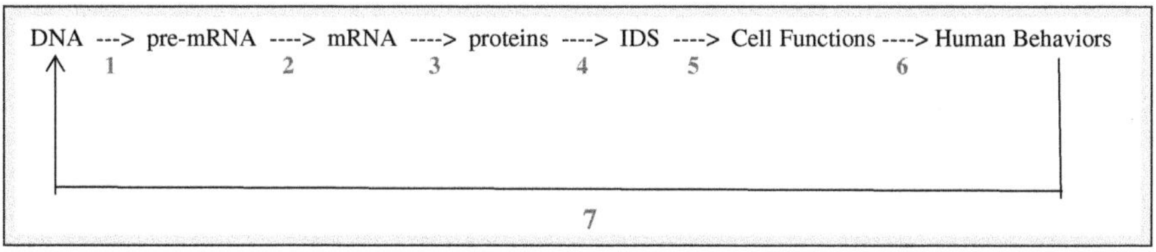

Figure 2. A simplified diagram of the flow of genetic and epigenetic information between DNA and the human behavior. IDS stands for the *Intracellular Dissipative Structures* (also called the *Dissipative Structures of Prigogine*) (Ji, 2012, Section 3.1) such as ion gradients across cell membranes and within the cytoplasm without any membrane barriers. According to the Bhopalator, a molecular model of the living cell proposed in 1985 in Bhopal, India, IDS's are postulated to be the immediate or the proximal causes for all cell functions (Ji, 1985). The seven steps in the scheme are as follows:
1 = transcription;
2 = splicing;
3 = translation (explained in Figure 3 in more detail);
4 = enzyme catalysis;
5 = cell motions;
6 = body motions;
7 = the effect of human behavior or emotion on gene expression, e.g., see the phenomenon of the *conserved transcriptional response to adversity* (CTRA) (Cole, 2014).

The causal link between DNA and our behaviors can be briefly summarized as in Figure 2.

It is hoped that the *information flow scheme* shown in Figure 2 can serve as a concrete example of information in action as information scientists strive to come up with a generally acceptable definition of information.

Unlike in Steps 1 and 2 where the same kinds of molecules, i.e., the nucleic acids, DNA and RNA, directly interact (or contact or touch each other) via the Watson–Crick base-paring mechanism, in Step 3 there is no such direct interaction between mRNA and amino acids, but rather their interactions are mediated by tRNA, which recognizes mRNA at its *anticodon arm* and amino acids at its *3′-acceptor stem*, about 60 angstroms away (see the blue region in Figure 3).

The universality of the wave–particle duality demonstrated in Ji (2015a, 2015b, 2018a) suggests that the tripartite coupling among codon, anticodon, and amino acid in the ribosome-mRNA-tRNA complex may be mediated by *resonant vibrations* or *standing waves* (also called *resonance* or *resonant waves*) generated within the tripartite complex, just as the vibrational patterns located at distant regions on the Chladni (1756–1827) plate (Patel, 2013) (see also "Amazing Resonance Experiment!" at https://www.

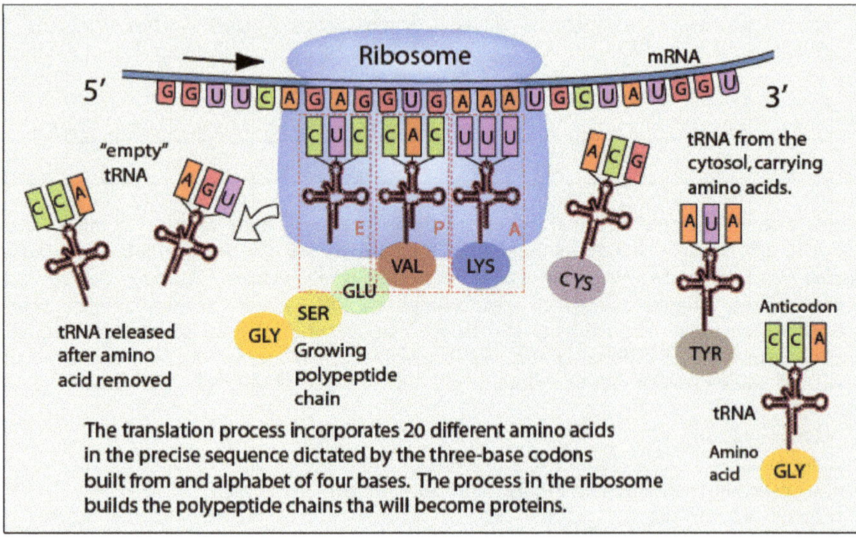

Figure 3. The molecular mechanism of decoding mRNA to synthesize proteins by combining amino acids catalyzed by the ribosome and tRNAs.
Source: Retrieved from https://www.quora.com/Why-are-ribosomes-so-important-in-plant-cells on 01/16/2018.

youtube.com/watch?v=wvJAgrUBF4w) are coordinated via resonance (Ji, 2018a, Figure 4.13).

The Chladni plate is an ideal model for illustrating the role of *resonance* in molecular biology (Ji, 2018a, Figure 4.13). At a given resonance frequency, the particles on remote regions of the Chaldni plate are coordinated without any direct interactions between them and yet form ordered patterns of particle distributions. To me, this is similar to what happens in the ribosome system when a peptide molecule is synthesized; i.e., different components of the ribosome–mRNA–tRNA complex execute their resonant vibrational motions that are so coordinated (as a result of biological evolution) as to achieve the peptide synthesis. The ribosome and the Chladni plate are compared at several levels in Table 7.

The idea of using the Chladni plate as a molecular model of protein synthesis in the ribosome occurred to me while preparing for the keynote lecture, "RASER model of single-molecule enzyme catalysis", that I presented at the 8th International Conference on Proteomics and Bioinformatics, held in Osaka, Japan, in May 2017 (see the right bottom panel in Figure 4) (Ji, 2018b). I formulated this idea before I realized that J. Franck and his group were working on the mechanism of protein synthesis in the

Table 7. The Ribosome–Chladni plate comparison.

	Chladni plate	Ribosome
1. Vibrating system	Metal plate	Ribosome + mRNA + tRNA complex
2. Objects being organized	Sand particles	Codons, anti-codons, aminoacyl residues
3. Scale	Macroscopic	Microscopic (or molecular)
4. Source of energy driving the organization	Electricity/sound waves	Thermal energy paid back by the free energy of chemical reactions without violating the second law (Ji, 1991)
5. Common principle obeyed	Fourier theorem	Fourier theorem

ribosome based on a similar approach (Tama *et al.*, 2003), i.e., the wave approach to molecular biology (Ji, 2015a, 2015b), for which Franck shared the Nobel Prize in Chemistry in 2017 (https://www.nobelprize.org/nobel_prizes/chemistry/laureates/2017/).

There are several components in what is here referred to as the "resonance model of the ribosome structure and function" (RMRSF):

(i) the ribosome and its associated RNAs and proteins are a system of molecular oscillators that obey the Fourier theorem (see the left panel in Figure 4), i.e., all the wave patterns, both traveling and standing (see the middle panel in Figure 4) of the system can be constructed from the linear combinations of the vibrational motions of all the chemical bonds (Herbert, 1987);

(ii) the standing waves in the ribosome system are determined by the geometrical shape (or topology) and the energy content of the system;

(iii) the chemical (i.e., peptide bond formation) and mechanical events (i.e., translocation of the tRNA from the A site to the P site, etc.) taking place at the anti-codon arm and the 3′-acceptor stem are "coupled" or "coordinated" through the resonance mechanism, just as are the motions of the sand particles located in different regions on the Chladni plate, or just as the vibrational motions of the tuning forks are coupled to the vibrational motions of the water column by resonance thereby exchanging both energy and information between them (see the right upper panel in Figure 4).

The resonance model of the *ribosome structure and function* depicted in Figure 4 is based on the similarities and differences between the ribosome

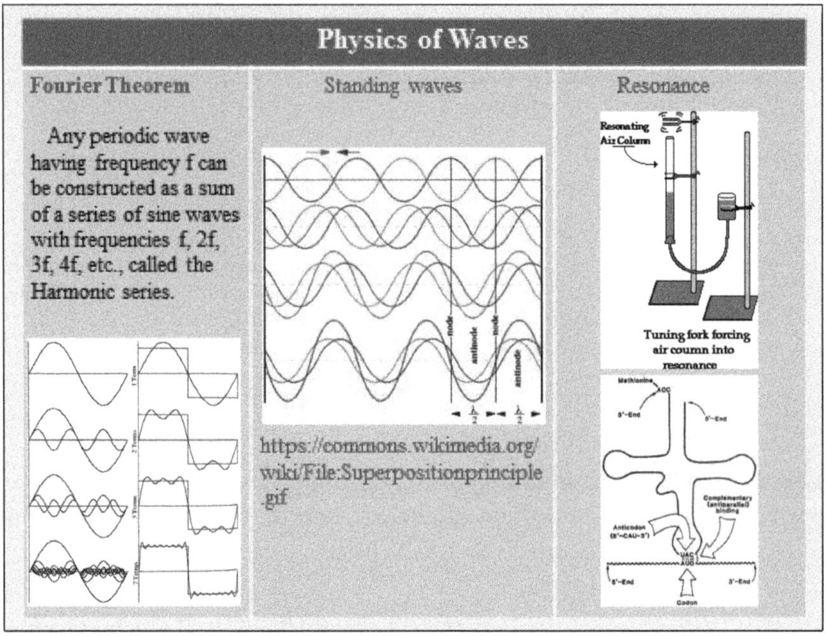

Figure 4. The wave approach to unraveling the molecular mechanisms underlying protein synthesis in the ribosome: The role of resonance.

and the Chaladni plate summarized in Table 7. It may be worth noting that my vibrational approach to enzymology/molecular biology dates back to 1974 (Ji, 1974) when I concluded as follows:

> "An ordinary enzyme possesses 10^3 to 10^4 vibrational degrees of freedom, as compared to 3 each for the translational and rotational degrees of freedom. It is therefore reasonable to assume that the vibrational motions of individual bonds in the enzyme will be far more important in enzyme catalysis than the translational or rotational motions of the enzyme as a whole. Given all the vibrational frequencies of the individual bonds in an enzyme, as well as their three-dimensional arrangements, we can in principle deduce the thermodynamic and catalytic properties of the enzyme under any conditions".

The resonance mechanism of coupling proposed in Figure 4 between the aminoacyl group at the 3′ acceptor stem and the anti-codon at the opposite side of the ribosome about 60 Å away is consistent with the conclusion reached by Petoukhov that *genes in DNA can be viewed as oscillators* that couple to one another via resonance (Petoukhov, 2016, 2017a, 2017b). One of the reasons that Petoukhov came to invoke the *resonance mechanism* is that many of the regularities seen in the genetic codes and DNA sequences (e.g., Chargaff's parity rules (Petoukhov, 2017b) obey the *tensor product rules of matrices* widely used in engineering and physics to analyze vibrational mechanics (Petoukhov, 2017a). Thus, it is clear that Petoukhov's resonance approach to modeling genetic structures indirectly supports the *resonance coupling mechanism* postulated for the transfer of genetic information from the codons of mRNA to the aminoacyl residues of tRNA as briefly summarized in Figure 4.

9. Fibonacci Numbers, Tensor Product of Matrices, and "Self-Replication with Variations" (SRV) in Biology

Fibonacci numbers, $\mathrm{F} = \{1, 1, 2, 3, 5, 8, 13, 21, \text{etc.}\}$, can be represented as

$$\mathrm{F}(n+1) = \mathrm{F}(n) + \mathrm{F}(n-1), \tag{6}$$

where $\mathrm{F}(n)$ is the nth Fibonacci number, counting from the second number of the first triplet of the Fibonacci numbers, i.e., 1, 1, 2, so that $\mathrm{F}(n-1) = 1$, $\mathrm{F}(n) = 1$, and $\mathrm{F}(n+1) = 2$. The basic unit of the Fibonacci series is the adjacent or contiguous triplet, $\mathrm{F}(n+1)$, $\mathrm{F}(n)$, and $\mathrm{F}(n-1)$, which is here referred to as the "Fibonacci triplet".

The Fibonacci triplet embodies an ITR (Irreducible Triadic Relation) [Ji, 2018a, Chapter 9] in the sense (i) that replacing any one of them with other integers will destroy the Fibonacci series and (ii) that the triplet constitutes a *commutative triangle* as shown in the upper panel in Figure 5 and hence a mathematical category (Ji, 2018a, Figure 2.2).

The Fibonacci series, in the form of rectangles, can be generated by iterating two simple operations of (i) squaring the longer side of a Fibonacci rectangle and (ii) adding to it the old Fibonacci rectangle. Repeated application of these two operations is referred to as the "replication with variation", or RV (see in (Ji, 2018a, Figure 9.7) and will be referred to here as "Self-Replication with Variations" (SRV). It can be readily demonstrated with a ruler and a pencil that the process of RV or SRV applied repeatedly can generate a series of Fibonacci rectangles that contain the *Fibonacci spirals* (e.g., see https://www.mathsisfun.com/numbers/fibonacci-sequence.html) found in nautilus shells, the composition of Mona Lisa, and the galactic spirals (see Ji, 2018a, Figures 9.3 (d), (h), and (l)). It is interesting to note that the ratio, $\mathrm{F}(n)/\mathrm{F}(n-1)$, approaches the Golden ratio of 1.618 when n is greater than 10.

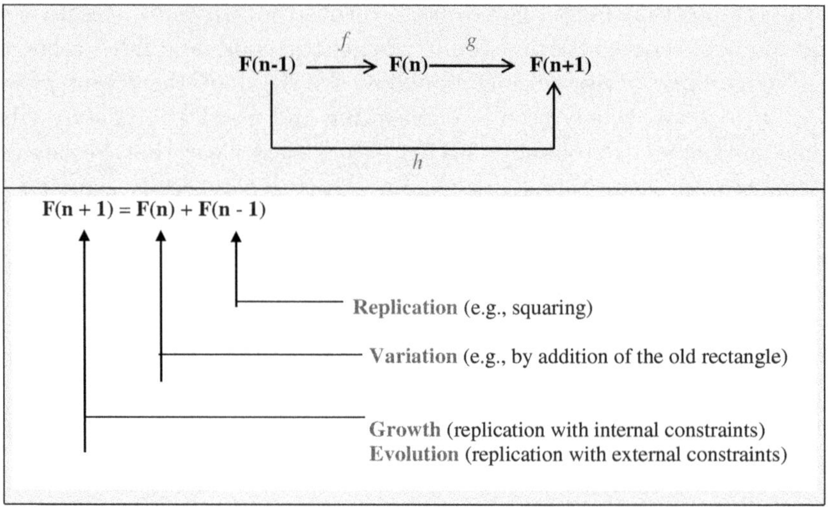

Figure 5. (*Upper panel*) The *Fibonacci triplet* as a mathematical category. f = growth by a factor of Φ; g = growth by a factor of Φ; h = growth by a factor of Φ^2. These mappings can be viewed as representing the mechanisms of "replication with variation (RV)" that seem to appear in biological development and evolution (Ji, 2018a). (*Lower panel*) A diagrammatic representation of the "theory" of "replication with variations (RV)" as the fundamental mechanism underlying the generation of the Fibonacci numbers, development/growth, and evolution. The Fibonacci numbers and hence RV are found in DNA (Yamagishi and Shimabukuro, 2008; Perez, 2010).

Source: Adapted from Ji (2018a, Figures 9.5 and 9.7).

The Fibonacci numbers can be thought of being generated via a series of three steps as indicated in the lower panel in Figure 5, i.e., replication, *variation*, and *growth/evolution*, which has been referred to as the "Replication with Variation" (RV) or "Self-Replication with Variations" (SRV). RV or SRV is thought to act as the Universal Mechanism underlying not only the Fibonacci series and the Golden ratio but also cell divisions in morphogenesis and epigenetic evolution, which may account for the wide occurrence of Fibonacci spirals and Golden ratios in biological structures and processes (Ji, 2018a, Chapter 9).

The process of tensor product of matrices that Petoukhov (Petoukhov, 2016, 2017a, 2017b) found useful in uncovering the regularities (or rules) embedded in long DNA sequences may be viewed as a mathematical representation of SRV as explained below.

The four bases of the genetic alphabet, i.e., A, C, G, and T (see the first row in Figure 6), divide into the well-known Watson–Crick base pairs,

Alphabet	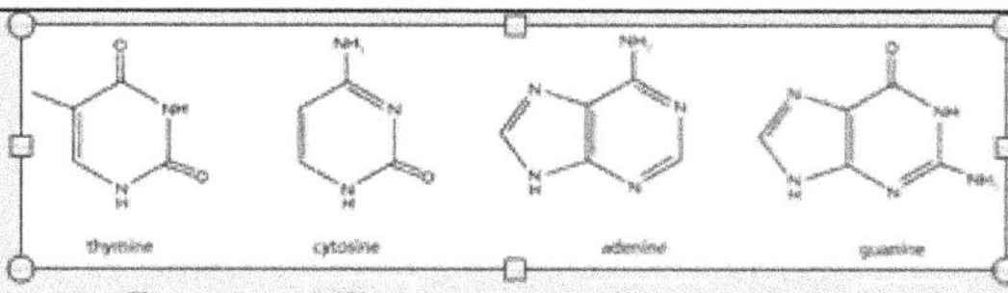 (T) (C) (A) (G)	
Watson-Crick pair	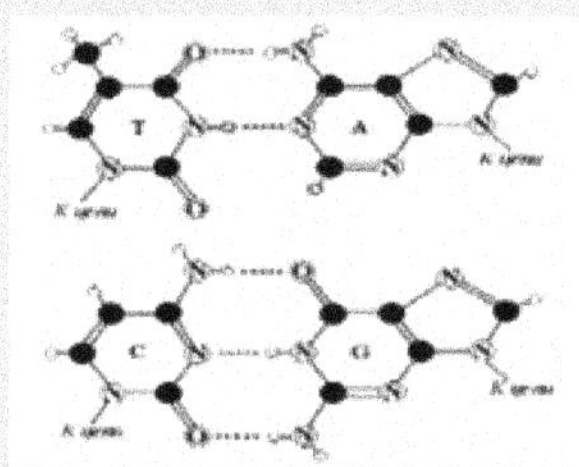	T vs. A C vs. G
Sub-Alphabet	2H-3H (also called Watson-Crick) pair	(T, A) vs. (C, G)
	Pyrimidine-purine pair	(C, T) vs. (A, G)
	Amino-keto pair	(C, T) vs. (A, G)
Petoukhov Coincidence [1]	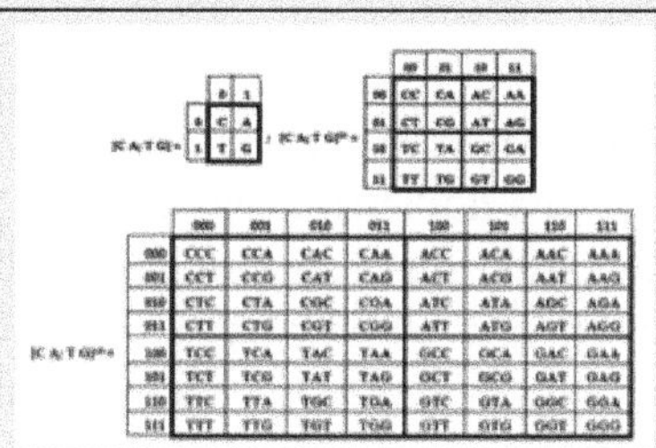	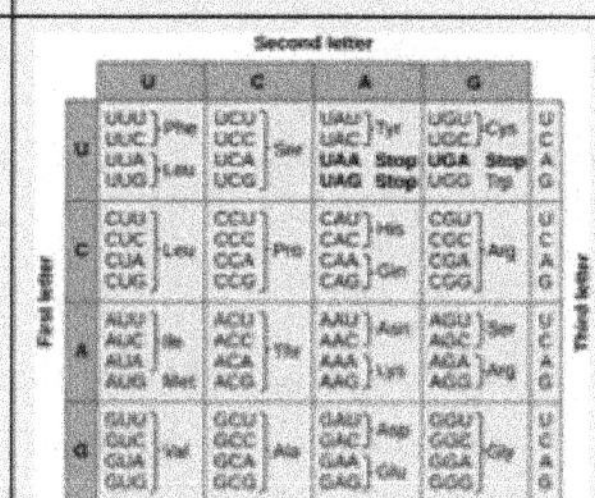

Figure 6. The nucleotide alphabet and subalphabet of genetics. [1] = Ji (2018a, p. 242).

Source: Adopted from (Ji, 2018a, Table 5.2); Petoukhov (2016, 2017a, 2017b).

i.e., the T–A pair characterized by two hydrogen bonds and the C–G pair characterized by three hydrogen bonds (see the second row). In addition, there are two other pairs less well known — the amino (A, C) and keto (G, T) pair, and the pyrimidine (C, T) and purine (A, G) pair, each being characterized by unique molecular features or traits (and hence most likely unique vibrational resonance properties as well) (see the third row). These three nucleobase pairs (which are three pairs of pairs, or three pairs of yin and yang) can be represented as a triadic set of *four* binary numbers as shown in the last four columns in the table in Figure 7 (omitting the subscripts), namely, 0011, 0101, and 1001. Stated in another way, each of

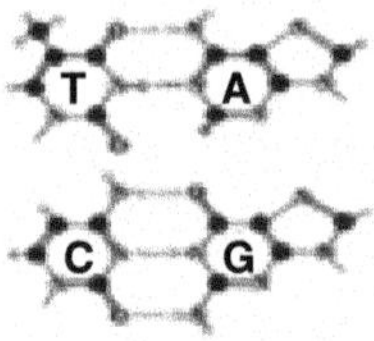

TRAITS	G	A	T	C
(1) pyrimidine (C,T), purine (A,G)	0_1	0_1	1_1	1_1
(2) amino (A,C), keto (G,T)	0_2	1_2	0_2	1_2
(3) complementarity (C,G) and (A,T) with 3 or 2 hydrogen bonds	1_3	0_3	0_3	1_3

Figure 7. The three *yin–yang pairs* (also called *binary opposite pairs* or *binary subalphabets* (Petoukhov, 2016, 2017a, 2017b)), i.e., (1), (2), and (3), that can be generated from the 4 letters of the genetic alphabet A, C, G, and T.

the four nucleobases can be represented either as 0 or 1 within a given pair of base pairs. That is, G and A can be assigned the same binary digit, 0, and T and C the same binary digit 1, within the pyrimidine–purine pair rule. On the other hand, the bases within each pair are assigned either 0 or 1. For example, within the 2H–3H pair, G and C are assigned 1 and A and T are assigned 0. In contrast, within the pyrimidine–purine pair rule, T and C are assigned 1 and G and A are assigned 0, etc. Two things are noteworthy:

(i) The addition of the first two rows of the binary digits in Figure 7 produces the third row, if the addition is carried out based on the rule of *modulo-2 addition*: i.e., addition of two identical digits = 1; addition of two different digits = 0. This in effect reduces the three base pairs into two. Thus, knowing any two of the 3 yin–yang pairs allows us to infer the third correctly. Nature may have utilized this fact in designing its error-immunity coding in genetics (Petoukhov, 2016, 2017a, 2017b).

(ii) Any sequence of nucleotides can carry two overlapping binary messages (Petoukhov, 2016), each message being decodable based on one of the two alternative base (or yin–yang) pair rules in Figure 7. An example is given below:

GCTCGCCTAA

0 1 1 1 0 1 1 1 0 0 according to the *pyrimidine–purine* pair rule

0 1 0 1 0 1 1 0 1 1 according to the *amino–keto* pair rule

1 1 0 1 1 1 1 0 0 0 according to the 2H–3H pair rule

In other words, the same set of 10 nucleotides above can encode two independent binary messages (i.e., the top two rows, the bottom row being their sum based on the rule of modulo-2 addition), leading to the conclusion

[C A; T G] =

	0	1
0	C	A
1	T	G

; [C A; T G]$^{(2)}$ =

	00	01	10	11
00	CC	CA	AC	AA
01	CT	CG	AT	AG
10	TC	TA	GC	GA
11	TT	TG	GT	GG

[C A; T G]$^{(3)}$ =

	000	001	010	011	100	101	110	111
000	CCC	CCA	CAC	CAA	ACC	ACA	AAC	AAA
001	CCT	CCG	CAT	CAG	ACT	ACG	AAT	AAG
010	CTC	CTA	CGC	CGA	ATC	ATA	AGC	AGA
011	CTT	CTG	CGT	CGG	ATT	ATG	AGT	AGG
100	TCC	TCA	TAC	TAA	GCC	GCA	GAC	GAA
101	TCT	TCG	TAT	TAG	GCT	GCG	GAT	GAG
110	TTC	TTA	TGC	TGA	GTC	GTA	GGC	GGA
111	TTT	TTG	TGT	TGG	GTT	GTG	GGT	GGG

Figure 8. The tensor product of 2×2 matrices, e.g., [C A; T G] (*the upper left table*), to generate a 4×4 (*the upper right table*) matrix and an 8×8 matrix (*the lower table*). *Source*: Reproduced from Petoukhov (2016).

that DNA is akin to a *broadband* communication channel, "broadband" indicating the ability of a communication channel to transfer information fast because of its ability to send multiple messages simultaneously (https://en.wikipedia.org/wiki/Broadband).

The rules of the tensor multiplication (or product) of matrices of nucleotides are illustrated in Figure 8. When the 2×2 matrix, denoted as [C A; T G], is tensor multiplied by itself once, the product is a 4×4 matrix, each quadrant having the original 2×2 matrix multiplied by the corresponding elements in the original matrix. For example, the quadrant, [CC CA; CT CG] is the product of C and [C A; T G], the quadrant on its right-hand side [AC AA; AT AG] is the product of A and the original matrix [C A; T G], etc. The same procedure applied twice, i.e., [C A; T G]^3, produces the 8×8 matrix shown in the lower panel in Figure 8. Evidently, the self-multiplication of matrices obeying the tensor product rule generates an enlarged matrix that retains the structure of the original matrix in a new environment. Hence, it would be logical to identify the tensor product of matrices as an example of what was referred to as the "self-replication with variations" (Ji, 2018a, Figure 9.7).

10. The Qualitative and Quantitative Similarities between *Cellese* and *Humanese*

There may be two kinds of similarities underlying the isomorphism between *cellese and humanese* — the *qualitative* and the *quantitative.* Tables 1 and 2 list the *qualitative* similarity (or qualitative functor, to borrow the category-theoretic idiom (Spivak, 2013; Brown and Porter, 1989) and Figure 9 lists the *quantitative* similarities (or quantitative functor). That is, both *cellese* and *humanese* obey the same pair of mathematical equations — PDE (Planckian Distribution Equation; see Section 11 below) was derived from physics (Ji, 2012) and the Menzerath–Altmann law was derived from linguistics (https://en.wikipedia.org/wiki/Menzerath%27s_law) but they are indistinguishable as far as their ability to fit long-tailed histograms generated from either biology (see the left column in Figure 9) or linguistics (the right column in Figure 9) is concerned. Therefore, it seems reasonable to conclude that Tables 1, 2 and Figure 9 provide both the *qualitative* and *quantitative* evidence supporting the validity of the isomorphism between *cellese* and *humanese.*

11. Planckian Distribution Equation (PDE) and *Planckons*

The Planckian Distribution Equation (PDE) refers to the mathematical equations (see Eqs. (8) and (9) in Figure 10) derived in 2008 at Rutgers from the Planck blackbody radiation equation (PBRE) (see Eq. (7) in Figure 10) discovered by Planck (1858–1947) in 1900. PDE replaces the universal constants and temperature in PBRE with three or four parameters (see Eqs. (8) and (10) in Figure 10) and has been found to fit almost any long-tailed histograms generated in many fields, including atomic physics, protein folding, single-molecule enzyme catalysis, RNA metabolism in cells and tissues, functional magnetic resonance imaging (fMRI), human decision-making process, glottometrics (the quantitative study of languages), econophysics, and cosmology (Ji, 2012, 2015a, 2015b, 2018a). It is convenient to define the physicochemical or formal processes that produce data that fit PDE as *Planckian processes*, or *Planckons* more briefly. Thus, it appears to us reasonable to assume that Planckons are the

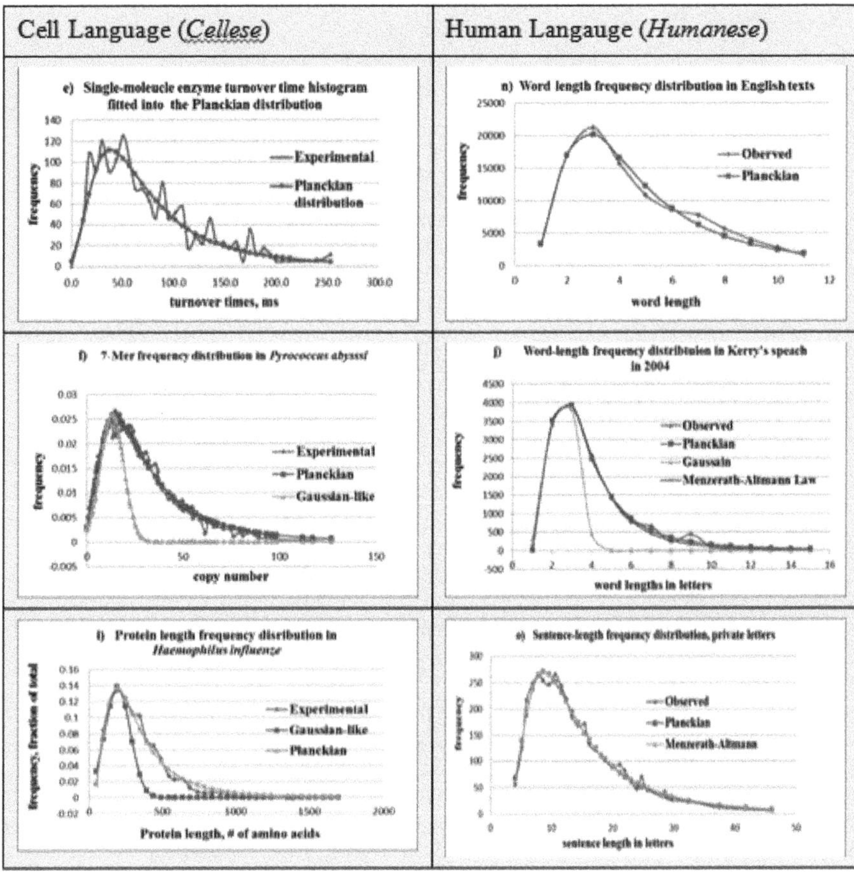

Figure 9. The Planckian Distribution Equation (PDE) and the Menzerath–Altmann Law (MAL) connect the *cell* and *human languages*.

Source: Adapted from Ji (2015b, 2018a).

units of organizations in the Universe. We have found over 100 *Planckons* in all fields of natural and human sciences as of December 2017.

Since the first term in PBRE is related to the number of the standing waves (synonymous with the modes of vibrations per unit frequency per unit volume) generated with the system of oscillations under consideration and the second term to the average energy of the standing waves (http://hyperphysics.phy-astr.gsu.edu/hbase/mod6.html), it is assumed that the first and the second terms of PDE can also be interpreted similarly. That is, whenever a system generates data that fit PDE, the system can be

(a)

$$E(\lambda, T) = \frac{2\pi hc^2}{\lambda^5\left(e^{\frac{hc}{\lambda kT}} - 1\right)} \quad (7)$$

λ = Wavelength
c = Speed of light
k = Boltzmann constant
h = Planck's constant
e = 2.71828182
[T] = Kelvin (Temperature)
[λ] = Meters
h = 6.626.1034 J.s
c = 2.998.108 m/s
k = 1.381.10-23 J/K

(b)

$$y = \frac{a}{(Ax+B)^5} \cdot \frac{1}{e^{b/(Ax+B)} - 1} \quad (8)$$

Number of modes of vibrations per unit frequency per unit volume

Average energy per mode

(c)

$$y = \frac{A}{(x+B)^5} \cdot \frac{1}{e^{C/(x+B)} - 1} \quad (9)$$

$$A = a/A^5 \quad (10)$$
$$B = B/A \quad (11)$$
$$C = b/A \quad (12)$$

Figure 10. (a) The Planck radiation equation. Reproduced from Blackbody radiation. http://hyperphysics.phy-astr.gsu.edu/hbase/mod6.html. (b) The 4-parameter blackbody radiation-like equation (BRE) (Ji, 2012, Chapters 11 and 12), also called the generalized Planck equation (GPE) or the Planckian Distribution equation (PDE) (Ji, 2018a, Chapter 8). (c) The 3-parameter (colored red) version of BRE/GPE/PDE. The relations between the 4- and 3-parameter versions of BRE/GPE/PDE are given in Eqs. (10)–(12).
Source: Adapted from Ji (2018a, Figure 8.1).

assumed to generate standing waves within it with a fixed average energy of the standing waves. We also assume that the observable function of the system of oscillations will be determined by the pattern of the standing waves and their average energy, as schematically represented in Figure 11.

12. What Is Planckian Information?

The Planckian information (I_P) is defined as the information produced (or used) by the so-called Planckian processes, which are in turn defined as follows:

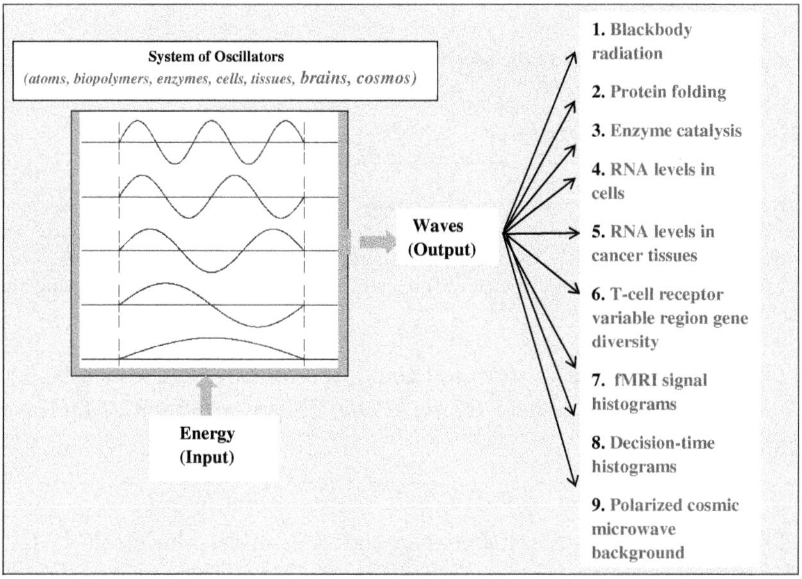

Figure 11. One possibility to account for the universality of the Planckian distribution in nature is to postulate that the *wave-particle duality* first discovered in atomic physics in the early 20th century operates at all scales of material systems, from atoms to the Universe.

Source: Reproduced from Ji (2015a, 2015b).

> "Planckian processes are the physicochemical, neurophysiological, biomedical, mental, linguistic, socioeconomic, cosmological, or any other processes that generate long-tailed histograms obeying the Planckian distribution equation (PDE)".

For convenience, Planckian processes will be referred to as *Planckons*. The relation between Gaussian processes and Planckian processes is schematically depicted in Figure 12.

12.1. *Planckian information of the first kind*

There are two ways of quantifying I_P:

(i) as the ratio of the area under the curve (AUC) of PDE to that of Gaussian-like symmetric curves (Ji, 2015a, 2015b, 2018a), and
(ii) as a function of the *skewedness* of the long-tailed histogram.

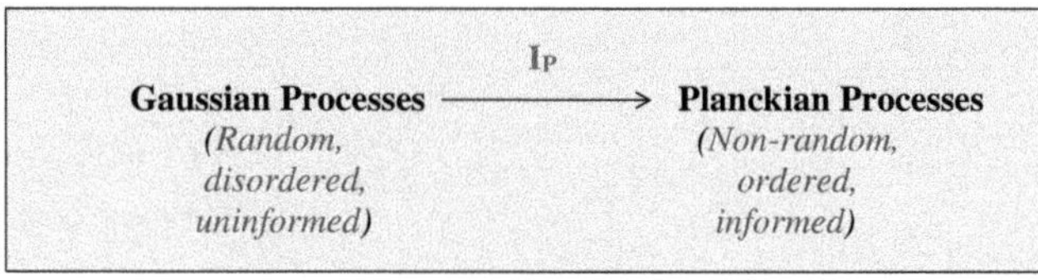

Figure 12. *A diagrammatic representation of the relation between Gaussian and Planckian processes.* Planckian processes (or *Planckons*) are thought to be the n ordered processes selected out of the N random Gaussian processes via some selection mechanism, thereby generating the Planckian information, $I_P = \log_2 (N/n)$, assuming that N processes are equiprobable.

The "Gaussian-like" curves are generated by Eq. (13), which is derived from the Gaussian equation by replacing its pre-exponential factor with free parameter A:

$$y = Ae^{-(x-\mu)\ \hat{}2/(2\sigma\ \hat{}2)}. \tag{13}$$

The degree of mismatch between the area under the curve (AUC) of PDE, Eq. (8) or (9), and that of GLE, Eq. (13), is postulated to be a measure of *non-randomness* (and hence *order* and *information*). GLE is associated with random processes, since it is symmetric with respect to the sign reversal of its exponential term, $(x - \mu)$. This *measure of order* is referred to as the Planckian Information (I_P) of the first kind defined quantitatively as shown in Eq. (14) or Eq. (15):

$$I_P = \log_2 (\text{AUC(PDE)}/\text{AUC(GLE)}) \quad \text{bits}, \tag{14}$$

or

$$I_P = \log_2 \left[\int P(x)dx \Big/ \int G(x)dx \right] \quad \text{bits}, \tag{15}$$

where $P(x)$ and $G(x)$ are the Plackian distribution equation and the Gaussian-like equation, respectively. The rising portion of GLE often completely overlaps with the rising portion of the right long-tailed PDE (see in Ji, 2015b, Figures 15, 16, 20, 22, and 23). It has been found that the greater the deviation of PDE from GLE, the greater is the I_P value, since GLE represents randomness and the deviation of PDE from GLE represents non-randomness, order, or information.

12.2. *Planckian information of the second kind*

Another mechanism of generating PDE from Gaussian distribution is what I call the "Rutgers University Admissions Mechanism" (RUAM). That is, if

RAUM does not take into account the students' heights in their admissions process, the height distribution of the RU students would be most likely Gaussian. However, if RUAM favors short students over tall ones, the RU students' height distribution will be skewed from the normal curve thus producing a long-tailed histogram that most likely fits PDE. The degree of skewness of PDE from its Gaussian counterpart (with an equal area under the curve) can be used as a measure of the information used by RAUM in selecting RU students. The information derived from PDE based on its skewness will be referred to as the Planckian information of the second kind defined by Eq. (16) to be distinguished from the Planckian information defined previously (see Eq. (15)) which is now called the Planckian information of the first kind, I_{PF}.

$$I_{PS} = -\log_2((\mu - \text{mode})/\sigma), \tag{16}$$

where $\propto$ and σ are the mean and the standard deviation of the long-tailed histogram under consideration.

We have found that some experimental data (e.g., digitized water wave patterns produced by the sonified Raman spectral bands measured from single cells) that fit PDE are better modeled with I_{PF} and some others (e.g., the mRNA levels measured from yeast cell ensembles) are better modeled with I_{PS}.

These observations clearly indicate the following:

(a) There can be more than one kind of information that can be defined based on the same empirically derived mathematical equation, probably depending on supporting physical *mechanisms*.
(b) The reasoning in (1) suggests that the mathematically defined "information" is *arbitrary* in the sense of Saussure's arbitrariness of signs.
(c) The mathematically defined "information" can be viewed as a sign in the Peircean sense and hence is irreducibly triadic as depicted in Figure 13.

12.3. *The Planckian Information is to* Order *What the Boltzmann–Gibbs–Shannon entropy is to* Disorder

The Planckian information represents the degree of organization/order of physical (or non-physical) systems in contrast to the Boltzmann or the Boltzmann–Gibbs entropy which represents the disorder/disorganization of a physical system, whether the system involved is atoms, enzymes, cells,

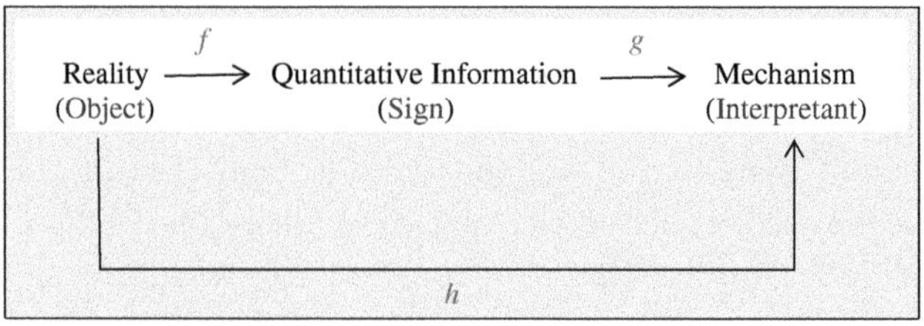

Figure 13. The irreducibly triadic nature of the "quantitative information" or the "mathematical information". f = measurement; g = mental process; h = correspondence, grounding. The terms in parentheses are those of the definition of the sign according to Peirce.

Source: (Ji, 2018a, Chapter 9).

Table 8. Two different views on the entropy–information relation.

	Schrödinger (1944) Schrödinger (1998)	Ji, 2015a, 2015b
Entropy (S)	S = k log D	S = k log D
Information (I)	– S = k log (1/D)	$I_P = \log_2$ [AUC(PDE)/AUC(GLE)]

Notes: I_P = the Planckian information, Eqs. (15) and (16). D = disorder. AUC = Area Under the Curve; PDE = Planckian Distribution Equations (8) or (9); GLE = Gaussian-like Equation (13).

brains, human societies, or the Universe (Ji, 2018a, Figure 8.6). I_P is related to the "organized complexity" and S is related to "disorganized complexity" of Weaver (1948). The organization represented by I_P results from *symmetry-breaking selection processes* applied to some randomly accessible (and hence symmetrically distributed) processes, whether the system involved is atoms, enzymes, cells, brains, languages, human societies, or the Universe (Ji, 2012, 2015a), as summarized in Table 8.

There is a great confusion in science and philosophy concerning the relation between the concepts of *information* and *entropy* as pointed out by Wicken (1987). A large part of this confusion may be traced back to the suggestions made by Schrödinger (1944) and others subsequently (e.g., von Neumann, Brillouin, etc.) that *order* can be measured as the *inverse of disorder* (D) and hence that information can be measured as negative entropy (see the second column in Table 8).

As I pointed out in Ji (2015, pp. 12–15), the concept of "negative entropy" violates the *Third Law of Thermodynamics* and hence cannot be used to define "order" nor "information". However, Planckian information,

I_P, can be positive, zero, or negative, depending on whether AUC (PDE) is greater than, equal to, or less than AUC (GLE), respectively, leading to the following conclusion:

"Information can be, but entropy cannot be, negative". (17)

Hence,

"Information is not entropy", (18)

as often claimed.

I recommended in Ji (2018a, Section 8.5.2) that Statement (17) or (18) be referred to as the *First Law of Informatics* (FLI). It is hoped that FLI will help clarify the decades-long confusions plaguing the fields of informatics, computer science, thermodynamics, biology, and philosophy of science (Wicken, 1987).

Another way of supporting the thesis that *information* and *entropy* are not equivalent is to invoke the notion of *irreducible triadic relations* (ITR) of Peirce (1839–1914) (Ji, 2018a, Chapter 9), according to whom the sign (i.e., anything that stands for something other than itself) is the irreducible triad of *object*, *representamen* (also called *sign or sign vehicle*), and *interpretant* (the effect that a sign has on the mind of an interpreter). The *irreducible triadic relation* (ITR) can be represented as a 3-node network shown in Figure 14. The *communication system* of Shannon is also irreducibly triadic, since it can be mapped onto the sign triad as indicated in the figure. Entropy (in the sense of Shannon's communication theory) is one of the three *nodes* and Information (in the sense of Peircean semiotics) is one of the three *edges*. Clearly, nodes and edges are two different classes of entities, consistent with the FLI Statement (Petoukhov, 2017b).

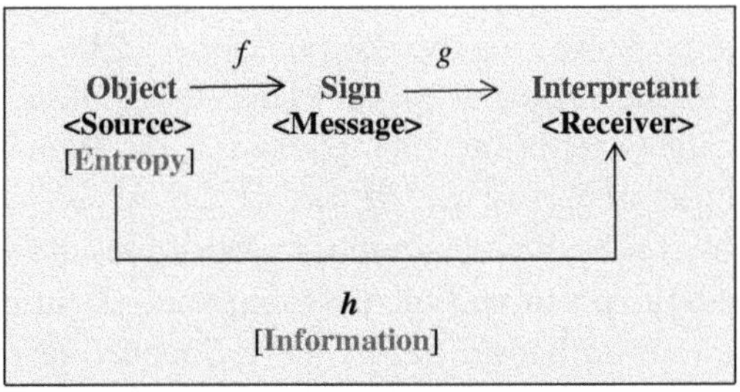

Figure 14. The isomorphism between the communication system (*the source–message–receiver triad*) and the semiotic system (*the object–sign–interpretant triad*), the "interpretant" being defined as the effect that a sign has on the mind of an interpreter. The arrows read "determines" or "constrains". f = sign/message production, g = sign/message interpretation; h = information flow, or correspondence.

13. The Planck–Shannon Space and the Semantic Space of Cell Language

Once a long-tailed histogram is fitted into PDE (see Figure 9 for examples), two numbers can be generated/computed from the PDE — (i) the *Planckain information* of the second kind, I_{PS}, Eq. (16), and (ii) the *Shannon entropy* (H) that can be calculated based on Eqs. (19) and (20):

$$H = -\Sigma p_i \log_2 p_i, \tag{19}$$

where p_i is the probability of observing the ith event or entity calculated as

$$p_i = y_i/\Sigma y_i, \tag{20}$$

where y_i is the frequency of the ith event of entity and the index i runs from 1 to n, the total number of events or entities.

Thus, the information encoded in PDE parameter values of a long-tailed histogram can be visualized as a point in what is referred to as the *Planck–Shannon space* or *plot* (see Figure 15).

When a group of 10 metabolic pathways (each having a varying number of open reading frames or genes as shown in Figure 15) is chosen from the budding yeast transcriptome measured over the 850 minute period of glucose–galactose shift experiments (Garcia-Martinez *et al.*, 2004) and their mRNA levels were represented as histograms which were then fitted into PDE, 10 pairs of numbers can be generated, each pair corresponding to the Shannon entropy and Planckian information of the second kind. When these 10 pairs of numbers are plotted on the Planckian–Shannon space, a reasonably linear correlation was found (see the bottom left-hand panel, Figure 15). However, when a similar set of 10 groups of mRNA levels are chosen that have no known metabolic functions, although they too generated long-tailed histograms that fitted PDE, the 10 pairs of numbers often did not produce any correlation when plotted in the Planck–Shannon space (see the right-hand panel, Figure 15).

Each point in the Planck–Shannon space represents a long-tailed histogram of a metabolic pathway (left bottom panel, Figure 15) or of a group of mRNA levels with no known functions (right bottom panel, Figure 15), but the linear correlation among the 10 points seems to occur frequently when they are functionally related. We analyzed 8 other pairs of 10 groups of mRNA levels of known and unknown metabolic functions and found that 6 out of the 8 sets of metabolic pathways showed linear correlations with correlation coefficients ranging from 0.60 to 0.75, while the 8 sets of mRNA

Pathway #	Biological Process	Number of Open Reading Frames
1	Cell cycle	72
2	Cell wall biogenesis	53
3	Chromatic structure	44
4	Cytoskeleton	71
5	DNA repair	32
6	rRNA processing	37
7	Nuclear protein targeting	43
8	Protein synthesis	156
9	Transport	129
10	Transcription	175

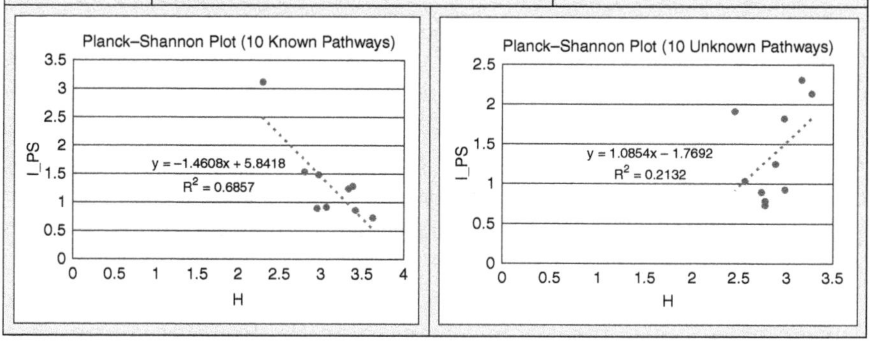

Figure 15. The 10 metabolic pathways analyzed.

levels with no known function showed relatively poor correlations, their correlation coefficients being less than 0.2. These observations support our following hypothesis:

> "Each point on the Planck–Shannon space represents a metabolic pathway and a linear correlation among 3 or more of these points represents the cell function supported by the linearly correlated metabolic pathways." (21)

Thus, we can identify the Planck–Shannon space as a *cell function space* or the *semantic space of cell language*.

The 10 points forming a correlated line in Figure 15 indicate that the *Planck–Shannon space* or *plot* can recognize the third level of metabolic organizations as predicted in the last row of Table 2. That is, the Planck–Shannon space can distinguish molecular sentences (or metabolic pathways) as individual points, regardless of whether or not they are correlated with one another and molecular texts as linearly correlated points three or more in number.

14. Applications of Ribonoscopy, the Cell Language Theory, and Planck–Shannon plots in Drug Discovery Research

Just as the study of electrons in atoms (called *atomic spectroscopy*) revolutionized physics and information technology in the last century, so it may be predicted that the study of RNA molecules in living cells (called "ribonoscopy", from "looking at RNA molecules", using the microarray technique and its equivalent (Ji, 2012, Chapters 18 and 19)) may revolutionize biology and medicine in the 21st century. *Quantum mechanics* that developed in physics between 1900 and 1925 provided the theoretical foundation for the study of electrons in atoms. Similarly, it seems to us that the *cell language theory* whose beginning may be traced at least to the time of Chargaff's discovery of his parity rules in the middle of the 20th century (Chargaff, 1971) may provide the theoretical foundation for the study of RNA molecules in living cells.

The practical applications of the cell language theory, especially the concept of molecular text, as implemented by the PDE-based analysis of microarray data, are illustrated in Figure 16.

If we define *molecular texts* as a *functionally related* set of 3 or more metabolic pathways within a patient or an identical pathway, e.g., the hypothetical protein pathway in Figure 16, distributed among 3 or more patients, we can study the effect of drugs on such molecular texts using the *Planck–Shannon plots* just as we can study the effect of drugs on ligand–receptor interactions using the Scatchard plot (https://en.wikipedia.org/wiki/Scatchard_equation) in biochemistry.

Since doxorubicin treatment induces the correlation of the hypothetical protein pathway among long survivors (see the upper two panels, Figure 16) but destroys the correlation among short survivors (see the lower two panels, Figure 16) as summarized in Table 9, it seems logical to conclude that the activation of this metabolic pathway (or molecular text) is beneficial for breast cancer tissues and hence the hypothetical protein pathways can serve as a biomarker for anti-breast cancer drug discovery.

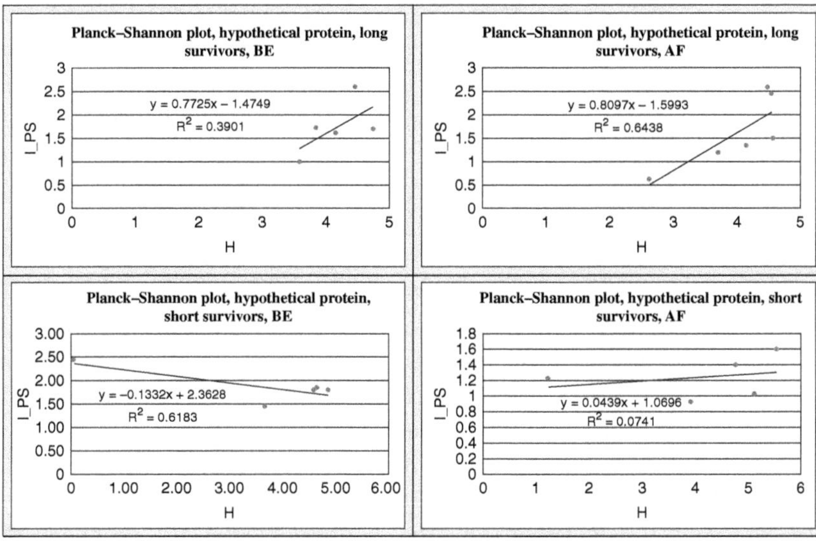

Figure 16. The Planck–Shannon plots of 5 or 6 sets of mRNA levels encoding hypothetical protein measured from breast cancer tissues of short (5 patients) and long (6 patients) survivors before (BE) and after (AF) treating with doxorubicin for 6 weeks. *Source*: Data from Perou *et al.* (2000).

Table 9. The effects of drug treatment on the *inter-metabolic pathways* (also called *molecular texts*, see Table 4) in short and long surviving cancer patients.

Drug effect	Short survivors	Long survivors
Type I (−, +)	−	+
Type II (+, −)	+	−

If one examines other metabolic pathways in the human genome using the Planck–Shannon plots, it may be possible to discover breast cancer biomarkers other than the hypothetical protein pathways. To discover potential anti-breast cancer drugs, it would be necessary to test them on cultured cells biopsied from short and long surviving breast cancer patients.

15. Conclusions

(1) There may be more than one genetic alphabet embedded in DNA, here denoted as the 1st order-, 2nd order-, 3rd order-, . . . , nth order-genetic

alphabet, each having 4^n letters and a unique biological function, the four letters of A, C, G, and T being the simplest and the first alphabets to be discovered whose function includes encoding 20 amino acids. But there are astronomically large numbers of heritable cell functions which may require more complex genetic alphabets than the 1st order one whose letters consist of n-plets of A, C, G, and T, e.g., $4^2 = 16$ doublets such as AG, CG, GT, AT, CT, $4^5 = 1024$ 5-plets such as ACTGC, CTAGT, TTCCA, etc.

(2) The DNA molecule may carry two kinds of information, variously called the *syntactic information* and the *semantic information, physical information* and *sematic information, quantitative* and *qualitative information*, etc. (see Table 3). The cell language may be considered primarily as a theory of qualitative information although it has some quantitative aspect as well, e.g., Chargaff's parity rules and Petoukhov's rules of tetra-groups (Petoukhov, 2016, 2017a, 2017b).

(3) The structural features (i.e., the syntactics) of DNA language obey the rules of the tensor product of genetic matrices. Since tensor multiplication of matrices is widely used in vibrational analysis in acoustic engineering and noise-immunity communication in computer engineering, it is possible that vibrations and resonance play an important role in mediating communication among genetic letters (of the 1st-, 2nd-, 3rd-, etc., types) in biocommunications catalyzed by biopolymers (proteins, RNAs, DNAs).

(4) The molecular language operating in living cells, called *cellese*, appears to obey the same principles of communication operating in human language, called *humanese*. Since we understand humanese better than cellese, it may be helpful to use humanese as a guide to understanding cellese. For example, since humanese is organized by four distinct levels, i.e., (i) letters, (ii) words, (iii) sentences, and (iv) texts, it may be inferred that cellese are also organized by 4 distinct levels, i.e., (i) n genetic alphabets, (ii) genes, (iii) metabolic pathways (MPs), and (iv) functionally related metabolic pathways (FRMPs) called hyper- or supermetabolic pathways.

(5) The Planck–Shannon plots of mRNA data may be the only currently available method for studying FRMPs.

(6) The FRMPs identified by the Planck–Shannon plots can be employed as a novel, higher-order (i.e., supermetabolic pathway) biomarkers for drug discovery research.

Appendix: Genetics and Human Language

(1) I listened to Berwick's lecture (Berwick, 2017) on what is unique about human language as compared to the languages used by other species of animals. Simply put, one of the the main conclusions he and Chomsky have arrived at after decades of intensive theoretical and experimental investigations is that the human language is unique because it is the only method of communication that involves the mental operation of what they call "merging", i.e., putting together A and B to produce C that is more than the sum of A and B. He illustrates this concept using the pencil as a simple example, where A = lead, B = eraser, and C = pencil with both lead and eraser as components of one unit. Apparently, Homo sapiens is the only species among animals who can externalize the mental operation of "merging" in this manner.

(2) The operation of "merging" that Berwick and Chomsky are talking about may have a mathematical counterpart in the algebra of tensor product of matrices that Sergey Petoukhov discussed extensively in connection with his analysis of long DNA sequences (Petoukhov, 2017). As you can see below, when two 2×2 matrices are tensor-multiplied, the results is a 4×4 matrix whose elements retain both the original feature and a variant feature. For example, C becomes CA, CT, or CG. To me, the same conclusion can be reached when two two different matrices are tensor-multiplied.

(3) In Ji (2018, Fig. 9.7), I defined the concept of "Replication with Variation" (RV) using the Fibonacci series as an example; $\mathrm{F}(n-1) + \mathrm{F}(n) = \mathrm{F}(n+1)$, where $\mathrm{F}(n)$ is the Fibonacci number at position n. Since $\mathrm{F}(n)/\mathrm{F}(n-1)$ approaches the Golden Ratio asymptotically after $n = 9$ (Ji, 2018, Table 9.5), it may be concluded that RV can generate the Fibonacci series under appropriate conditions. Cellular growth and differentiation in living systems, many of which show quantitative morphological features fitting the Fibonacci series, may be viewed as the biological example of the operation of RV as are, I suggest, the tensor product of matrices being the mathematical example of RV and the "merging operation" of Berwick and Chomsky the linguistic example of RV.

(4) If the above analyses are correct, it seems logical to conclude that

  (i) merging in linguistics,
  (ii) tensor product in mathematics, and
  (iii) Replication with variation in biology

are the different aspects/manifestations of the same universal principle which may be identified with the principle of the Irreducible Triadic Relation (ITR) of C. S. Peirce (1839–1914) (Ji, 2018, Chapter 9).

Acknowledgments

I thank the many students who took my Research in Pharmacology course over the past 10 years at the Ernest Mario School of Pharmacy, Rutgers University and carried out many of the computational work presented here. Special thanks go to Kenneth So, Vinay Vadali, Sun Yong Park, and Beum Jun Park for their careful and time-consuming computations and helpful discussions.

References

Babloyantz, A. (1986). *Molecules, Dynamics & Life: An Introduction to Self-Organization of Matter*, Wiley-Interscience, New York.

Berwick, R. (2017). Why Only Us: The Origin of Human Language. Available online: https://www.vofoundation.org/faith-and-science/robert-c-berwick-why-only-us-the-origin-of-human-language/.

Brenner, J. E. (2012). Mark Burgin's theory of information, *Information* 3, 224–228; doi:10.3390/9nf03020224.

Brown, R. and Porter, T. (1989). Category Theory: An Abstract Setting for Analogy and Comparison. Available online: http://groupoids.org.uk/pdffiles/Analogy-and-Comparison.pdf (accessed on 15 February 2017).

Burgin, M. (2010). *Theory of Information: Fundamentality, Diversity and Unification*, World Scientific, New Jersey.

Chargaff, E. (1971). Preface to a grammar of biology: A hundred years of nucleic acid research, *Science* 172, 637–642.

Chomsky, N. (1995). *The Minimalist Program*, The MIT Press, Cambridge, MA.

Cole, S. W. (2014). Human Social Genomics, *PLOS Genetics* 10(8), e1004601. https://doi.org/10.1371/journal.pgen.1004601

Diamant, E. (2015). Emanuel Diamant, The brain is processing information, not data. Does anybody care? *ISIS Summit Vienna 2015, Extended* Abstract. Available online: http://sciforum.net/conference/isis-summit-vienna-2015/paper/2842.

Form vs. Matter. Stanford Encyclopedia of Philosophy. Available online: https://plato.stanford.edu/entries/form-matter/

Garcia-Martinez, J., Aranda, A. and Perez-Ortin, J. E. (2004). Genomic run-on evaluates transcription rates for all yeast genes and identifies gene regulatory mechanisms, *Mol. Cell* 15, 303–313.

Herbert, N. (1987). *Quantum Reality: Beyond the New Physics, An Excursion into Metaphysics*, Anchor Books, Garden City, New York, p. 64.

Hockett, C. F. (1960). The origin of speech, *Sci. Am.* 203(3), 89–96.

Ji, S. (1974). Energy and negentropy in enzymic catalysis, *Ann. N. Y. Acad. Sci.* 227, 419–437.

Ji, S. (1985). The Bhopalator — A molecular model of the living cell based on the concepts of conformons and dissipative structures, *J. Theoret. Biol.* 116, 399–426.

Ji, S. (1988). Watson–crick and prigoginian forms of genetic information, *J. Theoret. Biol.* 130, 239–245.

Ji, S. (1991). Conformons and Molecular Machines, in Molecular Theory of Cell Life and Death, Rutgers University Press, New Brunswick. Avaible online: http://www.conformon.net/wp-content/uploads/2013/01/thermal_barrier_p31_p51.pdf.

Ji, S. (1997). Isomorphism between cell and human languages: Molecular biological, bioinformatic and linguistic implications, *BioSystems* 44, 17–39. Avaible online: http://www.conformon.net/wp-content/uploads/2012/05/Isomorphism1.pdf.

Ji, S. (1999). The linguistics of DNA: Words, sentences, grammar, phonetics, and semantics, *Ann. N. Y. Acad. Sci.* 870, 411–417.

Ji, S. (2001). Isomorphism between cell and human languages: Micro- and Macrosemiotics, in *Semiotics 2000*: "*Sebeok's Century*", S. Simpkins, J. Deely (eds.), Legas, Ottawa, pp. 357–374.

Ji, S. (2002). Microsemiotics of DNA, *Semiotica* 138(1/4), 15–42.

Ji, S. (2012). *Molecular Theory of the Living Cell: Concepts, Molecular Mechanisms, and Biomedical Applications*, Springer, New York.

Ji, S. (2015a). Planckian distributions in molecular machines, living cells, and brains: The wave-particle duality in biomedical sciences. In *Proceedings of the International Conference on Biology and Biomedical Engineering*, Vienna, Austria, 15–17 March 2015, pp. 115–137.

Ji, S. (2015b). Planckian information (I_P): A new measure of order in atoms, enzymes, cells, brains, human societies, and the cosmos, in *Unified Field Mechanics: Natural Science beyond the Veil of Spacetime*, Amoroso, R., Rowlands, P. and Kauffman, L. (eds.), World Scientific, New Jersey, 2015, pp. 579–589.

Ji, S. (2017). Neo-semiotics: Introducing zeroness into peircean semiotics may bridge the knowable and the unknowable, *Prog. Biophys. Mol. Biol.* 131, 387–401. Available at: http://www.sciencedirect.com/science/article/pii/S0079610717300858?via%3Dihub.

Ji, S. (2018a). *The Cell Language Theory: Connecting Mind and Matter*, World Scientific, New Jersey.

Ji, S. (2018b). RASER model of single-molecule enzyme catalysis and its application to the ribosome structure and function, *Arch. Mol. Med. Genetics* 1(1), 31–39. Available at: http://hendun.org/journals/AMMG/PDF/AMMG-18-1-104.pdf.

Ji, S. (2018c). Mathematical (quantitative) and cell linguistic (qualitative) evidence for hypermetabolic pathways as potential drug targets, *J. Mol. Genet. Medicine* 12(2), 1000343.

Ji, S. (2019). My email to Sean Nuallain, *et al.*, in the Foundations of Mind discussion group, on January 4, 2019, entitled "Genetics and human language". See Appendix.

Kondepudi, D. and Prigogine, I. (1998). *Modern Thermodynamics: From Heat Engine to Dissipative Structures*, John Wiley & Sons, Inc., Chichester.

Norris, V. *et al.* (1999). Hypothesis: Hyperstructures regulate bacterial structure and the cell cycle, *Biochimie* 81, 915–920.

Patel, K. (2013). Chladni Plates — Sand Vibration Patterns. Available online: https://physicsbridgecourse.blogspot.com/2013/11/chladni-plates-sand-vibration-patterns.html.

Perez, J.-C. (2010). Codon populations in single-stranded whole human genome DNA are fractal and fine-tuned by the Golden Ratio 1.618. *Interdisc. Sci: Comput. Life Sci.* 2, 1–13.

Perou, C. M., Sorlie, T., Eisen, M. B. *et al.* (2000). Molecular portraits of human breast Tumors, *Nature* 406(6797), 747–752.

Petoukhov, S. (2016). The system-resonance approach in modeling genetic structures, *BioSystems* 139, 1–11.

Petoukhov, S. (2017a). Genetic coding and united-hypercomplex systems in the models of algebraic biology, *BioSystems* 158, 31–46.

Petoukhov, S. (2017b). The rules of long DNA-sequences and tetra-groups of oligonucleotides, *arXiv:1709.04943v4 [q-bio.OT]*.

Piattelli-Palmarini, M. and Vitiello, G. (2015). Linguistics and some aspects of its underlying dynamics, *Biolinguistics* 9, 96–115.

Piattelli-Palmarini, M. and Vitiello, G. (2017). Quantum field theory and the linguistic minimalist program: A remarkable isomorphism, *J. Phys. Conf. Series* 880, 012016. DOI 10.1088/1742-6596/880/0112016.

Raanan, H., Pike, D. H., Moore, E. K., Falkowski, P. G., and Vikas Nanda, V. (2018). Modular origins of biological electron transfer chains, *Proc. Nat. Acad. Sci. USA*; 201714225 DOI: 10.1073/pnas.1714225115.

Ramsey, L. and Lee, S. (2016). Our DNA is 99.9% the Same as the Person Sitting Next to us — and we're Surprisingly Similar to a Bunch of Other Living Things. Available online: http://www.businessinsider.com/comparing-genetic-similarity-between-humans-and-other-things-2016-5.

Schrödinger, E. (1998). *What Is Life? With Mind and Matter and Autobiographical Sketches*, Cambridge University Press, Cambridge.

Spivak, D. I. (2013). Category Theory for Scientists. Available online: http://www.math.mit.edu/~dspivak/teaching/sp13/CT4S--static.pdf (accessed on 15 February 2017).

Tama, F., Valle, M., Frank, J. and Brooks, C. L. (2003). Dynamic reorganization of the functionally active ribosome explored by normal mode analysis and cryo-electron microscopy, *Pro. Nat. Acad. Sci. USA* 100(16), 9319–9323.

Trifonov, E. N. (1993). Available online: http://www.worldscientific.com/doi/abs/10.1142/9789814503655_0009.

Weaver, W. (1948). Science and complexity, *American Scientist* 36, 536–544.

Wicken, J. S. (1987). Entropy and information: Suggestions for common language, *Phil. Sci.* 54, 176–193.

Yamagishi, M. E. and Shimabukuro, A. I. (2008). Nucleotide frequencies in human genome and fibonacci numbers, *Bull. Math. Biol.* 70(3), 643–653. DOI 10.1007/s11538-007-9261-6.

Chapter 10

The Biology of Information

Robert K. Logan

Department of Physics, University of Toronto,
60 St. George St., Toronto, ON, Canada M5S 1A7
logan@physics.utoronto.ca

There are almost as many definitions of information as there are scholars writing about the philosophy of information. We shall narrow the definition of information for the purposes of this chapter and look at information from the perspective of the original meaning of the word as the formation of the mind. Mind within this context will include the way in which all living things become aware of their environment or umwelt and act accordingly to maintain themselves and propagate their organization. We will argue that information, as we define it, is that which guides all living organisms from the simplest single-cell organisms to fungi and plants to human beings and other animals to make choices that are in their own self-interest. We will argue that information as defined from Shannon's perspective or from the way physicists use the term refers to forms of information quite different from forms of what shall be defined as instructional or biotic information. The information we consider will be treated as a process and not a thing. The focus of this chapter will be on the biology of information and the information of biology (i.e., biological organisms).

1. Introduction

"To live effectively is to live with adequate information."

— Wiener (1950)

"Information is a distinction that makes a difference."

— MacKay (1969)

"Information is a difference that makes a difference."

— Bateson (1973)

Information ... arises ... as natural selection assembling the very constraints on the release of energy that then constitutes work and the propagation of the organization (Kauffman *et al.*, 2007).

We live in the Information Age and we are surrounded by information. Thanks to "new media" like the Internet, the Web, blogs, email, smart phones, iPods, iPads, computer notebooks, mainframe computers, servers and the Cloud we are blanketed in information — drowning in information according to some. Yet we do not have an agreement for a definition of information. There are almost as many definitions of information as there are scholars writing about the philosophy of information. I have been a participant for a number of years in the Foundations of Information Science (FIS) discussion list, which dates back to the early 1990s, and we have yet to agree on a definition of information. In one of my posts to the discussion list I likened the problem of defining information to the story of the three blind men describing an elephant that they encountered. It is a rope said the blind man holding the tail; no, it is a snake said the blind man holding the trunk; no, it is a tree said the blind man touching the elephant's leg. All words including the word "information" are metaphors and a metaphor cannot be right or wrong — it can only be illuminating if inspired or irrelevant if too narrow. Therefore, all definitions of information are both correct and incorrect, correct in that they catch some aspect of the notion and incorrect because they do not incorporate all the senses of what we mean by information and that includes this statement.

Information has a variety of meanings for researchers in different fields. Information for the engineer, the physicist and the biologist to mention a few is completely different. And even within the same field the definitions of information vary. I will not attempt to provide the definitive definition of information as I do not think it exists, but rather I will describe in this chapter information from a number of different perspectives including those of Shannon (1948), MacKay (1969), Bateson (1973), the biological points of view of Kauffman *et al.* (2007) and Deacon (1997, 2012), the cybernetic point of view of Wiener (1948, 1950) and Brillouin and Léon (2004), the way physicists use the term as well as researchers in the field of artificial intelligence (AI) and artificial life (AL). The main focus of this chapter will be on information from a biological point of view, which as I will argue in what follows is to me the most interesting, most relevant and the most productive. We shall therefore narrow the definition of information and look at information from the perspective of the original meaning of the word, as the formation of the mind. We will argue that biotic information, as we define it, is that which guides all living organisms from the simplest single-cell organisms to fungi and plants to human beings and other animals to make choices that are in their own self-interest.

Even though I will focus on biology, I will not ignore the work of Shannon (1948) and classical Information theory. I will examine Shannon's information theory and the complex nature of information and describe the historic development of Shannon's theory, which I believe is a very useful theory, but is basically a theory of signals and not information. I hope to demonstrate some of the limitations of Shannon's approach and show that there is more than one notion of information depending on the phenomena it is describing. I will examine the relationship of information to materiality, meaning and organization and show that Shannon's information is independent of these, which is not the case for biotic information. I will also show based on the work of Kauffman *et al.* (2007) that there exists a parallel between the information and organization of biotic systems as well as the elements of human culture and language.

In addition to the everyday interaction with information by the users of computer-based digital "new media", there is the role that information plays in the sciences of artificial intelligence (AI) and artificial life (AL). In AI, intelligence is posited to be a form of information that can be downloaded from the human brain onto a computer. In AL, life is posited to be a form of information that first became embodied in carbon-based chemicals but now can exist in silicon-based computers. Some AL scientists like Edward Fredkin insist that the universe is a computer and that life, including human life, is merely a program running on that computer (Hayles, 1999a,b).

The irony of our total immersion in information as well as the prominent role it plays in AI and AL is that for the most part we do not really have a clear understanding of exactly what information is. Information is not a simple straightforward idea but rather it is a very slippery concept used in many different ways in many different contexts. Linguistically and grammatically the word information is a noun, but in actuality it represents a process and hence is like a verb. In the consideration of the concept of information that I plan to examine, I will assert the following:

> There is not only one form of information but several kinds of information and they, for example, Shannon's information and biotic information, are quite different.

Information is not an invariant or a universal independent of the frame of reference or context in which it operates or in which it is invoked. Rather, I hope to demonstrate the notion of the relativity of information and show that the concept of information depends on the context of where and how it is being used.

That there is an important connection or relationship of information to meaning and organization as was first suggested by MacKay (1969) and Bateson (1973).

That despite the fact that the word information is a noun, information is more like a verb or a process.

That information is not a material thing nor is it a form of energy.

That there is some connection between information and patterning.

That information is not a uniquely human phenomenon, but that all living organisms experience the phenomenon of information and that it is vital to their well-being and their ability to survive, act in their own interest and propagate their organization.

These are some of the issues that I will address in this chapter as we try to flesh out our understanding of exactly what it is that we call information. We will consider the historic development of the concept of information to get a handle on the exact meaning of this process that permeates every aspect of our existence and is as fundamental to our economic well-being in the modern world as is energy.

We trace the development of the concept of information from the earliest uses of the word to the beginning of information theory as formulated by Shannon (1948) and Wiener (1948, 1950), to MacKay (1969) and Bateson's (1973) formulation of information as the distinction or difference that makes a difference to the inclusion of information in biotic systems. I wish to acknowledge that to address the issues that I have just formulated, I will make extensive use of the work of Kauffman *et al.* (2007), which I will review shortly.

2. The Origin of the Concept of Information in the English Language

The English word information first appears in the written record in 1386 by Chaucer: "Whanne Melibee hadde herd the grete skiles and resons of Dame Prudence, and hire wise informacions and techynges (Oxford English Dictionary, OED)". The word is derived from Latin through French by combining the word inform meaning giving a form to the mind with the ending "ation" denoting a noun of action.

The notion of information as a thing capable of storage in or the transfer or communication to something inanimate and the notion of information as a mathematically defined quantity do not arise until the 20th century. In the OED article on information, one reads the first use of information as something material, as follows: "The whole difficulty resides in the amount of definition in the [television] picture, or, as the engineers put it, the amount of information to be transmitted in a given time (1937 *Discovery* Nov. 329/1)". The first mathematization of information according to the OED appeared in articles by Fisher (1925) and Hartley (1928, p. 540).

3. Shannon and the Birth of Information Theory

Claude Shannon, working in telecommunication engineering for Bell Labs and building on the ideas of Fisher and Hartley, formulated information theory, where he defined information as a message in terms of bits sent by a sender to a receiver. Shannon was interested in how one could encode and accurately transmit data a sender wished to transmit to a receiver. Shannon gave information a numerical or mathematical value based on probability and measured it in terms of the decrease of uncertainty for a receiver. The amount of Shannon information is inversely proportional to the probability of the occurrence of that information. On the advice of John von Neumann, Shannon called his measure *information entropy.* This choice was criticized by Wicken (1987) who argued that in science a term should have only one meaning. Schneider and Sagan (2005) referring to the use of the term entropy in both thermodynamics and information theory also suggest that Shannon's use of the term is confusing when he wrote, "There is no simple correspondence between the two theories".

Both Wiener (1950) and Brillouin and Léon (2004) adopted Shannon's definition of information and its relation to entropy with the one exception of its sign, likely influenced by the arguments of Schrödinger (1944). Wiener (1950) wrote as follows:

> Messages are themselves a form of pattern and organization. Indeed, it is possible to treat sets of messages as having entropy like sets of states in the external world. Just as entropy is a measure of disorganization, the information carried by a set of messages is a measure of organization.

Brillouin and Léon (2004) also argued that a living system exports entropy in order to maintain its own entropy at a low level. Brillouin used the term negentropy to describe information rather than negative entropy.

According to Claude Shannon (1948) his definition of information is not connected to its meaning. However, as Shannon suggested, information in the form of a message often contains meaning, but that meaning is not a necessary condition for defining information. So, it is possible to have information without meaning, whatever that means.

4. MacKay's Counter Revolution: Where is the Meaning in Shannon Information?

With the support of Wiener (1950) and Brillouin and Léon (2004), Shannon's (1948) notion of information was readily adopted, but there were pockets of resistance. Not all of the members of the information science community were happy with Shannon's definition of information. Donald MacKay (1969) argued that he did not see too close a connection between the notion of information as we use it in communications engineering and what [we] are doing here: the problem here is not so much finding the best encoding of symbols, but, rather, the determination of the semantic question of what to send and to whom to send it, [namely,] "the change in a receiver's mind-set, and thus with meaning" and not just the sender's signal (Hayles, 1999a,b).

In 1951, MacKay (1969) proposed that "information is a distinction that makes a difference" and Bateson (1973) shortly thereafter proposed something similar, namely, that "information is a difference that makes a difference". Marshall McLuhan, influenced by his professor's (Richards, 1952) notion of feedforward, came up with a formula similar to MacKay and Bateson. Richards claimed that in order to be understood, the sender of a message had to feedforward the context of the information. McLuhan reformulated this idea with his one liner, "the user is the content," meaning that the interpretation of the message sent by the sender is a key component of the meaning of that message. Information is a metaphor that sometimes describes the signal and sometimes the interpretation. The pheromone trails that ants leave are information for the members of their nest, but they are not information for other organisms that might encounter these chemical signals.

Since interpretation is a key element of the process of informing, there is an intimate relationship between agency and information. An agent has choice and a non-agent has no choice. An agent must be a living organism because it has the choice to act in its own self-interest unlike abiotic matter or objects. Abiotic non-agents do not have information because they have no choice. We humans can have information about abiotic objects but those objects themselves do not have that information as they have no mind to be informed. We are using the word *mind* in the metaphoric sense of sentience no matter how primitive even at the level of single-cell organisms. A black hole makes no choices — it is ruled by the laws of physics, which once again illustrates that what is information for a physicist is quite different from what it is for a biologist. Deacon (1997) defines information as "about something for something towards some end" and therefore only living organisms that have ends, namely to survive and to propagate their organization, make choices and only they possess information.

Shannon's position defining information as a signal and not the meaning prevailed because MacKay and Bateson's definitions of meaning could not be measured or quantified. Shannon's motivation for his definition of information was to create a tool to analyze how to increase the ratio of signal to noise within telecommunications, but the problem with it was it did not fully describe communication as Shannon (1953) admitted:

> The word "information" has been given many different meanings by various writers in the general field of information theory. It is likely that at least a number of these will prove sufficiently useful in certain applications to deserve further study and permanent recognition. It is hardly to be expected that a single concept of information would satisfactorily account for the numerous possible applications of this general field. The present note outlines a new approach to information theory which is aimed specifically at the analysis of certain communication problems in which there exist a number of information sources simultaneously in operation.

It is proposed that Shannon information theory is really signal theory, or as McLuhan (1988) suggested a theory of transportation. Langefors (1968) has made a similar suggestion describing Shannon's information theory as "signal transmission theory". What I find extraordinary is that Shannon's definition of information, limited in scope by his own admission, has become the standard by which almost all forms of information are gauged.

In a certain sense, Shannon's information theory is more about data than it is about information, where the following definitions of data, knowledge, information and wisdom are taken into account:

- Data are the pure and simple facts without any particular structure or organization, the basic atoms of information.

- Information is structured data, which adds meaning to the data and gives it context and significance.
- Knowledge is the ability to use information strategically to achieve one's objectives.
- Wisdom is the capacity to choose objectives consistent with one's values and within a larger social context (Logan and Stokes, 2004).

T. S. Eliot has a more poetic way of expressing a similar thought:

> "Where is the wisdom we have lost in knowledge? Where is the knowledge we have lost in information?"

5. Biotic Information

Shannon information works fine for telecommunication engineering and many other applications, but it has a problem with biological systems because Shannon specified that the number of possible messages he needed to consider had to be finite. Darwinian preadaptations, however, cannot be predicted and are possibly infinite in number and therefore as a consequence "the ensemble of possibilities and their entropy cannot be calculated" (Kauffman *et al.*, 2007). A definition of information as reducing uncertainty does not make sense and therefore we defined biotic or instructional information not with Shannon, but with constraints or boundary conditions. The amount of information will be related to the diversity of constraints and the diversity of processes that they can partially cause to occur. By taking this step, we embed the concept of information in the ongoing processes of the biosphere, for they are causally relevant to that which happens in the unfolding of the biosphere.

> We therefore conclude that constraints are information and ... information is constraints.... We use the term **"instructional information"** because of the instructional function this information performs and we sometimes call it **"biotic information"** because this is the domain it acts in, as opposed to human telecommunication or computer information systems where Shannon information operates (Kauffman *et al.*, 2007).

There is an amusing story of how we came to this definition of constraints in biological organisms as information. Kauffman asked me as we were relaxing over a beer after a systems biology conference in Ottawa, Ontario, how was it that living cells were able to do the work to build constraints that allowed them to convert external energy from their environment into work if the constraints did not exist at first (a chicken and egg problem). I told him I could not answer his question but that the constraints are the information which I came up with by channeling Marshall McLuhan and his idea that "the medium is the message". That exchange led to our paper "Propagating Organization: An Enquiry (Kauffman *et al.*, 2007)" and the idea of biotic information where we argued that constraints acting as instructional information are essential to the operation of a cell and the propagation of its organization.

> The working of a cell is, in part, a complex web of constraints, or boundary conditions, which partially direct or cause the events which happen. Importantly, the propagating organization in the cell is the structural union of constraints as instructional information, the constrained release of energy as work, the use of work in the construction of copies of information, the use of work in the construction of other structures, and the construction of further constraints as instructional information. This instructional information further constrains the further release of energy *in diverse specific ways,* all of which propagates organization of process that completes a closure of tasks whereby the cell reproduces (Kauffman *et al.*, 2004).

Biotic or instructional information allows an organism to propagate its organization. This contradicts Shannon's definition of information and the notion that a random set or soup of organic chemicals has more Shannon information than a structured and organized set of organic chemicals found in a living organism.

> The biotic agent has more meaning than the soup, however. The living organism with more structure and more organization has less Shannon information. This is counterintuitive to a biologist's understanding of a living organism. We therefore conclude that the use of Shannon information to describe a biotic system would not be valid. Shannon information for a biotic system is simply a category error. A living organism has meaning because it is an autonomous agent acting on its own behalf. A random soup of organic chemicals has no meaning and no organization (Kauffman *et al.*, 2004).

Shannon information can be defined independent of meaning, whereas biotic or instructional information is intimately connected to the meaning of the organism's information, namely its organization. Thus, we see that the organization of an organism is a form of information. This is a much more dynamic form of information than the string of symbols or bits that define Shannon information.

According to Shannon's definition of information, a set of random biochemicals would have more information than a living organism which has the structure of an organized set of biochemicals. This totally contradicts the conclusion a systems biologist would reach. It is clear that biotic and Shannon are two different forms of information with utility in different domains and support the notion of the relativity of information, the idea that different domains require different definitions of information, which we now address in the next section.

6. The Relativity of Information

In an article titled "A Discipline Independent Definition of Information" published in the *Journal of the American Society for Information Science*, Losee (1997) defined information in the following terms:

> Information may be defined as the characteristics of the output of a process, these being informative about the process and the input. This discipline-independent definition may be applied to all domains, from physics to epistemology.

Losee's suggestion that the term *information* is the same for all the domains and phenomena and is therefore an invariant, like the speed of light for all frames of reference, is at odds with what has been reported so far. Biotic and Shannon forms of information are quite different. Information as defined by MacKay and Bateson is all about meaning whereas Shannon information is agnostic with respect to meaning. Both MacKay (1969) and Bateson (1973) indicate that information must make a difference (Information is a distinction/difference that makes a difference — MacKay/Bateson) and that difference must be the difference registered by the receiver of the information. Shannon's measure of information is a measure of what the sender sends and has nothing to do with the distinction or difference for the receiver, and therefore, Shannon information is quite different from that of MacKay and Bateson. I therefore conclude that information is not an invariant like the speed of light, but depends on the frame of reference or context in which it is used. Burgin (2003) reached a similar conclusion, "Traditionally, researchers tried to define information as some absolute essence that exists autonomously. Here, we suggest that this may be considered only as a theoretical abstraction, while on the empirical level information is subjected to dynamical relativity".

Shannon information works for telecommunication channel engineering. Kolmogorov information, closely related to Shannon information, is defined as the minimum computational resources needed to describe a program or a text. It is useful for the study of information compression with respect to

Turing machines (Burgin, 2003). Shannon and Kolmogorov forms of information are quite different from notions of information of MacKay, Bateson and the biotic information described above.

Information can be thought of as a tool that comes in different forms depending on the job that needs to be done. MacKay identified two main categories of information: structural information, specifically linked to meaning, and selective information, in terms of signals like Shannon and Kolmogorov forms of information. Kolmogorov information was intended to measure information content as the complexity of an algorithm on a Turing Machine. Shannon information was formulated to deal with the signal-to-noise ratio in telecommunications. Shannon and Kolmogorov forms of information are what MacKay termed selective information. Biotic or instructional information, on the other hand, is a form of structural information. The information of DNA is like MacKay structural information that depends on context so that identical genotypes can give rise to different phenotypes depending on the environment or context. It is not fixed like Shannon selective information. According to Hayles (1999a,b), "Shannon and Wiener define information in terms of what it is, MacKay defines it in terms of what it does". MacKay, Bateson and the biotic forms of information include a verb or process. DNA, RNA and proteins are not informational "things" as such but rather they catalyze "processes" and actions that give rise to the propagation of organization, and hence, the transmission of information — information with meaning at that. Put simply, instructional information is structural information as the root of the word instructional reveals. Both Shannon and Wiener forms of information include a noun or a thing.

Another distinction between Shannon information and biotic or instructional information as defined in Kauffman *et al.* (2007) is that with Shannon there is no explanation as to where information comes from and how it came into being. Information in Shannon's theory arrives *deus ex machina*, whereas biotic information as described above arises from the constraints that allow a living organism to harness free energy and turn it into work so that it can carry out its metabolism and replicate its organization. Kauffman (1995, 2000) has described how this organization emerges through autocatalysis as an emergent phenomenon with properties that cannot be derived from, predicted from or reduced to the properties of the biomolecules of which the living organism is composed. This, therefore, provides an explanation of where biotic information comes from.

7. Information and Materiality

Information is information, not matter or energy. No materialism which does not admit this can survive at the present day — Wiener (1948).

Shannon's theory defines information as a probability function with no dimension, no materiality, and no necessary connection with meaning. It is a pattern, not a presence — Hayles (1999a,b)

A simple argument that information is not material is the observation that identical forms of information can be in multiple places at the same time as when I sent an email to multiple recipients. Although information is not material, the relationship of Shannon information and biotic information to the material media in which they are embedded is quite different and is another way to demonstrate the difference between these two forms of information.

As Hayles suggests, Shannon information is a pattern, not a presence. Biotic information, on the other hand, is both a pattern and a presence. One cannot disassociate the pattern of information in DNA or RNA from the biochemical molecules in which they reside. DNA and RNA catalyze chemical interactions and cannot be moved from their molecular medium to a computer, for example. A receptor for food is not just a pattern — it is also a presence. A biological system is both an information pattern and a material object, or more accurately, information patterns instantiated in a material presence. Schrödinger (1944), long ago before the discovery of DNA, described this dual aspect of chromosomal material metaphorically. "The chromosome structures are at the same time instrumental in bringing about the development they foreshadow. They are law-code and executive power — or, to use another simile, they are architect's plan and builder's craft — in one". It is the dynamic of the interaction between the patterns of information and the material composition of the biotic agents that determines their behavior.

Shannon information is not as dependent on the physical medium in which it is embedded as is the case with biotic information such as DNA and RNA. A text can be replicated or duplicated exactly. A text can also be transmitted and reformatted from one medium to another, for example, from a computer file to a text printed on paper or from a live performance to a YouTube video. One cannot reformat a person from a biological medium to a file on a computer despite the plans of some AI enthusiasts who think they can upload their brain to a computer some day in the future.

8. Information and Organization

Is organization a form of information? Kauffman (2000) in his book *At Home in the Universe: The Search for the Laws of Self-Organization and Complexity* posits that self-organization in the form of the auto-catalysis of biomolecules is at the root of the emergence of life and that the evolution of life is a combination of both natural selection and self-organization. When this idea is paired with the notion that constraints that allow cells to convert free energy into work represent information as developed in Kauffman *et al.* (2007), we can simply conclude that in biotic systems organization and information are basically equivalent. Organization is a form of information. Hayles (1999a,b) is in agreement with this conclusion when she writes, "Information does not exist in this paradigm [referring to self-organization] or that it has sunk so deeply into the system as to become indistinguishable from the organizational properties defining the system as such".

Wiener like Shannon related information to entropy but, unlike Shannon, Wiener (1948) saw a connection between organization and information, "The notion of the amount of information attaches itself very naturally to a classical notion in statistical mechanics: that of entropy. Just as the amount of information in a system is a measure of its degree of organization, so the entropy of a system is a measure of its degree of disorganization".

Kauffman *et al.* (2007) made a similar claim when it was asserted that the constraints that allow the propagation of organization in a living organism represent the information content of that organism. In other words, the propagating organization of a living organism is its information content. Our position in a certain sense recapitulates similar sentiments expressed by Wiener (1950) when he wrote "We are not stuff that abides but patterns that perpetuate themselves". I have great respect for Professor Weiner, with whom I had some conversations as an undergraduate at MIT in the late 1950s, but in this instance, I would beg to differ with him. We are both stuff and patterns and it is the patterns of the stuff we are made of that give us life. As Deacon (2012) expressed it in his book *Incomplete Nature*, "There is more here than stuff. There is how this stuff is organized and related to other stuff". So, *contra* Wiener, we are stuff but stuff that is patterned in such a way as to perpetuate itself and propagate its organization as Kauffman *et al.* (2007) have suggested. The information in a biological system is not symbolic, but rather chemical. One cannot separate the information that is life from its biochemical medium and upload it on a computer. As we have already asserted, the medium of the flesh is both the message and the content of a biotic system.

The organic chemicals of which we are composed are continually replaced so that after seven years, there is a completely new set of molecules. So, we are not flesh of a particular set of molecules but the organization of the molecules of which we are composed. So, in this sense, Professor Wiener was correct to a certain extent: we are not just stuff, but without that stuff we would not be. In a certain sense, we are a process of the biochemicals of which we are composed.

9. Human Language and Culture as Forms of Propagation

9.1. *Organization and information*

The focus of this chapter is biological information, but now we will consider the following forms of information that are biological in the sense that they are created by human biological creatures as they organize their activities in terms of language and culture. I find support for this idea in the following quote by Katherine Hayles (1999a,b): "I take informatics to mean the technologies of information as well as the biological, social, linguistic and cultural changes that initiate, accompany and complicate their development." Not only are language, culture and the various forms of culture products of human biology, they also parallel the behavior of living organisms in the way in which they have emerged and evolved. We (myself and the scholars I am about to reference) now make this case.

We begin with the suggestion made by Christiansen (1994) that language could be considered as an organism that evolved through natural selection to be easily learned. He rejected Noam Chomsky's notion that language emerged as the result of a fortuitous genetic mutation that provided the human brain with Universal Grammar (UG) and a Language Acquisition Device (LAD). This is how Chomsky explains that all the languages of the world possess a similar grammatical structure and the fact that young children seem to automatically pick up the language of their parents despite what he calls the poverty of stimulus. Chomsky makes no attempt to connect the existence of this cognitive endowment with the origin of speech or any other cognitive ability for that matter. Rather than embracing a Darwinian explanation of the UG and the LAD, he disavows himself of such an explanation and merely suggests that this facility arose or emerged as a consequence of a genetic mutation and the fact that so many brain cells are packed into the small space of the human skull.

Deacon (1997) critiques Chomsky's theory:

> as a 'hopeful monster' theory of human language: the evolutionary theorist's counterpart to divine intervention, in which a freak mutation just happens to produce a radically different and serendipitously better-equipped organism.... The accidental language organ theory politely begs us to ignore the messy details of language origins.... It rephrases the problem by giving it a new name.

Tomasello (1999) also criticizes Chomsky on the grounds that he provides no explanation for UG or the LAD other than a lucky genetic mutation.

Morten Christiansen (1994) proposed another alternative to Chomsky's lucky genetic mutation hardwired hypothesis by suggesting that verbal language could be considered as an organism that evolved to be easily learned, especially by children despite the poverty of stimulus problem. As for the explanation of UG, that can be easily explained as due to the universality of human cognition. In a later paper with his colleagues, Christiansen *et al.* (2001) wrote the following:

> Language exists only because humans can learn, produce, and process them. Without humans there would be no language. It therefore makes sense to construe languages as organisms that have had to adapt themselves through natural selection to fit a particular ecological niche: the human brain. In order for languages to 'survive', they must adapt to the properties of the human learning and processing mechanisms. This is not to say that having a language does not confer selective advantages onto humans. It seems clear that humans with superior language abilities are likely to have a selective advantage over other humans (and other organisms) with lesser communicative powers. This is an uncontroversial point, forming the basic premise of many of the adaptationist theories of language evolution. However, what is often not appreciated is that the selection forces working on language to fit humans are significantly stronger than the selection pressures on humans to be able to use language. In the case of the former, a language can only survive if it is learnable and processable by humans. On the other hand, adaptation toward language use is merely one out of many selective pressures working on humans (such as, for example, being able to avoid predators and find food). Whereas humans can survive without language, the opposite is not the case. Thus, language is more likely to have adapted itself to its human hosts than the other way around.
>
> Languages that are hard for humans to learn simply die out, or more likely, do not come into existence at all.

This hypothesis at once explains why grammars are universal and are easily learned by children without the need of a hardwired LAD. Languages simply evolved to have these properties.

Culture is another important adaptive mental tool that is dependent on and intimately connected to verbal language and symbolic information. It is the mechanism whereby the learning of previous generations is passed on to the next generation through communication and social interactions. Just as language behaves as an organism that evolved to be easily learned, the same can be said of culture. It, too, evolved to be easily learned and can be considered to be an organism. This is a simple extension of the ideas that Christiansen developed for language.

10. Conclusion

We have demonstrated the relativity of information by showing that information is not an invariant concept independent of the phenomena it is describing or the frame of reference with respect to which it is defined. In particular, we have shown that Shannon information cannot properly describe the information associated with living organisms. We have examined the relationship of information to materiality, meaning and organization and showed that Shannon information is independent of meaning, organization and its material instantiation, which is just the opposite for biotic information, and the information associated with language and culture. We have also shown that there exists an intimate relationship between information and organization for living organisms and the elements of human culture.

References

Bateson, G. (1973). *Steps to an Ecology of Mind*, Paladin, Frogmore: St. Albans.

Brillouin, L. and Léon (2004). *Science and Information Theory*, Dover, Mineola, New York.

Burgin, M. (2003). Information: Problems, paradoxes, and solutions, *tripleC* 1(1), 53–70.

Christiansen, M. (1994). Infinite Languages Finite Minds: Connectionism, Learning and Linguistic Structure, Unpublished Doctoral Dissertation, Centre for Cognitive Studies, University of Edinburgh, Edinburgh.

Christiansen, M., Dale, R., Ellefson, M. and Conway, C. (2001). The role of sequential learning in language evolution: Computational and experimental

studies, in *Simulating the Evolution of Language*, Cangelosi, A. and Parisi, D. (eds.), Springer-Verlag, London.
Deacon, T. (1997). *The Symbolic Species: The Co-evolution of the Brain and Language*, W. Norton & Co., New York.
Deacon, T. (2012). *Incomplete Nature: How Mind Emerged from Matter*, W. W. Norton & Co., New York.
Fisher, R. A. (1925). Theory of statistical estimation, *Proc. Cambridge Philos. Soc.* XXII, 709.
Hartley, R. V. L. (1928). Transmission of information, *Bell Sys. Tech. J.* VII, 535–563.
Hayles, K. (1999a). *How We Became Posthuman*, University of Chicago Press, Chicago.
Hayles, K. (1999b). The condition of virtuality, in *The Digital Dialectic*, Lunenfeld, P. (ed.), MIT Press, Cambridge MA.
Kauffman, S. (1995). *At Home in the Universe: The Search for the Laws of Self-Organization and Complexity*, Oxford University Press, Oxford.
Kauffman, S. (2000). *Investigations.* Oxford University Press, Oxford.
Kauffman, S. Stuart, Logan, R. K., Este, R., Goebel, R., Hobill, D. and Shmulevich, I. (2007). Propagating organization: An enquiry, *Biol. Philos.* 23, 27–45.
Langefors, B. (1968). *System för Företagsstyrning*, Studentlitteratur, Lund.
Logan, R. K. and Stokes, L. W. (2004). *Collaborate to Compete: Driving Profitability in the Knowledge Economy*, Wiley, Toronto and New York.
Losee, R. M. (1997). A discipline independent definition of information, *J. Am. Soc. Inform. Sci.* 48(3), 254–269.
MacKay, D. (1969). *Information, Mechanism and Meaning*, MIT Press, Cambridge MA.
McLuhan, M. and McLuhan, E. (1988). *Laws of Media: The New Science*, University of Toronto Press, Toronto.
Richards, I. A. (1952). Communication between men: The meaning of language, in *Transactions of 8th Macy Conference — Cybernetics: Circular Causal and Feedback Mechanisms in Biological and Social System*, Foerster, H. (ed.), Josiah Macy, Jr. Foundation, New York.
Schneider, E. and Sagan, D. (2005). *Into the Cool: Energy, Flow, Thermodynamics and Life*, University of Chicago Press, Chicago.
Schrödinger, E. (1944). *What is Life*? Cambridge University Press, Cambridge.
Shannon, C. E. (1948). A mathematical theory of communication, *Bell Syst. Tech. J.* 27, 379–423 and 623–656.
Shannon, C. E. (1953). The lattice theory of information, *IEEE Trans. Inform. Theory* 1(1), 105–107.
Shiryayev, A. N. (1993). (ed.), *Selected Works of A.N. Kolmogorov: Volume III: Information Theory and the Theory of Algorithms* (*Mathematics and its Applications*), Kluwer Academic Publishing, New York.
Tomasello, M. (1999). *The Cultural Origins of Human Cognition*, Harvard University Press, Cambridge, MA.

Wicken, J. (1987). Entropy and information: Suggestion for a common language, *Philos. Sci.* 54, 176–193.
Wiener, N. (1948). *Cybernetics: Or Control and Communication in the Animal and the Machine*, MIT Press, Cambridge MA.
Wiener, N. (1950). *The Human Use of Human Beings*, Houghton Mifflin, Boston.

Chapter 11

Towards a New Information Thinking — Biologically Inspired

Jorge Navarro* and Pedro C. Marijuán†

Bioinformation Group,
Aragon Health Sciences Institute (IACS),
CIBA Building San Juan Bosco 13, Zaragoza 50009, Spain
**jnavarro.iacs@aragon.es*
†*pcmarijuan.iacs@aragon.es*

An increasing variety of research practices could potentially converge into renewed information science, as astounding new social uses of knowledge and even more astounding social transformations are revolving around information technologies. Whether a robust information science will finally emerge may depend not only on successful discussions about the definition of information and its related taxonomies but also on discussions about the accelerated technological development or about more potent theorizations. Perhaps the most important adjustment to make is about framing a new "mode of thought" within information science itself: an inner thinking system interconnecting the research practices in the fundamental empirical–theoretical areas that can contribute better to crystallize the new science. Transcending the different empirical approaches to living systems, and in general to biological–social informational entities, should be one of the primary goals. Indeed, the "invisible hand" of information has been the great shaper of all the complexity evolved within the different biological and social informational realms around. But information is never detached; it can only be detected through the transformations it provokes, symbolically flowing amidst the different subjects or agency realms and contributing to modify their material structures and self-production processes adaptively. Like in the historical birth of other major sciences, the empirical exploration and the comparative understanding of informational entities, understood as those that self-produce and communicate with their environment, should take place on a priority basis. It is along this direct reflection from the empirical, coupled with the theoretical and philosophical, comprehension that crafting a great scientific domain around information science could be feasible, becoming a common goal for all the scholars and researchers involved in information studies.

1. Introduction: The Transformations of Science

Although the sciences are often seen as a compact stock of knowledge that every generation receives and changes rather parsimoniously, except for the occurrence of momentous "revolutionary" transformations (Kuhn, 1962), they are in fact one of the fastest adaptive systems of society, continuously evolving, putting to trial their contents, and transforming them according to the ongoing discoveries and changes and the rising and falling social needs of the times. After the industrial revolution, a "fact" repeated almost regularly has been the duplication of essential social structures, both economic and scientific, that has taken place within every passing generation. That is to say, the whole economic wealth, the social effort devoted to research, the number of scientific disciplines, and the number of people devoted to scientific research have approximately doubled along each generational period — unfortunately the wealth disparity between the haves and the have-nots of societies has also been doubled (Landes, 1998; Klein, 2004; Marijuán *et al.*, 2012). This curious economic–scientific "law", which has taken hold during most of the last two centuries at least, resembles Moore's law on the doubling capacity of information circuits every few years. It might be interpreted as if every generation would have created by itself a whole extra world, enlarging the old one it has inherited, but recreating the new portion more attuned to its own vision and experience of the world — and usually, revolving around a preferred fundamental discipline that takes the intellectual lead.

That's the gist of the present chapter: the fresh opportunity for information science in our times demands a new way of thinking or mode of thought that, as we will argue, should receive a fundamental biological inspiration.

The tight relationship between knowledge and the economic structures of societies that we have commented on is hardly a surprise. As said, new generations spurred by their new necessities and new visions of the world explore new phenomena and transform the system of knowledge and the map of the sciences giving birth to ad hoc new disciplines — would it be a surprise that there was a rise and boom of classical information science in the 1960s, and nowadays there is the renewed emergence of information science along the stunning informational uprising of contemporary societies?

The historical parallel with the emergence of thermodynamics in the generations closer to the social–technological experimentation with steam-engines fortifies the case. Engineering, economics, and theoretical science advanced hand in hand in the whole fields studying energy during the first and second industrial revolutions. Drawing thus a parallel, most of the technological and scientific growth during the last decades has been produced in information-related fields: bioinformatics and omics revolution, neuroinformatics, robotics, artificial intelligence, quantum information, supercomputing, virtual economy, virtual learning, social networks, and so on. Thus, along this parallel, would the consolidation of a robust information science in our times be surprising?

Although the efforts to build a coherent discipline around the manifold scientific manifestations of information have not succeeded yet, they are gaining momentum and might be contemplated with increasing hope and intellectual ambition. Rather than attempting the construction of another average or standard discipline, information science is about the creation of one of the "great scientific domains" of contemporary knowledge (Rosenbloom, 2013).

The great scientific domains and their mutual interrelationships, as well as the role that information science should play there, will be discussed later on. Let us however emphasize that it is information science — rather than computer science and technology as argued by Rosenbloom — that is the strategic domain that has the mandate to perform the greatest cognitive interrelationships: the natural sciences with the social sciences and humanities; the individual cognitive limitations with the social recombination of knowledge; the generation of scientific knowledge with its systematic use in social organizations and management structures; the human needs of communication with the endless growth of social complexity; the closed views of political–economic systems with the ecological sustainability and planetary biogeochemical flows, and so on. Providing the next generation with a more coherent and workable system of knowledge would be a valuable legacy from our generation indeed — perhaps by having launched in the chaotic sea of disciplines an information hub of fresh design.

But the challenges are multitudinal, coming from multiple sources and compounding each other. Probably, developing a new way of thinking within Information Science is needed to substantially advance in the enterprise. Physical, engineering, computing, biological, or philosophical approaches are not going to provide by themselves, out from their well-trodden paths, the essentials of the new perspective. In the present chapter, some basic ideas will be presented that could (hopefully) help to reorient the ongoing explorations.

2. The Difficulties: Why Information Science is Lingering?

A number of recent works have addressed the topic of information and its scientific history (see among others, Hobart and Schiffman, 1998; Lanham, 2006, Wright, 2007; Gleick, 2011; Poe, 2011; Deacon, 2012; Hidalgo, 2015; Hustved, 2016; Volk, 2017). Depending on each author's perspective, the pioneers become figures close to the humanities such as W. Ong, E. Havelock, and M. McLuhan; to communication and control, C. Shannon and N. Wiener; to computer science, C. Babbage, V. Bush, J. von Neumann, and A. Turing; or closer to natural science, S. Dancoff, Henry Quastler, G. Bateson, H. von Foerster, C.H. Waddington, B. Mandelbrot, and I. Prigogine among others. They configure an amazing plurality of disciplinary fields which extend across the whole spectrum of knowledge. Indeed, high level research, bold syntheses, and integrative attempts on disparaging theories and conceptions of the world have been in the information science scope — from the very beginning.

Integrative problems were badly felt already in the Macy conference's series, during 1946–1953, which gathered many of the above luminaries (Gleick, 2011; Poe, 2011). Notwithstanding the shared intellectual excitation, the difficult communication between such diverse perspectives — engineering, cybernetics, psychology, anthropology, linguistics, neuroscience, economics — gradually dissolved the "golden age" attempt and damped the expectation of fast progress in the multidisciplinary convergence. Arguably, that occasion was one of the most important foundational attempts, although it was "informational" only in an implicit way, as explicitly these conferences were contemplating a common science "on the workings of human mind". Another conference's series attempt, during the early 1990s, has been the Foundations of Information Science (FIS) initiative, with a similar broad scope if not wider, though it started with the explicit goal of developing a "vertical" or "transdisciplinary" science connecting the different threads and scales of informational processes (Conrad, 1996; Marijuán, 1996). The five FIS conferences held until now have carried the torch of a multidisciplinary synthesis across information fields until the present times.

It is not easy summarizing the major difficulties of the informational integrative effort. Compounding the heterogeneity of the main contributing fields, the strange cross-scale and all-inclusive nature of information processes — from the quantum, to molecules, cells, nervous systems, and societies — makes for an unwieldy conceptual framework. When almost all scales of nature are involved in a phenomenon, the prospect of a smooth synthesis may be discounted in advance. The complexity burden not only impinges on the workings of the practitioner scholar but the limitations of the "detached" scientific observer in the handling of all the related contents also become a problematic circumstance that cannot be ignored by scientific idealization — it somehow brings to mind the thermodynamic cost of information processing in the conjecture of Maxwell's demon. We will return to the observer theme (meta-observer, actually) later on.

From the point of view of the practicing researcher, there are two interrelated disciplinary "propensities": physicalism and computationalism. On the one hand, we don't understand very well the mixing of disciplines, their processes of knowledge recombination, and in the absence of well-founded inter-multidisciplinary methodology there is still a hegemony of the physically oriented way of thinking. In the multidisciplinary obscurity, finally being able to connect with physical tenets is taken as a guarantee of rightness. On the other hand, the stupendous success of computing, in spite of its (relative) theoretical shallowness, has also contributed to displace the research and scholarly interest out from the apparently "murky" information camp. Actually, during the 1960s and 1970s, it became clear that computer science was winning the day against the incipient information science, and the latter was relegated as an enlarged or elegant version of "library science". The paradox nowadays is that the stunning transformation of many branches of science and technology fuelled by computers, including physics itself, is leading one to reconsider the "natural" structures and processes of information, not only in the quantum–cosmological realms or in the biomolecular, or in the human-to-machine interaction, but particularly in the human-to-human interrelationship and in the bonding structures of societies as the great drivers of global communication networks.

A frequent informational surrogate is the anthropocentric shortcut, when information is directly established around the human subject. It should not be confused with the informational analysis of human neurophysiological processes, which in general involve the use of information theory and network tools applied to sensory perception, connectomics, cognitive schemes, and consciousness (Berthoz, 2000; Edelman and

Tononi, 2000; Fuster, 2003; Buzsáki, 2006; Dehaene, 2009; Tononi, 2012; Ledoux, 2015; Damasio, 2018). These studies are of a limited span at the present time, notwithstanding their acute necessity and all their future potential — there is an essential role that this neural approach has to fill in. But far more frequently, information has been directly linked to the conceptual constructions of language as the essential vehicle of human communication. In this case, information turns out to be easily amenable to philosophical analysis: no wonder that numerous schools of thought and great philosophical figures have drafted implicitly or explicitly their own informational version. In this sense, information can be easily connected with logics, semiotics, artificial intelligence, library science, etc. Meaning and knowledge may be appended too, but they become plagued with the real-world problems of formal systems: brittleness, shallowness, self-reference, etc. (Dupré, 1995). Precisely transcending those limitations has been one of the central goals of new fields in artificial life, advanced artificial intelligence, natural computing, and so on (Brooks, 2001; Zhong, 2011). Another consequence of anthropocentric information is that the connection with physical information and biological information, as well as with the roots of social information, is completely lost, or left in a penumbra. Freed from constraints, half-baked surrogates of anthropocentric information have proliferated in multiple fields: economics, business management, urban planning, industrial organization, social engineering, the military, etc.

Although the anthropocentric factor can never be eliminated totally — as the late Yates put, "our science smells human, and we cannot get it free of that odor" (Yates, 1987) — at least it should be subordinated within a robust neurocognitive scheme that explains "ecologically" the way human individuals confront information and create knowledge. That cognitive scheme should give biological sense to the limitations of the individual, both cognitive and communicational, and illuminate the ways that knowledge is maintained, applied, and recombined — and that includes the organizational practices upon which the complexity of social structures has been developed historically.

3. The Object: Can Information be Defined?

Having established the necessary background, we should advance the new conceptualizations. Rather than discussing the hundreds of information definitions (Lenski, 2010), let us definitely discard that game.

Information cannot have a universal definition because it corresponds to the open-ended interactions that a subject can engage in with whatsoever elements or objects are present in its environment. Information is relative, it pertains to an unmediated coupling between objects–subjects, always in mutual contact. We demand either that a physical contact be established between the pair object–subject, or that a "channel" be arranged to bring to the adjacency of the subject those objects or their very direct effects.

Thereafter, if we advocate undefinability, it does not mean the ineffability of the term; rather what it means from a naturalistic perspective is that the plurality of subjects (can't we talk about information in cells, in organisms, in individuals, or in social bodies as legitimate subjects?), plus the vastness of possible coupled objects, and the multiplicity of "coupling" modes between subjects and objects do not permit any universalistic definition. It does not follow that a consensus notion (or notions) could not be established. The achievement of a "winning" notion getting the highest marks in most fields is quite possible and highly recommended (actually we are going to propose an interesting candidate in the next section). Implicitly, that's what most of the proponents of information definitions have in mind. Although a good working notion may be crafted, universality is precluded. There will always be some other fields in which even the best established notion will not work appropriately; all the more when explicit conditions on what should count as subject, object, and coupling have to enter into the game too — and may be changed on the spot due to the highly volatile nature of communicative interactions.

That a phenomenon does not yield to rigorous conceptual definition does not imply its scientific lack of value or that it cannot be appropriately quantified. That's precisely what happens with two of the most fundamental categories of physics: space and time. Couldn't we paraphrase St. Augustine's famous question about time? What then is information? If no one asks of me, I know; if I wish to explain to him who asks, I know not... Actually, in the conceptual constructions of science, time and space are two fundamental forms of information, involved in all the inter-subjective frameworks.

An essential point in this alternative discussion hinges on what being informational means when applied to a subject. The main analytical focus is changing from the "name" information as an object in itself, or as an emergent property of an object in the relationship with a subject, to the "adjective" informational. What entities might be attributed the quality or characteristic of being informational? The response is concise: those capable of both communication and self-production. The first trait is well known in many different fields related to information science — apparently. From our perspective, however, the communication theme demands full clarification. In the same way that the energetic-metabolic aspect of biological self-production was better understood once the energy flow was properly recognized and characterized (Morowitz, 1968), the information flow of communication demands its own general conceptualization, beyond current conceptions restricted to specialized uses.

We need to transcend the subject–object coupling scheme. The central point to discuss is how the informational entity organizes its openness towards the environment in the communication exchanges. In principle, what we find in the surroundings of an informational entity is not an isolated object or single communication item, but continuous streams or flows of them. Even when a single object is examined or brought into communication by the informational entity, the observation actually becomes a "flow" of signals. In general, there will be many signaling objects flowing at once around the informational entity. The concrete ways these objects or signals flow around the subject is an important concern. The capture of the surrounding signaling flows is propitiated by ad hoc designs that tend to cover increased regions of space throughout special sensing structures, incorporating fractal distributions, for instance. How these signaling flows will be processed and the effects they produce inside do not concern us right now — there are very different processing structures and degrees of complexity in all of those "natural" entities we term as informational. In general, the communication trait has been evolved adaptively: exploiting the maximum surrounding flows with the highest possible diversity of objects/events, using the lowest capture energy, and applying the faster processing rhythms compatible with the own structures. This can be observed in cellular signaling systems, in nervous systems, and in the communication systems of human organizations and in societies at large. The external communication flow, or better, the information flow, antedates and permeates the evolution of whatsoever self-producing structures of complexity.

Looking at the other essential trait of the informational entity, self-production, something similar would happen. We see again a stream or flow of objects, destined to fuel the self-production processes, which are introjected or channelled inwards to be consumed — also appearing as a parallel flow of extrojected material outputs. The ingoing and outgoing flows constitute the metabolism of the self-producing informational entity. For the analysis of these matter and energy flows, we use well-developed thermodynamic and bioenergetic approaches: dissipative processes, nonlinear systems, self-organization, etc. As said, the energy flow has been well-characterized for the biosphere and ecosystems (Morowitz, 1968); technically it is also a crucial ingredient of industrial engineering, urban planning, economic management, etc. From this point of view, the informational entity appears as an open system, out from equilibrium, endowed with the required energy–entropy–matter flows, and often involving the generation of well-ordered dissipative structures. Depending on the different classes of self-production processes, the informational entity will require highly differentiated flows of stuff — water, oxygen, photons, ions, metals, molecular nutrients, affordances of the ecological niches, goods and services, energetic products, etc. — which in most cases do not flow spontaneously to encounter the entity; rather, they have to be searched, channelled, and internalized by following optimized behavioural and search strategies (Holland, 2012; Bejan and Peder, 2012; West, 2017).

Although the information flow and the energy flow may partially overlap in their constitutive components, they are treated in a highly different way: "reading" the environment becomes utterly different from, and prior to, "eating" it. The living cell is the clearest instance, as we will discuss in Section 4. In general, the high-energy, highly valuable self-production flows will be anticipated, detected, and captured by the faster and cheaper communication flows with the surrounding environment. The mutual congruence in functional terms between both kinds of flows is crucial for the viability of the informational entity. A frequent commonality of forms occurs too, manifested in supporting structures that often display fractal forms derived from the necessity to cover a region of space and to transport the affordances of both the material "stuff" and the communicational "fluff" to a center (Bejan and Peder, 2012; West, 2017).

Around the information flow, a new conceptual framework might be established. At its core, information science should deal with informational entities that exist "in the flow", that both communicate and self-produce. The intertwining of the two operational realms begets a seamless existential unity. As Lanham (2006) has stated about our societies, it is the almost immaterial fluff of communication which provides guidance to the constitutive material stuff; and vice versa, it is social life that provides the functions and goals of communication systems — always in service of social self-production. Hence, an "economy of attention" emerges: how the networks of self-production processes have to look at the world for their survival.

4. The Subject: On Being Informational

We have just substituted the information flow for the object. Now, in order to explicate the innards of the informational entity itself, the subject, we need a notion of information capable of making functional sense of how communication and self-production dovetail. Herein we will propose information as "distinction on the adjacent" (Marijuán, 2004, 2010; del Moral *et al.*, 2013; Marijuán *et al.*, 2015).

The distinction term refers to the processing capabilities of the informational entity, or the subject, engaged in the communicative exchanges — what "receiving the information" implies, what internalizing a portion of the information flow involves. Contrary to metabolic items, signaling objects themselves are not internalized: only some special relationships extracted in the reception process are circulated inwards. The informational entity interacts with the impinging signaling flows — via specialized receptors, sensory surfaces, or dedicated "agents" situated at its real or symbolic surface — and then it simultaneously creates preliminary streams of relationships, "distinctions", derived from the incoming signals. These distinctional relationships are successively transduced or translated onto the operational order of the entity as they circulate throughout the interior.

As an example of how distinctions could be extracted systematically, a simplified scheme of distinctional logic based on multidimensional partitions was discussed elsewhere (Marijuán *et al.*, 1998; García-Olivares *et al.*, 1998). It was based on three simplifying principles: economy, parsimony, and symmetry. Together, those three logical principles (equivalent to discarding redundant items, discarding differences already taken into account, and discarding any positional difference among the items) provide an "attentional window" that considerably simplifies the received messages. Finally a consistent skeleton of logical distinctions is obtained among the surviving parts within the overlapping elements of the message. The procedure was inspired in the workings of K. Javorszky (2000, 2003) about "multidimensional partitions". The guiding idea is that, by following this communicational logic, not far away from our customary simplifications in linguistic exchanges, we could apply multidimensional partitions as formal descriptors of the messages in different qualitative dimensions.

The adjacent term of the proposed notion of information refers to the physical contact to be achieved, and the need of counting with sensory elements or excitatory surfaces to be impinged upon by the incoming communicative signals. Given that the signaling objects are not internalized, and only their distinctional consequences are circulated, increasing the potential reach of adjacency becomes one of the most strategic and cheapest assets for the informational entity. Extending the portion of the environment potentially controlled through communication flows is a formidable drive of biological evolution — cellular pili, flagella, cilia, arborization of axons and dendrites, the neuronal multiplicity of sensors and receptors, specialized neuronal maps, sensing modalities, etc. — not to speak about the cultural artifacts, means of communication, and scientific–technological apparatuses that systematically transcend the immediate adjacency of subjects in the hyper-complex information flows of contemporary societies. By transcending the limits of immediate space-time adjacency (and memory), subjects may perform a myriad of further distinctions and cognitive operations.

The internalization of the communicative information flow, now transformed onto all the distinctions performed at the reception stage plus those distinctions generated along further transduction processes, provides adaptive guidance to the inner generative (and degradative) self-production processes of the informational entity. It happens in living cells, organisms, individuals, enterprises, and institutions, and in societies at large. In all of them, the information derived from the communication processes widely circulates and gets mixed with the inner self-production processes, adaptively changing the ongoing operations that constitute the "metabolic life" of the entity. That's the existential fate of all informational entities: they are adaptive, structurally always in the making and in the dismantling, subject to limited life spans.

Therefore, two different information flows, situated at different poles of the hierarchical scale, have been mixed together. The communicative information flow collected from the environment, once internalized and properly transduced, becomes irretrievably mixed with the "productive" information flow that circulates among the lower-level constitutive entities, themselves again informational entities endowed with their own spheres of communication and self-production. Looking from the low-level signaling flow, the consequence of receiving the new portion of transduced information implies a cascade of alterations in the ongoing internal self-production processes. Populations of these lower-level agents will solve through their own adaptive mechanisms the corresponding portion of the higher-level adaptive problems. An upward direction is established too, as the autonomous changes of the low-level signaling flows are also capable of influencing and guiding the self-production and communication processes of the whole informational entity, changing its adaptive behaviour. Indeed, the verb changes the flesh, but the flesh may change the verb as well.

Conversely to the "hardness" of artificial systems for information processing, and their strict separation between hardware and software, the mixing assimilation of the communicative information flow takes place inside a unitary "wetware" all along the *scala naturae* of life. The collective action of self-producing entities, connected in multiple, flexible ways by a permeating information flow, makes possible the unfathomable complexity and adaptability evolved along the Great Chain of Being. As we have written elsewhere (Marijuán, 2010) — a homogeneous description is out of hand — these informational entities belong to conceptual disciplines worlds apart. A parsimonious strategy may consist in attempting a relatively independent description for each one of the informational scales, to be followed then by some tentative abstractions/conclusions interrelating them "vertically" (Conrad, 1996; Marijuán, 1996). In what follows, we will make a brief description of the main communication processes of the living cell — the simplest ones, prokaryotic cells.

5. The Living Cell Observed: Towards a New Observational Methodology

Prokaryotic cells will be taken as a practical case (almost the single one in this brief chapter) to discuss the problems that an information science's observer has to confront. These cells contain perhaps one of the most interesting panoramas for informational analysis, where at least nearly all the intervening phenomena can be molecularly described. Around the examination of these cells, we will start to discuss the methodological and observational troubles plaguing the information studies. Notwithstanding that, in the description which follows, the language has to switch necessarily towards the molecular biological.

How different looks the animate from the inanimate! Just count the number of different molecular species teeming at the interior of a prokaryotic cell (the multiple classes of peptides, enzymes, receptors, phospholipids, RNAs, DNA, metals, nutrients, ions, etc.); the figure is staggering, in the order of 10,000 species, unthinkable of any regular physical system of that magnitude — a mere cubic micron. The special properties of water, the polymerization strategies, the organization in "architectures", the coded correspondence between triplets and amino acids, the folding process, the catalytic properties of enzymes, the semi-permeable membrane, etc. should be invoked as molecular builders of the cellular order, providers of "processing power" to this basic informational entity.

From our informational perspective, it is interesting that the vocabulary developed by molecular biologists was crafted, almost from the very beginning, around the communication metaphor: molecular recognition, genetic and epigenetic codes, transcription, translation, processors, messengers, signaling systems, effectors, transducers, second messengers, regulators, interferences, complexes, networks, modules, etc. It is quite revealing that the founding fathers of molecular biology so heavily relied on the communication metaphor for naming the molecular events they were uncovering. Advertently or inadvertently, they were fundamentally right.

5.1. *Cellular coupling of self-production and communication: how signaling systematically guides transcription*

Briefly analyzing the two essential informational traits, self-production is basically performed by a "network society" of specialized enzyme and protein agents that are coded onto the DNA "sequential architecture" and are continuously exchanging information about their specific activities thanks to the special solvent properties of the water matrix. In response to signals of the environment, to be described in what follows, enzymes and proteins are synthesized (and also degraded) from the sequential information of DNA and RNA, which are themselves incessantly subject to evolutionary combinatory games (Marijuán, 2002, 2010). The whole productive process culminates in the regularity of a specific cell cycle that is open to the environment. Nevertheless, by itself the transcription network expressing the DNA genes is closed, "blind". In other words, the coupling between the sequential genetic architecture and the diluted architecture of enzymes and proteins needs the injection of further adaptive capability to respond to environmental demands. This is done by means of signaling guidance, so as to partially deploy the genetic circuits in response to relevant happenstances of the environment and also of the cellular interior. Most of the topological governance of the transcription regulatory network, the decision of what parts should be activated or what particular circuits should be inhibited, is achieved by means of the cellular signaling system that activates the transcription factors (TFs) needed (Navarro, 2010).

Going on then to discuss communication, a variety of signaling systems may be found in prokaryotic cells, ranging from simple transcription-sensory regulators (a single protein comprising two domains), to those systems of multiple components and interconnected pathways that regulate key stages of the cell cycle, such as latency, pathogenesis, replication, and dispersion. A basic taxonomy of bacterial signaling systems was proposed elsewhere (Marijuán *et al.*, 2010), which was centered on "the 1–2–3 scheme". In a specific bacterium, for instance *E. coli* K-12, the number of different signaling pathways is almost two hundred, the majority belonging to the "one component systems" class, with around two dozen members of the "two component systems" class, and a few members of miscellaneous classes ("three component systems" and others). The flows of information crossing throughout the different receptors and further transduced along the signaling pathways are informing in prokaryotic cells about the presence of metabolic items in the environment, while in eukaryotic cells these very flows mostly relate to direct cell-to-cell communication about developmental and physiological matters. In both cases, the information flows become systematically transformed into variations of the self-producing structure.

A more detailed view of the whole signaling and transcription processes and the roles of transcription factors appears in Figure 1; it also represents in different colours the distinct sensing realms that impinge upon gene transcription: the externally sensed substances, the external substances transported inside, the hybrid external–internal substances, the purely internal ones, and the sensing of DNA states. See Figure 1 from RegulonDB (Gama-Castro *et al.*, 2015).

The transcriptional scheme of *E. coli* K-12 constitutes, as we have reiterated, the essence of the way this bacterium is in the world: how it "feels", stays, reproduces, associates, and transforms the world around. This cell is far more complex than the average prokaryote as it must survive occasionally in inanimate harsh environments, but it mainly proliferates inside the gut of animals, being involved in dense exchanges with other microbial communities and with the immune defenses of the host as well. The numerous two-component systems (29 actually) present in the bacterium reveal an active, vigilant sensing of its environment, including the social exchange with congeners by means of quorum sensing (Qsec-QseB). However, the far more numerous one-component TFs devoted to inner sensing — around 140, from an estimated total of 300 TFs (Ishihama *et al.*, 2014) — reveal a rich metabolic network and a vast transportation system of external fluxes to be carefully controlled. All this complexity has been acquired adaptively,

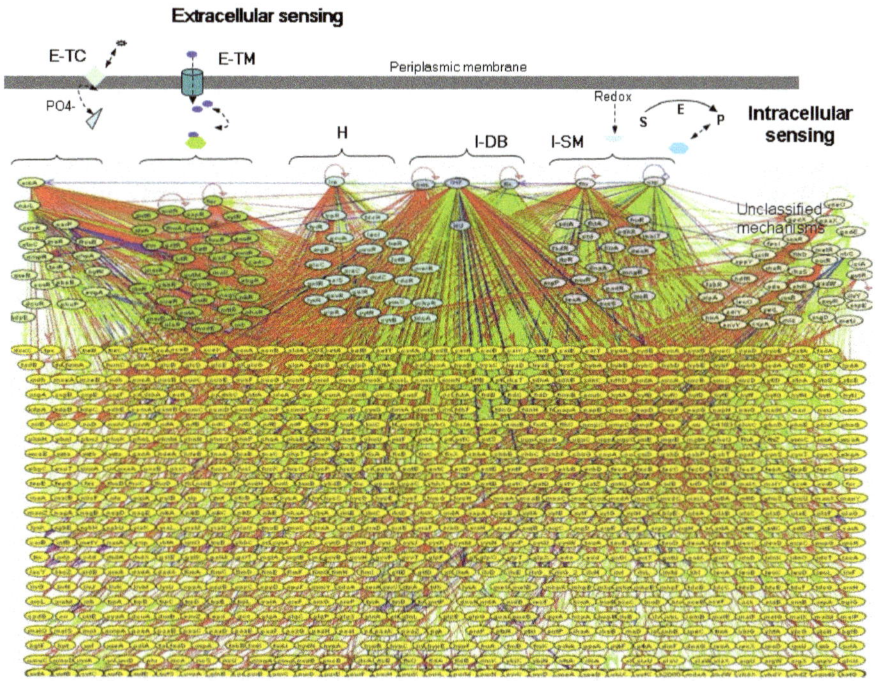

Figure 1. The *Escherichia coli* transcriptional regulatory network for sensing the extracellular and intracellular environment. In the upper part and from left to right, in green, are those transcription factors (TFs) corresponding to the extracellular class of sensing; in light green are those TFs from two-component systems (E-TC) and in dark green are those TFs using exogenous metabolites transported into the cells by transport systems (E-TM). In light blue are those TFs corresponding to hybrid system of sensing (H), i.e., those TFs using metabolites synthesized inside the cell and incorporated from the milieu. In dark blue are those TFs for DNA-bending or chromatin architectural TFs, they do not sense metabolites directly. In pink are those TFs for sensing intracellular conditions or sensing the internal cellular redox-state. Finally, in light orange are those TFs without metabolites or unknown mechanisms to modulate their activities. Global TFs (ArcA, Lrp, Hns, IHF, FIS, FNR and CRP) are at the top level. The connections: green lines represent activation; red, repression; blue, dual (activation and repression). In yellow (low) are those genes which do not code for TF products. Abbreviations: S, substrate; E, enzyme; and P, product.

Source: From RegulonDB, with permission.

along the evolutionary co-construction of microbial niches with the other interrelated species.

All this wealth of biomolecular information may be connected with the general ideas sketched in Section 4. Perhaps the most general statement about this bacterial way of life consists in the continual adjustment that the ongoing life cycle performs with respect to the medium, the continuous "reading" on the external and the internal environments it performs in. Depending on the external signals impinging on its external and internal receptors, on the transported flows of metabolites and nutrients, and on the inner DNA-sensing modalities, *E. coli* will express cohorts of genes that will be translated into the ribosomes. In Figure 1, we can clearly appreciate these different realms of transcriptional sensibility. In line with what we have stated in Section 4, the communication flow (1CS, 2CS, 3CS), the energy flow through transporters, and the productive information flow of multiple activators and inhibitors operating on enzyme networks, including protein degradation by proteasomes, harmoniously contribute to obtain a life cycle that culminates in reproduction. Perturbing in whatever manner, this scheme of flows directly affects the multitude of sensors related to protein synthesis and protein degradation, informing in different ways the ongoing self-production process.

5.2. *Cellular elaboration of meaning: Observer problems*

In particular, if we want to ascertain the effect that a given signal produces, we must count the new molecular presences and absences derived from the gene expression consequences of the signaling event. The meaning of a particular signal is thus established through "molecular mining". But there is no fixed reference there: the life cycle itself, in all its enormous multiplicity of possible "moods" and trajectories, can only be established, in retrospect, by "freezing" it; at every instant, we might look behind, the reference that provides, generates, and fabricates the meaning has changed. The whole life cycle is but a temporal sequence of instantaneous meanings continuously churned out from the entire self-production processes and apparatuses of life. Looking from the angle of semiotic conventions, signals appear as compositional structures of the objects themselves, quite indistinguishable and inseparable from them and from the outer world as well; only with the advent of quorum signals and interspecies communication, we may partially distinguish between signals that denote the presence of a very important "animate" object. As for the subjects, they appear themselves as life cycles in progress, and only that which pertains to the advancement of the life cycle has been evolutionarily incorporated as being part of the subject's own communication and energy flows. The relevance or value of a signal can subsequently be considered and gauged — this in general corresponds to second messengers and the cell cycle "checkpoints". Completion of the cell cycle always appears as the fundamental reference. The phenomenon of knowledge may be appended too, once the generative codes of the protein agents implementing successful responses have been evolutionarily selected, refined, and cohered within the life cycle. The recombination strategy performed upon the DNA stretches emerges as an abstract problem-solving tool of far-reaching evolutionary consequences, basically understood as "domain recombination" (Marijuán and del Moral, 2007; del Moral *et al.*, 2011).

Let us emphasize the difference between signals and metabolites. Essential molecular components of the environment are continuously scanned by the signaling system: nutrients, ions, metals, peptides, amino acids, toxics, signaling hormones, etc. Once important molecular presences are detected, the system activates gene expression programs or directly induces changes in the cytoplasm and membrane. But signals themselves are left untouched: the molecules participating as "messengers" are merely recognized as signals and do not suffer further transformations as happens with metabolites along metabolic pathways. The case of lactose may illustrate what we mean. François Jacob and Jacques Monod were the first scientists to discover the basic mechanisms involved in lactose metabolism and regulation, what is called the "lac operon" (Alberts *et al.*, 2007). These genes are turned on only when lactose is present in the environment or if glucose concentration decreases. The essential point is that lactose is treated first as a signal, specifically received by a "one component system" (LacI), and then, only after expression of the lac operon, it is treated metabolically by some other metabolic enzymes. So, lactose is "detected" or "read" first, thereafter it is "eaten". Similar happenstances may occur for several dozen substances that later on could enter into the metabolic networks (nutrients, metals, signaling peptides, etc.).

Although the above short description makes sense in molecular biological terms, it betrays the complexity involved. In actuality, most of the concepts involved in cellular communication have been coined recently (signaling system, 1–2–3 component systems, etc.). In addition, the transcriptional networks are difficult to frame experimentally, to analyze formally, and to cohere into macroscopic functions and ontologies. Further, achieving the right coupling between signaling from the environment, often involving simultaneous activation of several signaling paths, and the too complex programs of gene expression — for example, cAMP signaling involves more than 400 genes in *E. coli* — cannot be systematically realized in general, only in simplified instances. If this is what happens regarding "simple" prokaryotes, in eukaryotes the degree of structural complexity may be one or two orders of magnitude higher for each differentiated cell type, plus the emergence of a new layer of interactive complexity related to development and physiology. The point is that the real observation of a bacterium — or the surrogate observation through scientific literature — involves some previous choices about what is the expertise that the scientific observer must count with. For instance, following the description of a leading bioengineer (Kirckpatrick, 2009), up to 32 different scientific and technological disciplines may be involved in modern tissue engineering and biomaterial research.

Therefore, if the observation of this simplest informational entity has to be made from a rigorous scientific background, a meta-observer is needed in order to establish what of the endless biomolecular, biophysical, bioenergetic, bioinformatic, etc. aspects are relevant. All the other complexities will have to be discarded given the limited conceptual scope and the even more limited capability of the written discourse. This "methodological" problem of the information scientist becomes far worse in multicellular organisms, when force fields, cytoskeletons, electric fields, etc. have to enter in the description of the communication game (Marijuán *et al.*, 2013).

From another angle, it is the unsolved methodological problem of inter-, multi-, pluri-, transdisciplinarity, which very often can be safely ignored in specialized disciplines, and which can also be pragmatically tackled in multidisciplinary groups focused on concrete research problems. But that's not the case in information science, when the potential examination of the whole panoply of communication tools of the informational entity may be needed, as well as the specific impact that these tools may have in the advancement of the life cycle. It is not a rhetorical problem, it is a real one. The observer is too limited regarding the multidisciplinary "mix" needed for the necessarily holistic informational analysis — and this limitation cannot be surmounted by mere pragmatism in research expertise. A glance at some of the scarce reviews on signaling systems may be illustrative of the magnitude of the problem (e.g., Marijuán *et al.*, 2010, 2013, respectively, on prokaryotic and eukaryotic signaling "intelligence"). Even in current systems of biology, the general discourse interrelating communication and life cycle is absent. The endless chase, nowadays on global networks and "omic" lists, and yesterday on individual molecules and pathways, has barred the necessary biomolecular participation in the informational, holistic debate.

To conclude this brief description of the simplest informational entity (for the time being the only one we can describe molecularly with some reasonable approximation), the methodological problems found are not insurmountable. Like in other more mature disciplines, pragmatic solutions may also be found for the bioinformational analysis: those safe guidelines under which an efficient reduction of complexity may be achieved. In that regard, it is interesting discussing the growth of the complexity itself.

6. Complexity Growth: How Information Flows Orchestrate Complex Structures

The informational approach we have drafted from the simplest cells can somehow be generalized upwards. The information flow — the communication sphere — exists around the successive compartments of cells and organisms, accompanied always by a parallel flow of nutrients and maintenance and repair components. Actually, every new organization realm develops some specific information flow; at the frontiers or junctures between compartments, the previous, more basic information flow is maintained — as happens for instance with "old" molecular diffusion in between "modern" synaptic contacts. This successive composition of information layers seems valid for organisms, individuals, and societies too.

Is there any "natural" information flow we can point at in human societies? Anthropologically, the most distinguishing trait of our species is the use of language for communication among individuals. Evolutionarily, these communication activities, basically face-to-face conversation, have taken a genuine evolutionary category along the "social brain hypothesis" that was recently framed (Dunbar, 2004). It means that the most plausible evolutionary hypothesis on the hypertrophic growth of the human brain relates to the adaptation to bigger social groups and to the use of language as a fundamental vehicle of communication and as a new form of social "grooming" (Allman, 1999; Silk, 2007). Oral exchanges become the way to share group experiences, techniques, knowledge, culture, etc. far beyond any previous "cultural repertoires" of other anthropoids. When the open-ended combinatory structure of language is finally pictured into written "permanent" form (the parallel with the "permanent" DNA of the cell is unavoidable), a whole new world for the circulation of information and knowledge is generated in human societies. Without entering into far-reaching discussions, the historical fact is that the development of social complexity becomes irreversibly linked to a chain of inventions for communication and knowledge generation. There is a succession of fundamental inventions that dramatically alter the "infostructure" of societies: numbers, writing, alphabet, codices, universities, printing press, books, steam engines, means of communication, computers, etc. (Stonier, 1990; Hobart and Schiffman, 1998). However, the natural flow of human information — conversation — always remains present at the interstices of the new information and communication technologies and as the essential instance of knowledge creation and dissemination.

Another crucial aspect of human cognition is the recombination of knowledge. It is a phenomenon which has passed almost unnoticed in traditional philosophy of science, notwithstanding its massive presence in contemporary scientific–technological societies (Scott, 1998; Hapke, 2008; Arthur, 2009). The term is not exclusively related to the social or to the biomolecular fields (DNA domain recombination, as we have mentioned before). It is quite relevant that contemporary neuroscience has also recognized the importance of the recombinatory dynamics in the "neuronal workspace" of individuals, as argued by distinguished neuroscientists: Edelman, Changeux, Allman, Arbib, etc. (Dehaene, 2009). Very briefly, the main function of the neuronal workspace is to "assemble, confront, recombine, and synthesize" knowledge. The social creation of knowledge, and obviously the growth of science, derives from knowledge recombination processes taking place in the cerebral workspace of individuals. The strict conditions put by the scientific method would represent the ways and means to directly interconnect standardized individual perceptions and actions beyond time, space, and cultural limitations, allowing the social decomposability of problems and the occurrence of knowledge recombination dynamics at a global and intergenerational scale (Marijuán *et al.*, 2012).

Within the abstract procedures of information science, we may legitimately contemplate the complexity growth of cells, nervous systems, and societies along their respective evolutionary, ontogenetic, and historic trajectories. To what extent has this complexity growth been based on cumulative consequences of knowledge recombination phenomena and remodelling of the information flows? Understanding the common abstract processes of adaptability behind the vertical convergence of cell systems, nervous systems, and social systems on similar knowledge recombination strategies should be one of the theoretical goals of information science (Conrad, 1996; Kirby, 2000; del Moral *et al.*, 2011; Marijuán *et al.*, 2013, 2015).

Summing up, the "invisible hand" of information has been the great shaper of all the complexity evolved by the different informational realms around, either in the biomolecular and neuronal, or in the social and economical contexts. Information cannot be detected but in action, communicatively flowing amidst the different agency realms, and contributing to modify their self-producing structures in an adaptive way. The empirical, comparative understanding of these phenomena should take place first. Like in the historical birth of other major sciences, the hands-on experience should lead the theoretical and philosophical comprehension. The utmost contribution of information philosophy could be, in our opinion, adumbrating the research philosophy needed in information science.

7. Conclusions: Building a Great Scientific Domain

To conclude these rough explorations, at the risk of being too reiterative, a few recommendations follow that are addressed to promote a new mode of thinking among information scientists and information philosophers:

- Rather than continuing with the discussions on how to define the information concept, a conceptual shift from information to "informational entities" is advocated. It implies a new type of scientific observer, actually a non-ideal "meta-observer", a consensus notion on information, and an essentially empirical orientation for the new science.
- Achieving a new understanding of the natural flows of information, basically making sense of how flows of both "real stuff" (self-production materials) and "signaling fluff" (communication) become intertwined in the workings of the informational entity. It can be extended to a set of canonical informational entities: cells, multicell organisms, nervous systems, and societies. The simplest cells — prokaryotes — appear as an immediate model system.
- Placing the emphasis on applied informational themes, trying to cooperate in useful research topics with other disciplines and paying attention to relevant social problems; and also contributing to analyze the social pitfalls of contemporary information technologies, both on the mental side by the excess derived from the overwhelming information flows (stress, depression, diminished learning, weakened social bonds, tunnel vision) and on the material side (energy burden, e-waste, pollution, pilfering of resources).
- Producing a new vision on the social dynamics of knowledge and the combinatoric interrelationship between thousands of scientific and

technological disciplines, an essential problem of today's "knowledge societies" (see Coda).

- Contributing to more sophisticated cultural syntheses, needed in these global "instantaneous" times of continuous information overload and critical demands. New structures of social bonding, of social problem-solving, and of collective intelligence should be delineated for the sustainable societies of tomorrow.

7.1. *Coda*

Providing a new vantage point about the whole social dynamics of scientific knowledge is an essential problem of today's "information societies" — our sciences have become a Babel Tower of more than 6,000 scientific and technological disciplines, where classical information (library) science is helpless to provide any interesting guidance to society (Marijuán *et al.*, 2012).

In order to make sense of our system of knowledge, we need a clearer idea of how sciences recombine their contents and how new disciplines are born. Every discipline provides reliable partial knowledge about the external world, but in order to cope with the (non-restricted) real-world problems, it needs integration and overlapping with the extra contents provided by the other disciplines. The sciences are continuously mixing and rearranging their contents for the sake of the problems they have to solve, and also as the result of communities of dedicated scientists in continuous interaction. The realization of this socio-integrative process, in the double interdisciplinary and intradisciplinary dimension, becomes the essential engine for the recombination of knowledge in the praxis of science (the reductionist approach only characterizes a very narrow aspect). Scientific problem-solving means a continuous necessity of face-to-face meetings, collective gatherings, communication exchanges, interdisciplinary explorations, creation of new specialties, and so on. Along these permanent exchanges, whole domains of knowledge are "assembled, confronted, recombined, and synthesized". This is the way new disciplines appear (and others disappear), increasing their number as the whole scientific system grows generationally — at least during the last two centuries.

The knowledge recombination hypothesis might be applied to the historical evolution of science too. The combinatory activity of disciplinary modules or domains of theoretical–practical knowledge traveling to other disciplines will change there the local textures of knowledge, altering the regional maps of science and the whole complexion of the world of knowledge at large. Highly influential domains of knowledge, such as

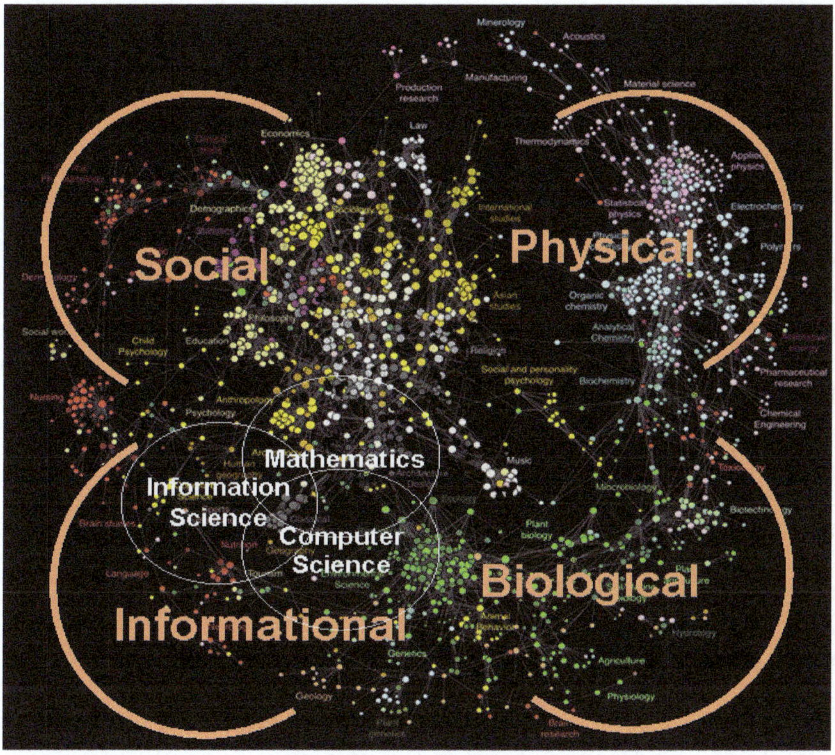

Figure 2. The four great domains of science. The graphic shows the network of contemporary disciplines in the background (following Bollen *et al.*, 2009); while the superimposed "four-leaf clover" represents the four great scientific domains: physical, biological, social, and informational.

Euclidean geometry, Newtonian mechanics, differential equations, genetics, and so on (and a multitude of other minor modules), would have generated the history of sciences, not only "developmentally" inside their own fields but even more in a "combinatorial" way, propelling the multidisciplinary evolution and cross-fertilization among scientific disciplines (Marijuán *et al.*, 2012).

Indeed, information science should promote a new vision to help make sense of the historical expansion of science, and of the requirements of human knowledge in action. Information science, properly developed and linked with computer science and mathematics, should constitute one of the Great Domains of contemporary science. The informational would go together with the physical, the biological, and the social: constituting the four great domains of science (see Figure 2).

The idea has been recently proposed by Rosenbloom (2013) for the computational domain — not without some doubts from his side on whether the computational or the informational should be given priority for the name of the whole domain. However, the doubts dissipate if we want to connect meaningfully the biological with the social and the humanities beyond the use of computational tools and models; if we want to deal with meaning, knowledge, and intelligence in a new way; if we want to understand the collective dynamics of the sciences themselves and illuminate on the new social relationships to establish with knowledge for a sustainable future — then there is no doubt that the priority should be given to the informational. Finally, as we were saying in the Introduction (Section 1), rather than attempting the construction of another average or standard discipline, information science is about the creation of one of the "great scientific domains" of contemporary knowledge.

Sources of Support

None declared.

Acknowledgments

To the colleagues of FIS Discussion List (Foundations of Information Science), for the many years of company, inspiration, and endurance.

References

Alberts, B., Johnson, A., Lewis, J., Raff, M., Keith, R. and Walter P. (2007). *Molecular Biology of the Cell.*, 5th edn., B. Alberts *et al.* (eds.), Garland Science, New York.

Allman, J. M. (1999). *Evolving Brains*, Scientific American Library, New York, NY, USA.

Arthur, B. W. (2009). *The Nature of Technology: What It Is and How it Evolves*, The Free Press, New York.

Bejan, A. and Peder, J. (2012). *Design in Nature*, Doubleday, New York.

Bollen, J., Van de Sompel, H., Hagberg, A., Bettencourt, L., Chute, R., Rodríguez, M. A. and Balakireva, L. (2009). Clickstream Data Yields High-Resolution Maps of Science. *PLoS ONE*, 4(3): e4803, doi: 10.1371/journal.pone.0004803.

Brooks, R. A. (2001). The relationship between matter and life, *Nature*, 409, 409–411.

Buzsáki, G. (2006). *Rhythms of the Brain*, Oxford University Press, New York.

Conrad, M. (1996). Cross-scale information processing in evolution, development and intelligence, *BioSystems*. 38, 97–109.

Damasio, A. (2018). *The Strange Order of Things: Life, Feeling and the Making of Cultures*, Penguin, New York.

Deacon, T. W. (2012). *Incomplete Nature: How Mind Emerged from Matter*, W. W. Norton & Company, Inc. New York.

Dehaene, S. (2009). *Reading in the Brain*, Penguin, New York.

del Moral, R. Navarro, J. and Marijuán, P. C. (2013). New times and new challenges for information science. *Proceedings of the 5th Foundations of Information Science. Information.*

del Moral, R. González, M. Navarro, J. and Marijuán, P. C. (2011). From genomics to scientomics: Expanding the bioinformation paradigm, *Information* 2(4), 651–671.

Dunbar, R. (2004). *The Human Story: A New History of Mankind's Evolution*, Faber & Faber Ltd., London.

Dupré, J. (1995). *The Disorder of Things*, Harvard University Press, Cambridge (MA).

Edelman, G. M. and Tononi, G. (2000). *A Universe of Consciousness: How Matter Becomes Imagination*, Basic Books, New York, NY, USA.

Fuster, J. (2003). *Cortex and Mind: Unifying Cognition*, Oxford University Press, New York.

Gama-Castro, S. Salgado, H., Santos-Zavaleta, A., Ledezma-Tejeida, D. *et al.*, (2015). RegulonDB version 9.0: High-level integration of gene regulation, coexpression, motif clustering and beyond, *Nucleic Acids Res* Jan 4;44.

García-Olivares, A., Villarroel, M. and Marijuán, P. C. (2000). Enzymes as molecular automata: A stochastic model of self-oscillatory glycolitic cycles in cellular metabolism, *BioSystems* 56, 121–129.

Gleick, J. (2011). *The Information: A History, A Theory*, A Flood Pantheon Books, New York, NY, USA.

Hapke, T. (2008). Combinatorics and order as a foundation of creativity, information organization and art in the work of Wilhelm Ostwald. Poster at the conference: Analogous spaces — architecture and the space of information, intellect and action, May 15–17, Ghent University, Belgium.

Hidalgo, C. (2015). *Why Information Grows: The Evolution of Order, from Atoms to Economies*, Basic Books.

Hobart, M. E. and Schiffman, Z. S. (1998). *Information Ages*, The John Hopkins University, Baltimore, MD, USA.

Holland, J. H. (2012). *Signals and Boundaries: Building Blocks for Complex Adaptive Systems*, The MIT Press, Cambridge, MA, USA.

Hustvedt, S. (2016). *The Delusions of Certainty*, Hachette UK, London.

Ishihama, A., Kori, A., Koshio, E., Yamada, K., Maeda, H., Shimada, T., Makinoshima, H., Iwata, A. and Fujita, N. (2014). Intracellular concentrations of 65 species of transcription factors with known regulatory functions in *Escherichia coli*, *J. Bact.* 196(15), 2718e2727.

Javorszky, K. (2000). Interaction between sequences and mixtures, *J. Theor. Biol.* 205, 663–665.

Javorszky, K. (2003). Information processing in auto-regulated systems, *Entropy* 5, 161–192.

Kirby, K. G. (2002). Biological adaptabilities and quantum entropies, *BioSystems* 64, 33–41.

Kirkpatrick, J. (2009). Multidisciplinarity and interdisciplinarity in bioengineering, biomaterials & nanomedicine, in *Proceedings of the CIBER-BBN: Networking Biomedical Research Centre for Bioengineering, Biomaterials and Nanomedicine*, February, 17–18. Zaragoza, Spain.

Klein, J. T. (2004). Interdisciplinary and complexity: An evolving relationship, *Complexity* 6(1–2), 2–10.

Kuhn, T. (1962). *The Structure of Scientific Revolutions*, University of Chicago Press, Chicago.

Landes, D. S. (1998). *The Wealth and Poverty of Nations: Why Some Are So Rich and Some So Poor*, W. W. Norton, Brooklyn, NY, USA.

Lanham, R. A. (2006). *The Economics of Attention*, The University Chicago Press, Chicago, IL, USA.

LeDoux, J. E. (2015). *Anxious: Using the Brain to Understand and Treat Fear and Anxiety*, Viking, New York.

Lenski, W. (2010). Information: A conceptual investigation, *Information* 1(2), 74–118.

Marijuán, P. C. (1996). First conference on foundations of information science: From computers and quantum physics, to cells, nervous systems, and societies, *BioSystems* 38, 87–96.

Marijuán, P. C. (2002). Bioinformation: Untangling the networks of life, *BioSystems* 64, 11–118.

Marijuán, P. C. (2004). Information and life: Towards a biological understanding of informational phenomena, *TripleC* 2(1), 9–19.

Marijuán, P. C. (2010). Knowledge recombination: On the informational adaptability of cells, nervous systems, and societies, *International Journal Information Theories and Applications* 8(1), 3–15.

Marijuán, P. C. and del Moral, R. (2007). The informational architectures of biological complexity, in *Computation, Information, Cognition — The Nexus and The Liminal*, Dodig-Crnković G. and Stuart S. (eds.), Cambridge University Press, Cambridge.

Marijuán, P. C., del Moral, R. and Navarro, J. (2012). Scientomics: An emergent perspective in knowledge organization, *Knowledge Organization* 39(3), 153–164.

Marijuán, P. C., del Moral, R. and Navarro, J. (2013). On eukaryotic intelligence: Signaling system's guidance in the evolution of multicellular organization, *BioSystems* 114(1), 8–24.

Marijuán, P. C. Navarro, J. and del Moral, R. (2010). On prokaryotic intelligence: Strategies for sensing the environment, *BioSystems* 99, 94–103.

Marijuán, P. C., Navarro, J. and del Moral, R. (2015). How the living is in the world: An inquiry into the informational choreographies of life, *Progress in Biophysics and Molecular Biology* 119(3), 469–480.

Marijuán, P. C., Pastor, J. and Villarroel, M. (1998). The language of cells: A partitional approach to cell-signaling, *Symmetry: Culture and Science* 9, 383–392.

Morowitz, H. J. (1968). *Energy Flow in Biology: Biological Organization as a Problem in Thermal Physics*, Academic Press, New York, London.

Navarro, J. (2010). Transcriptional regulatory network of *M. tuberculosis*: Functional and signaling aspects, Master Thesis, Universidad de Zaragoza, Zaragoza, Spain.

Poe, M. T. (2011). *A History of Communications. Media and Society from the Evolution of Speech to the Internet*, Cambridge University Press, New York, USA.

Rosenbloom, P. S. (2013). *On Computing, The Fourth Great Scientific Domain*, The MIT Press, Cambridge, MA.

Scott, J. C. (1998). *Seeing Like a State*, Yale University Press, New Haven, CT, USA.

Silk, J. B. (2007). Social components of fitness in primate groups, *Science* 317, 1347–1351.

Stonier, T. (1990). *Information and the Internal Structure of the Universe*, Springer Verlag (UK), London.

Tononi, G. (2012). *Phi: A Voyage from the Brain to the Soul*, Pantheon Books.

West, G. (2017). *Scale: The Universal Laws of Growth, Innovation, Sustainability, and the Pace of Life in Organisms, Cities, Economies, and Companies*, Penguin, New York.

Wright, A. (2007). *Glut: Mastering Information through the Ages*, Joseph Henry Press, Washington DC, USA.

Yates, F. E. (1987). Quanstuff and Biostuff, A view of Patterns of Convergence in Contemporary Science, in *Self-organization Systems, The Emergence of Order*, Yates (ed.), Plenum Press, New York.

Zhong, Y. X. (2010). On information science: An introduction to "Principles of Information Science", in *Proceedings of Fourth International Conference on the Foundations of Information Science*, FIS 2010, August, Beijing, China.

Chapter 12

Studies on Molecular Signaling Network Devices

Walter Riofrío

Faculty of Science and Philosophy, Universidad Peruana Cayetano Heredia, Av. Honorio Delgado 430, Urbanización Ingeniería, Lima, Peru

walter.riofrio.r@upch.pe

It is interesting to note that the community of biologists is progressively accepting the concept of information vis-à-vis the case of living systems. The results of research into the deepest intracellular level and upward through the rest of the intracellular, physiological, systemic, interspecies, and even ecological levels, more often than not, discuss such concepts as information, codes, signals, language, meaning, and other related terms. Most of us know the importance of understanding molecular components and their interactions. However, it is also vital to find out how these chemical compounds were able to produce a dynamic self-organization that is minimally complex. On the contrary, it seems that information has become a key concept in many areas of biology such as evolution, ecology, and molecular biology, among others. Likewise, it is suggested that many of the most fundamental biochemical processes, within living systems, involve the transfer and processing of information. In this chapter, I will discuss works on the design and development of new synthetic biology devices that may mimic the transmission and processing of biological information.

1. Introduction

For nearly two decades, research into synthetic biology has proven to be a successful approach to understanding the set of properties that proceed from the dynamics of biological phenomena. The primary goals of this new form of biological research are the design, construction, and development of artificial biological systems that do not exist in nature.

Of particular note is the impact brought about by the design and development of certain molecular circuits a few years ago.

To begin with, we must spotlight the research work that concretely demonstrated the essential spirit of synthetic biology, and that is to create a synthetic biological circuit.

We are talking about the work published in *Nature* in 2000 by Tim Gardner, Charles Cantor, and Jim Collins: "The Toggle Switch" (Gardner *et al.*, 2000). For the first time, they showed that it was possible to use parts within a cell to create designs that could develop logical functions and, in that way, imitate those that are found in electronic circuits.

This artificial biological circuit is a bistable synthetic gene regulatory network that can be used to control the on/off switch of gene expression. From a mathematical point of view, a biological switch is a bistable biological system: it can be in one of two possible states.

A toggle switch can be engineered from any two repressible promoters that are organized in a mutually inhibitory network. Said network is formed by two genes that mutually repress each other, i.e., gene X expresses a protein, PrX, which represses gene Y, while gene Y expresses a protein, PrY, which represses gene X.

The toggle consists of (A) promoters, (B) repressors, and (C) inducers:

(1) The promoters stimulate expression of a gene.
(2) The repressors bind to the promotors to inhibit gene expression.

The inducers bind to the repressors to prevent a repressor from binding to the promoters, thereby stimulating gene expression.

All these components are arranged to be mutually inhibiting. The bistability of the toggle switch stems from the mutually inhibiting design of the repressor genes (Gardner *et al.*, 2000, p. 339).

Another molecular circuit, the repressilator, is a milestone in synthetic biology and confirmed that gene regulatory networks, which perform a newly desired function, can be designed and implemented. As it happens, Michael Elowitz and Stanislas Leibler also published the results of this oscillating network in *Nature* in 2000 (Elowitz and Leibler, 2000).

The network was programmed into *Escherichia coli* by way of standard methods in molecular biology and manifested the anticipated oscillating behavior. It is basically a negative feedback loop created by three repressors which, after binding with a green fluorescent protein (GFP), cause the

E. coli cells to flash periodically, thereby demonstrating that oscillations can be genetically programmed.

The design employs three genes that constitute a feedback loop, in which each gene represses the next one in the loop and that is, in turn, repressed by the previous gene. In addition, the green fluorescent protein is used as the reporter, and the network operates in the following manner (Elowitz and Leibler, 2000, p. 337):

- Promoter 1 stimulates the expression of repressor 2.
- Promoter 2 stimulates the expression of repressor 1.

Assuming that repressor 2 is in excess of repressor 1, then

(a) Promoter 1 is not active.
(b) Introducing the inducer to repressor 2,
   - Inducer 2 binds to repressor 2.
   - Repressor 2 can no longer bind to promoter 2.
   - Promoter 2 stimulates the expression of promoter 1.

Thus, the toggle has been made!

In this chapter, I will present recent lines of thought in works about the design and development of new synthetic biology devices that might have components which mimic the transmission and processing of biological information in living systems.

The chapter is divided into three sections: Section 1 which serves as a brief introduction to the subject, Section 2, which shows approaches of research in bacterial Quorum Sensing and its possible applications on the design of synthetic biology devices, and Section 3 which shows a very recent line of research together with its potential to address important, timely scientific and technological issues that are of broad interest.

2. Cell-to-Cell Communication

How do cells process information? Although it is easy to pose this question, it is extremely difficult to provide a generic answer that can encompass the plethora of ways in which biological information has diversified itself over the 4 billion years of evolution on Earth.

In spite of this, we would like to underscore the fact that it is perhaps in the context of unicellular systems that we can begin to uncover the initial evidence and the first ways in which life originated the most elemental forms of processing information.

It is vital to remember that the survival of these living beings necessitated that they were able to respond, adequately and in very short periods of time, to the signs and signals that emanated from their surroundings (Di Roberto *et al.*, 2017).

Not very long ago, researchers were not aware of the fact that bacteria could be capable of exchanging information in a cell-to-cell level. It was mainly believed that they could only develop some type of sensors to recognize changes in temperature, pressure, oxygen, extracellular pH, as well as the level and availability of nutrients.

So let us talk on to the subject of microbial Quorum Sensing Systems (QS). They are microbial systems in charge of regulating microbial physiology based on the following details: synthesis, and the release and intake of small signaling molecules or autoinducers.

Detection of signal molecules by bacteria acts as stimulation which leads to altered gene expression and also the increase of its own biosynthesis once the minimal threshold is reached. This often leads to autoinduction of the signal, and so a rapid increase of signal concentration in the surrounding environment is observed. At critical cell densities, the binding of a regulator protein to the signal leads to the switching on of genes controlled by QS and therefore a coordinated population response.

The tasks carried out by QS are diverse. They basically prove that each kind of bacteria presents an array of needs, according to their own surroundings.

There are a number of important characteristics that must be shown for a molecule to be classed as a QS signal:

(a) the production of the QS signal takes place during specific stages of growth or in response to particular environmental changes;
(b) the QS signal accumulates in the extracellular environment and is recognized by a specific bacterial receptor;
(c) the accumulation of a critical threshold concentration of the QS signal generates a concerted response;
(d) the cellular response extends beyond the physiological changes required to metabolize the molecule.

Now, let us focus our attention on gram-positive bacteria. This bacterium uses modified oligopeptides to work as intercellular signaling molecules. Most gram-positive bacteria, holding the characteristics described above, release autoinducer molecules outside the cell. These molecules belong to the family of acyl homoserine lactones (AHLs) and many AHLs cross membranes freely and are detected in the cytoplasm by LuxR-type proteins (Waters and Bassler, 2005; Keller and Surette, 2006; Reading and Sperandio, 2006; Sifri, 2008).

On the contrary, in the gram-negative bacteria the process of signaling takes place at the intracellular level, and it is regulated by receptor proteins in the cytoplasm. Generally speaking, the interaction of the nearby AHL molecules with those receptors fuels the latter to bind with DNA, thereby modulating the expression of response genes.

The most well-known example of QS with an autoinducer is the *Vibrio fischeri*, which produces a diffusible compound (AHL) that only collects in the surrounding environment during the phase of exponential growth of the bacteria and triggers the expression of the luciferase gen.

The system is regulated by the luxL/luxR operon that constitutively expresses basal levels of the protein, luxL, and the receptor protein, luxR. When the concentration of the bacteria is quite low, the production of the autoinducing molecule (luxL) is also rather limited, and it is simply diffused into the extracellular environment where it accumulates but does not attain significant concentrations. In these conditions, the bacteria do not produce light (Case *et al.*, 2008; von Bodman *et al.*, 2008; Galloway *et al.*, 2011).

Additionally, it is possible that the principles that underlie bacterial quorum sensing networks are also crucial for collective behaviors in higher organisms. Moreover, new discoveries raise the plausible possibility that quorum sensing is not restricted to microorganisms, but rather is a general mechanism that functions throughout the tree of life (Papenfort and Bassler, 2016).

While research has been done in the recent past on understanding the multiple pathways of QS, few studies have focused on creating artificial modules that include extracellular components, like signaling molecules, and that allow for the examination of cellular communication and synchronization, which could prove to be tremendously important to the functioning of cellular groups, like organs, tissues, or bacterial populations.

Nevertheless, one recent work clearly emphasizes the huge potential that is buried in this line of research and that entails the creation of artificial modules (we will discuss another set of synthetic networks in the following section).

Synchronization is one of the fundamental nonlinear phenomena of the universe. It has been shown to be critical for the correct functioning of several biological systems, for instance, the existence of the so-called circadian clock (Yamaguchi *et al.*, 2003; Honma *et al.*, 2012).

This leads us to ask the next question: How do autonomous oscillators in living systems operate in a coherent state of oscillation?

The research we are presenting shows that QS could serve as a coupling mechanism for autonomous genetic clocks, thus establishing a well-defined study framework for analyzing the properties of synchronization in natural systems (Fernández-Niño *et al.*, 2017).

The purpose of the work was to create a synthetic bacterial communication module using synthetic biological techniques that enabled the modeling and study of QS circuits in bacterial populations.

Constructing the synthetic module required genes from the *Vibrio fisheri* and *Agrobacterium tumefaciens* QS systems as well as promotor elements that are regulated by external stimuli like lac operon promoter activated by IPTG, an analog of lactose (Fernández-Niño *et al.*, 2017, pp. 3–4).

This system was made of a synthetic network of bacterial communication that is capable of synchronization since it uses an Acyl Homoserine Lactone synthase (AHL) which produces an autoinducing AI molecule.

Even in the model, this AI molecule can pass through the cellular membrane and wall and bind to the luxR receptor when exceeding a threshold concentration. The AI-luxR complex acts like a transcription factor, thus stimulating the expression of a group of genes regulated by the lux box promoter. In the model, this promoter regulates the expression of two genes: (a) the green fluorescent protein (GFP) gene that is the reporter of synchronization and (b) the aiiB lactonase-encoding gene capable of hydrolyzing the concentrations of AHL. Generating the synchronization consists of coupling this synthetic bacterial communication module to a Repressilator (Fernández-Niño *et al.*, 2017, p. 3).

The two independent self-sustained synthetic units are able to autonomously oscillate in individual cells in order to produce bulk oscillations. It is the modularity of the system that could serve additionally as a basis to construct more complex populations of synthetic units or for the construction of cooperative and competitive synthetic microbial systems (Fernández-Niño *et al.*, 2017, p. 7).

3. Artificial Signaling Networks

Another research subject that could, in a not so distant future, enter the picture as a collection of possible developments in synthetic biology is the study of new artificial molecular devices that pose questions about how we can create molecular circuits that might simulate relatively basic signaling networks.

In other words, these are small molecular signaling networks that would have very specific purposes, that would outperform their natural counterparts, that would have enhancements which might be applicable to industry and biomedical science, and/or that may contribute to reducing some of the major environmental and health issues plaguing the world.

Microorganisms are capable of detecting a large number of external and internal stimuli and can process information rather efficiently; therefore, they generate a wide range of physical and chemical responses.

We know that, in general, all cells are able to respond to external stimuli and to produce specific physiological responses using their signal transduction systems.

As a result, it would seem to us that research into the next stage in the design of artificial circuits in synthetic biology would target, in the near future, molecular signaling networks.

Let us briefly review some of the studies that have called attention to this line of research, i.e., as a topic that should be seriously considered for the agendas of future synthetic biological investigation.

As a prior step, I would like to quickly point out some of the results from other lines of research that are dedicated to discovering how to specifically monitor the detection of pollutants that are introduced into our environment.

If living systems are constantly monitoring their surroundings to develop responses that will allow them to manage their adaptation strategies, then a good idea would seem to be improving those systems or at least a part of them.

So, there are studies that have attempted to computationally redesign periplasmic binding proteins (PBPs), which are normally involved in bacterial chemotaxis, in order to produce highly specific and responsive biosensing capacities for substances like explosives, chemical agents, and metals (Looger *et al.*, 2003; Dwyer and Hellinga, 2004; Dwyer *et al.*, 2004).

However, these computer redesigned proteins were unstable *in vitro*, making it impossible to use this powerful technology in electronic detectors (de Lorimier *et al.*, 2006).

Besides, in more practical terms, it would not appear to be very realistic to use bacteria with computer-modified receptors in possible detection systems applied to large environments.

After pondering these questions and the constraints associated with the required scaling, other researchers decided to focus on different living organisms. If we are talking about possible scaling to detect broad sectors of the environment, then why not think about the plant kingdom?

The detection ability of plants has developed through their evolution, and they are found naturally in most environments in which humans live. In addition, they can cover broad areas. These aforementioned qualities turn plants into low-cost detectors. The task, then, is to engineer them into detectors of specific substances whose response is easily seen (Antunes *et al.*, 2006).

Along these lines, Antunes *et al.* (2011) did some research into developing a synthetic signal transduction system in plants. To create it, they used histidine kinase (HK) signal transduction components, which are evolutionarily conserved.

This artificial signaling pathway prototype in plants with detection capability also had computationally redesigned bacterial receptors linked to the HK.

The importance of this study that we just summarized rests on the fact that transgenic plants with a fully synthetic signaling pathway in a higher eukaryotic organism were created for the first time. Their construction used a set of modular domains of bacterial and plant proteins. Additionally, it was shown that the computationally redesigned receptors could provide a modular platform for designing highly specific receptors for the majority of small molecules.

The Medford Laboratory was the developer of a synthetic signal transduction system that links exogenous perception of a particular ligand to a transcriptional response. Engineering of plants that detect and respond to a specific ligand is one area of application of this system.

The system was designed on the basis of evolutionary conservation of histidine kinase signaling and uses bacterial components adapted to function in plants. The synthetic signaling system is responsive to extracellular ligand perception by wild-type or modified ribose binding protein (RBP) scaffold.

After the ligand binding, the RBP binds and activates a synthetic fusion histidine kinase made from the extracellular portion of the bacterial chemotactic receptor Trg and the cytoplasmic portion of the bacterial phosphate sensor PhoR. Activated Trg-PhoR transmits a phosphate signal to the bacterial response regulator PhoB. After phosphorylation, PhoB translocates into the nucleus and activates transcription of certain response genes (Morey *et al.*, 2011).

Other research in the vein of synthetic biology is centered on "reshaping" a known signaling pathway (Bashor *et al.*, 2008).

Cellular behavior has been the target of multiple reshapings throughout evolution, such that the spatial and temporal organization of the molecules in the coordination of different cellular activities has played a critical and fundamental role.

We know that cells use a host of signaling proteins to modify cellular behavior in response to changes in their external and internal contexts. We have also discovered that scaffold proteins are the protagonists of the physical arrangement of the relevant molecular components in most of the biological signaling processes. These proteins direct the flow of information in the cell, thereby activating, coordinating, and regulating the signaling processes in the regulatory networks (Buday and Tompa, 2010).

The function of scaffolding proteins is to bring together two or more proteins in a relatively stable configuration, as the name so aptly describes. Numerous scaffolding proteins are found in nature, and many of them have multiple protein–protein interaction modules. They can, furthermore, exercise complex allosteric control over their associated molecules, or they can become the object of regulation (Garbett and Bretscher, 2014).

Scaffold proteins can also have a higher-order role, that is as signal processing hubs, serving as the target of feedback loops that optimize signaling amplitude and timing.

The Bashor *et al.* (2008) study that we discussed above is an example of what we just mentioned. As such, it showed that the Ste5 scaffold protein can be used as a platform to systematically reshape output of the yeast mating MAP kinase pathway.

The researchers constructed synthetic positive and negative feedback loops by dynamically regulating recruitment of pathway modulators to an artificial binding site on Ste5. These modified circuits produced a variety of behaviors: ultrasensitive dose response, response times that were either accelerated or delayed, and tunable adaptation. Protein scaffolds provide a flexible and robust platform for reprogramming cellular responses; hence, they might also be used to design cells with new therapeutic and biotechnological functions.

Other examples of the promise of intensively engineering signal transduction systems are the outcomes of studies which look to control the spatial and temporal aspects of signaling processes (Zhang and Cui, 2015; Kolar and Weber, 2017).

In order to attain those results, the investigators used a recent and innovative biological technique, optogenetics, which uses light to modulate molecular events in a targeted manner in living cells or organisms (Knöpfel and Boyden, 2012; Guglielmi *et al.*, 2016; Zhao *et al.*, 2018).

An interesting question that continues to intrigue researchers is as follows: What is the information dynamic in cellular signaling systems? The current methods have uncovered, to a certain degree, the components of these systems.

Yet, everything involved in the understanding and the dynamic information flow control, from its incorporation to the execution of a determined task or biological function, remains out of reach of our conventional techniques and methods.

Initial studies into optogenetics used light to regulate neuroelectric activities and, since then, it has definitely transformed experimental neurobiology (Boyden *et al.*, 2005; Deisseroth, 2011).

It has furthermore been used for the optochemical control of cell signaling, using small molecules (Shestopalov *et al.*, 2007; Liu and Deiters, 2014) or non-proteinogenic amino acids (Arbely *et al.*, 2012; Nguyen *et al.*, 2014) to trigger the activation states of signaling molecules.

An important step for optogenetic control of intracellular signal transduction was the discovery of photoactivatable proteins (Levskaya *et al.*, 2009; Wu *et al.*, 2009; Crefcoeur *et al.*, 2013).

Essentially, the mechanisms of these photoactivatable systems feature the following behaviors: they absorb energy from photons of light of a certain wavelength and the photoactivatable proteins go through conformational changes; therefore, they reorganize the inter- or intraprotein contacts and/or modulate the inter- or intraprotein interactions (Tischer and Weiner, 2014).

As is widely accepted, cells constantly receive external signals and then dynamically transmit and process them in space and time.

Optogenetics, as a new field of study, may entail not only the control of intracellular signal transduction but it could also monitor how the temporal signaling dynamic would have an impact on cellular functioning, how signals would be integrated into or filtered through certain key hubs along the signaling pathway, and also how the highly localized activation or repression in proteins could direct cellular behavior in biological systems (Repina *et al.*, 2017).

To wrap up, I will succinctly propose a possible alternative to the current lines of research in synthetic biology.

I will focus on a particularly important problem, the possible solution of which in the not so distant future would contribute greatly to the advancement of scientific knowledge as well as provide a wide range of possible industrial applications and improvements in current environmental conditions.

We are talking about actual research being done for the primary purpose of finding scientifically and economically viable ways of capturing atmospheric CO_2.

One of the major climate change issues is fundamentally associated with the increased concentration of atmospheric CO_2. Estimates place said concentration at close to 32% since the Industrial Revolution (De Silva *et al.*, 2015).

Moreover, the International Energy Agency (IEA) has raised the alarm on the danger of unchecked CO_2 emission and on the continued use of fossil fuel-dependent technologies.

Should humans hold to this course, then the agency warns that CO_2 emissions by 2030 will be 63% greater than they are now and 90% greater than what they were in 1990 (Ganesh, 2014).

The quest to discover methods of capturing CO_2 from the environment is, thus, a strategic and urgent topic of research.

One possible solution, in keeping with the discussions above, might be the practical creation of a signaling network design, in which an important process along that pathway would be the capture of atmospheric CO_2 into cellular metabolism.

It is a working hypothesis that is still in the theoretical–conceptual phase, but the general idea can be summed up in the following manner: designing an artificial intracellular, photoactivatable signaling network that would include certain processes for capturing atmospheric CO_2, a MAP kinase process, some scaffold proteins, and photoactivatable proteins, all of which are incorporated into the genome of a bacteria or yeast through CRISPR technology.

This is not a simple task, but I put it forth basically as a possible subject of research that could begin to materialize in the not so distant future.

4. Concluding Remarks

It is still unclear which concept of information is the most relevant to characterize biological information. More precisely, what would be the more salient core aspects that determine the transmission and processing of biological information which could differentiate it from a simple correlation between the different states of two molecular mechanisms involved in a given biological process?

There is an interesting conceptual difference between say, for instance,

(1) "X is correlated with Y",
(2) "X carries a message to (of) Y".

When we are talking about biological information, the first sentence could be interpreted in terms of Shannon's concept of information, while the second sentence could be interpreted as if there is a transmission of information that would imply a richer conception than Shannon's.

From a philosophical point of view, conceptualizing the role of genetic material expressed in terms of information flow from DNA to RNA to protein using Shannon's concept of information as correlations between the states of two systems, independent of the meaning and any semantic content of those states does not present interesting conceptual problems because we are only dealing with statistical properties of a given system.

This leads to the more contentious question of whether or not biology needs another, richer concept of information, sometimes called "semantic" or "meaningful" information. Several researchers claim it seems there are still some open questions in the studies on biological information which are somehow addressed by defenders of a richer concept than Shannon's information, from a naturalistic point of view.

More recently, another approach is centered on the nature of signs, as carriers of biological information. Focusing on the sender–receiver signaling systems, it is trying to show that there is an important role for a notion of content that goes beyond purely informational content. With this, it is supposed that the understanding of communication and its evolution could advance significantly through the study of simple models of interacting senders and receivers of signals.

Notwithstanding, there are still many open problems in this approach of signaling systems. For instance, in this chapter I sketched some new lines of research in synthetic biology which address these questions.

References

Antunes, M. S., Ha, S. B., Tewari-Singh, N., Morey, K. J., Trofka, A. M., Kugrens, P., Deyholos, M. and Medford, J. I. (2006). A synthetic de-greening gene circuit provides a reporting system that is remotely detectable and has a re-set capacity, *Plant Biotechnol J.* 4, 605–622.

Antunes, M. S., Morey, K. J., Smith, J. J., Albrecht, K. D., Bowen, T. A., Zdunek, J. K., Troupe, J. F., Cuneo, M. J., Webb, C. T., Hellinga, H. W. and Medford, J. I. (2011). Programmable ligand detection system in plants through a synthetic signal transduction pathway. *PLoS ONE* 6(1), e16292.

Arbely, E., Torres-Kolbus, J., Deiters, A. and Chin, J. W. (2012). Photocontrol of tyrosine phosphorylation in mammalian cells via genetic encoding of photocaged tyrosine, *J. Am. Chem. Soc.* 134, 11912–11915.

Boyden, E. S., Zhang, F., Bamberg, E., Nagel, G. and Deisseroth, K. (2005). Millisecond-timescale, genetically targeted optical control of neural activity, *Nat. Neurosci.* 8, 1263–1268.

Buday, L. and Tompa, P. (2010). Functional classification of scaffold proteins and related molecules, *FEBS* 277(21), 4348–4355.

Case, R. J., Labbate, M. and Kjelleberg, S. (2008). AHL-driven quorum-sensing circuits: Their frequency and function among the Proteobacteria, *ISME J.* 2, 345–349.

Crefcoeur, R. P., Yin, R. H., Ulm, R. and Halazonetis, T. D. (2013). Ultraviolet-B-mediated induction of protein-protein interactions in mammalian cells, *Nat. Commun.* 4, 1779. doi:10.1038/ncomms2800.

de Lorimier, R. M., Tian, Y. and Hellinga, H. W. (2006). Binding and signaling of surfaceimmobilized reagentless fluorescent biosensors derived from periplasmic binding proteins, *Protein Sci.* 15, 1936–1944.

De Silva, G., Ranjith, P. and Perera, M. (2015). Geochemical aspects of CO_2 sequestration in deep saline aquifers: A review, *Fuel* 155, 128–143.

Deisseroth, K. (2011). Optogenetics, *Nat. Meth.* 8, 26–29.

Di Roberto, R. B., Scott, B. M. and Peisajovich, S. G. (2017). Directed evolution methods to rewire signaling networks, in *Synthetic Protein Switches, Methods in Molecular Biology*, Stein, V. (ed.), Vol. 1596, Humana Press, New York, NY.

Dwyer, M. A. and Hellinga, H. W. (2004). Periplasmic binding proteins: A versatile superfamily for protein engineering, *Curr. Opin. Struct. Biol.* 14, 495–504.

Dwyer, M. A., Looger, L. L. and Hellinga, H. W. (2004). Computational design of a biologically active enzyme, *Science* 304, 1967–1971.

Elowitz, M. B. and Leibler, S. (2000). A synthetic oscillatory network of transcriptional regulators, *Nature* 403(6767), 335–338.

Fernández-Niño, M., Giraldo, D., Gomez-Porras, J. L., Dreyer, I., González Barrios, A. F. and Arevalo-Ferro, C. (2017). A synthetic multi-cellular network of coupled self-sustained oscillators, *PLoS ONE* 12(6); e0180155, https://doi.org/10.1371/journal.pone.0180155.

Furukawa, K. and Hohmann, S. (2015). A fungicide-responsive kinase as a tool for synthetic cell fate regulation, *Nucleic Acids Res.* 43, 7162–7170.

Galloway, W. R., Hodgkinson, J. T., Bowden, S. D., Welch, M. and Spring, D. R. (2011). Quorum sensing in gramnegative bacteria: Small-molecule modulation of AHL and AI-2 quorum sensing pathways, *Chem. Rev.* 111, 28–67.

Ganesh, I. (2014). Conversion of carbon dioxide into methanol — a potential liquid fuel: Fundamental challenges and opportunities (a review), *Renew. Sustain. Energy Rev.* 31, 221–257.

Garbett, D., Bretscher, A. (2014). The surprising dynamics of scaffolding proteins, *Mol. Biol. Cell* 25(16), 2315–2319.

Gardner, T., Cantor, C. R. and Collins, J. (2000). Construction of a genetic toggle switch in *Escherichia coli*, *Nature* 403, 339–342.

Guglielmi, G., Falk, H. J. and De Renzis, S. (2016). Optogenetic control of protein function: From intracellular processes to tissue morphogenesis, *Trends Cell. Biol.* 26(11), 864–874.

Honma, S., Ono, D., Suzuki, Y., Inagaki, N., Yoshikawa, T., Nakamura, W. and Honma, K. (2012). Suprachiasmatic nucleus: Cellular clocks and networks, *Prog. Brain Res.* 199, 129–141.

Keller, L. and Surette, M. G. (2006). Communication in bacteria and ecological and evolutionary perspective, *Nat. Rev. Microbiol.* 4, 249–258.

Knöpfel, T. and Boyden, E. S. (Eds.) (2012). Optogenetics: Tools for controlling and monitoring neuronal activity, *Prog. Brain Res.* 196, 2–278.

Kolar, K. and Weber, W. (2017). Synthetic biological approaches to optogenetically control cell signaling, *Curr. Opin. Biotechnol.* 47, 112–119.

Levskaya, A., Weiner, O. D., Lim, W. A. and Voigt, C. A. (2009). Spatiotemporal control of cell signalling using a light-switchable protein interaction, *Nature* 461, 997–1001.

Liu, Q. Y. and Deiters, A. (2014). Optochemical control of deoxyoligonucleotide function via a nucleobase-caging approach, *Acc. Chem. Res.* 47, 45–55.

Looger, L. L., Dwyer, M. A., Smith, J. J. and Hellinga, H. W. (2003). Computational design of receptor and sensor proteins with novel functions, *Nature* 423, 185–190.

Morey, K. J., Antunes, M. S., Albrecht, K. D., Bowen, T. A., Troupe, J. F., Havens, K. L. and Medford, J. I. (2011). Developing a synthetic signal transduction system in plants. In synthetic biology, (Part A, Chapter Twenty-Five), *Meth. Enzymol.* 497, 581–602.

Papenfort, K. and Bassler, B. L. (2016). Quorum sensing signal-response systems in Gram-negative bacteria, *Nat Rev Microbiol.* 14(9), 576–588.

Pikovsky, A., Rosenblum, M. and Kurths, J. (2003). *Synchronization A Universal Concept in Nonlinear Sciences, Cambridge Nonlinear Science Series*, Cambridge University Press.

Reading, N. C. and Sperandio, V. (2006). Quorum sensing: The many languages of bacteria, *FEMS Microbiol Lett.* 254(1), 1–11.

Repina, N. A., Rosenbloom, A., Mukherjee, A., Schaffer, D. V. and Kane, R. S. (2017). At light speed: Advances in optogenetic systems for regulating cell signaling and behavior, *Annu. Rev. Chem. Biomol. Eng.* 8, 13–39.

Shestopalov, I. A., Sinha, S. and Chen, J. K. (2007). Light-controlled gene silencing in zebrafish embryos, *Nat. Chem. Biol.* 3, 650–651.

Sifri, C. D. (2008). Quorum sensing: Bacteria talk sense, *Clin. Infect Dis.* 47(8), 1070–1076.

Tischer, D. and Weiner, O. D. (2014). Illuminating cell signalling with optogenetic tools, *Nat. Rev. Mol. Cell. Bio.* 15, 551–558.

von Bodman, S. B., Willey, J. M. and Diggle, S. P. (2008). Cell–cell communication in bacteria: United we stand, *J. Bacteriol.* 190, 4377–4391.

Waters, C. and Bassler, B. L. (2005). Quorum sensing: Cell to cell communication in bacteria, *Ann. Revi. Cell. Develop. Biol.* 21, 319–346.

Wu, Y. I., Frey, D., Lungu, O. I., Jaehrig, A., Schlichting, I., Kuhlman, B. and Hahn, K. M. (2009). A genetically encoded photoactivatable Rac controls the motility of living cells, *Nature* 461, 104–108.

Yamaguchi, S., Isejima, H., Matsuo, T., Okura, R., Yagita, K., Kobayashi, M. and Okamura, H. (2003). Synchronization of cellular clocks in the suprachiasmatic nucleus, *Science* 302, 1408–1412.

Zhang, K. and Cui, B. (2015). Optogenetic control of intracellular signaling pathways, *Trends Biotechnol.* 33(2), 92–100.

Zhao, E. M., Zhang, Y., Mehl, J., Park, H., Lalwani, M. A., Toettcher, J. E. and Avalos, J. L. (2018). Optogenetic regulation of engineered cellular metabolism for microbial chemical production, *Nature* 555, 683–687.

Chapter 13

Evolution of Genetic Information without Error Replication

Guenther Witzany

Telos-Philosophische Praxis, 5111 Buermoos, Austria
witzany@sbg.at

Darwinian evolutionary theory has two key terms, variations and biological selection, which finally lead to survival of the fittest variant. With the rise of molecular genetics, variations were explained as results of error replications out of the genetic master templates. For more than half a century, it has been accepted that new genetic information is mostly derived from random error-based events. But the error replication narrative has problems explaining the sudden emergence of new species, new phenotypic traits, and genome innovations as a sudden single event. Meanwhile, it is recognized that errors cannot explain the evolution of genetic information, genetic novelty, and complexity. Now, empirical evidence establishes the crucial role of non-random genetic content editors, such as viruses, diversity generating retroelements, and other RNA networks, to produce new genetic information, complex regulatory control, inheritance vectors, genetic identity, immunity, new sequence space, evolution of complex organisms, and evolutionary transitions.

1. Introduction

In the decades where physics and chemistry dominated biology, especially molecular biology and genetics, it was a common assumption that evolutionary novelty is a result of cumulation of beneficial replication errors ("mutations"), i.e., changes and variations of a master sequence structure.

The new synthesis in biology was the mainstream evolution narrative, in that genetic novelties and genomic variations could only be viewed as errors of previous original sequences. Although it was known early that all of the empirically measured and documented mutations cause deformed variants, most of them leading to disease or even death, very few seemed to represent neutral or even better variants. It was assumed that, in the long run of evolutionary times, such beneficial variants became an object to biological selection processes leading to better inheritable variants.

The lasting, dominant, new synthesis was constructed at a time before new and abundant data from (a) RNA biology, (b) comparative genomics, (c) phylogenetic analyses, and the (d) comeback of virology were available. Additionally, core paradigms such as (1) one gene–one protein concept, (2) central dogma of molecular biology (DNA–RNA–Protein), and (3) non-coding DNA representing "junk" were falsified in the meanwhile (Shapiro, 2009; Witzany, 2017). Concerning a coherent update of evolutionary theory, these are good reasons to look at alternatives to the error replication narrative. What are these alternatives? Let's start with a look at crucial deficits of the former "mutation" concept and the theory of quasispecies.

2. The Error Replication Narrative: Quasispecies Mutant Spectra

In a series of articles in the early 1970s, Manfred Eigen and Peter Schuster constructed a formalizable (mathematically based) explanation of how basic features of life and their complexities can be explained solely by physical properties of matter: If certain chemical properties exist on a planet with certain physical conditions, life will start by two self-reproducing macromolecular cycles in a complementary way. First, there are information-carrying nucleic acids building one reproductive cycle. Second, there are functional amino acids building the other cycle. Together they build a catalytic hypercycle, which is the basis of self-reproductivity of life (Eigen, 1971; Eigen and Schuster, 1978).

Both parts can be reconstructed physically in a formalizable depiction. The system of nucleic acids (information) and proteins (function) represents a closed loop, because no function is available without information, and information represents meaning by function. A clear hierarchy of reaction cycles leads to all empirically known features of life. The resulting evolutionary dynamics is an optimizing process, in that Darwinian selection values the fittest results of error replications (mutations). Mutations with selective advantage are instabilities in this system that can be explained sufficiently by irreversible thermodynamic processes.

A series of such mutations of master copies lead to quasispecies, i.e., mutant distributions of primitive replicating RNA entities. Such dynamic distributions of genomes that share genetic variation, competition, and selection generate fittest types (i.e., master copies) or, in the extreme case, "error tresholds", i.e., too high mutation rates/variations, in which information can't further reproduce in the case of too much mutational load.

Eigen followed the opinion of his time that a natural language as the genetic code is determined by a universal and context-free grammar as outlined in great detail by Noam Chomsky. Such a context-free grammar underlies natural laws strictly because it represents the universal logic of the material reality. The core functions of natural languages or codes are formalizable, predicable, and computable. The only scientific tool needed to depict the material reality is mathematics. Therefore, the molecular genetic code can be investigated and described sufficiently in a formalizable way (Schuster, 2011).

Today, we know that Eigen's concept of natural codes is not coherent with current knowledge on natural language or code use (Witzany, 1995, 2014b, 2015). First, no natural language speaks itself as no natural code codes itself (Witzany and Baluska, 2012; Witzany, 2016). The sequence structures of natural languages or codes are not a statistical, random assembly. Empirical data of all known observations indicate that all the natural languages or codes are the result of competent usage by groups of living agents. Second, natural languages or codes are not the inventions of single biotic agents, but they exclusively emerge in biotic interacting groups. This means, emergence of natural languages or codes is essentially a social event. Additionally, it is important not to forget the following: All known agents using natural code follow three kinds of semiotic rules (not laws), such as combinatorial (syntax), content-related (semantics), and context-dependent (pragmatics) rules. As proved by Charles Morris, if one level of rules is missing, nobody can reliably speak of a natural code or language.

In Eigen's concept, the syntax of the master copy determines the semantics (meaning). This is not coherent with the empirical fact that, in natural languages and codes, pragmatics (context) determine semantics. Pragmatic rules are completely absent in Eigen's concept. Several other derivatives of mathematical theories of language at that time, such as systems theory, information theory, game theory, and later on, synthetic biology, and even biolinguistics, all share this deficit.

3. Non-coding RNAs are Derived from Invading Genetic Parasites

Current knowledge indicates that DNA is not solely a genetic storage medium but also the main ecological niche for invading RNA species. Could it replace the error replication narrative to more coherently explain genetic novelty and genomic variations?

Endogenous viruses and their defectives, transposons, retrotransposons, long terminal repeats, non-long terminal repeats, long interspersed nuclear elements, short interspersed nuclear elements, ALUs, group I introns, group II introns, phages, and plasmids are the currently investigated examples that use DNA genomes as their preferred live habitat (Witzany, 2015).

Some of these genome-invading agents have been termed as mobile genetic elements and an abundance of such agents has been identified during the last 40 years as obligate inhabitants of all genomes (Frost *et al.*, 2005; Villarreal and Witzany, 2018). It is important to note that all of them share behavioral motifs that cannot be found on abiotic planets: They rely on a natural code; they infect, insert, delete, cut and paste, or copy and paste, and spread within the genome. They change host genetic identities either by insertion/deletion and recombination, or the epigenetic markings of genetic content, co-evolve with the host, and interact in a typically module-like manner (Villarreal 2009a; Shapiro, 2004; Koonin *et al.*, 2015; Koonin and Krupovic, 2017).

Together with non-coding RNAs, these agents shape genome architecture and regulation. In this respect, they are agents of change not only in evolutionary time but also in real time (Mattick, 2009).

Many of the currently known non-coding RNAs are derived from such invading genetic parasites (Villarreal, 2005). Non-coding RNAs interact with DNA, RNA, and proteins and play important roles in nuclear organization, transcription, post-transcriptional modifications (RNA editing, splicing), repair, immunity, and epigenetic modifications (Mattick 2001, 2003; Chuong *et al.*, 2016). Non-coding RNAs transcribe in the sense and antisense directions and are expressed with relation to cell type, subcellular compartment, developmental stage, and environmental stimuli (Zinad *et al.*, 2017). This means their function is context dependent (Witzany, 2009).

Additionally, RNA polymerases overlap in transcriptional contents, which means that each nucleotide can participate in varying transcriptional content arrangements according to varying contexts.

Non-coding RNAs regulate in a varying manner, coordinated or independent, autonomously or functionally interrelated, and can regulate individual genes as well as large genetic networks; they can precisely control the spatiotemporal deployment of genes that are executing neuronal processes with extreme cell specificity. Various classes of non-coding RNAs target each other for post-transcriptional regulation via alternative splicing, polyadenylation, 5′ capping, non-templated modifications, and RNA editing. In particular, RNA editing can transmit environmental information to the epigenome (Yablonovitch *et al.*, 2017). Additionally, it enables neuronal plasticity, which is essential for learning and memory (Qureshi and Mehler, 2012; Mercer and Mattick, 2013).

Non-coding RNAs can undergo nuclear-cytoplasmic, nuclear-mitochondrial, axodendritic trafficking via ribonucleprotein complexes that promote spatiotemporal distribution and function of various combinations of ncRNAs, mRNAs, and RNA-binding proteins (Spadoro and Bredy, 2012). Interestingly, the non-coding RNAs in host genomes are more conserved than the protein-coding sequences (Villarreal, 2004). Does this indicate that they are older in terms of evolution than coding sequences?

Non-coding RNAs are not the result of error replications, but in most cases derived from consortial interactions of RNA groups to stabilize or develop new identity traits embedded in a global network of the current RNA world, embedded also into a global virosphere of incredible density (Tycowski *et al.*, 2015). Infection and counter-defence are constantly changing according to the environmental context of these highly interactive groups that integrate to or preclude new, infectious agents. Let us never forget that 1 ml of sea water contains 1 million bacteria, but ten times more viruses and viral parts. If we take solely the number of phages present on this planet placed side by side, we would get 42 million light-years spanned by 10^{31} phage virions (Rohwer *et al.*, 2014).

4. From Physics to Biology: Emergence of Biological Selection

Single RNA stem loops react strictly according to physical and chemical laws, without any biotic characteristics. If such RNA stem loops built groups which can compete and cooperate, biological selection emerges

(Hayden and Lehman, 2006; Briones *et al.*, 2009; Vaidya *et al.*, 2012; Higgs and Lehman, 2015). How do agents emerge from chemicals to form biological identity and then form groups that learn membership (to belong or not to belong)? Single RNA stem-loop generation occurs by physical, chemical properties solely as demonstrated by natural and randomized RNA experiments. As mentioned above, if stem loops build complex consortia, they initiate social interactions not present in a pure chemical world, that is, biological selection emerges. This designates the crucial step from inanimate world to life.

In contrast to inanimate nature, RNA stem-loop groups actively generate behavioral motifs and patterns of interaction, that is, coordinate common behavior according to rules that lead to consortia of groups that differentiate self and non-self by integrating certain RNA stem-loop structures, which are beneficial for the stem-loop group, warding off others that do not fit into the consortia. We can find *de novo* initiation of behavior that cannot be deduced from former behavioral patterns, such as highly adaptive processes, retaining a contextual history, optimal energy costs (smart), solving problems beyond the capacity of its individual members, and fast-changing reactions against non-members (Villarreal and Witzany, 2013a, 2013b). These capabilities are all absent in inanimate nature.

5. RNA Group Interactions: Networking, Immune Reaction, Membership Roles

Interestingly, an RNA group's membership can never be fully specified, since it can always be further parasitized by as yet unknown parasites. This essential feature renders the ability to absolutely specify membership as basically undeniable. An RNA stem-loop group can never be fully secure from as yet unidentified genetic parasites. But the most important consequence from this insecurity is that it provides the inherent capacity for novelty, that is, the precondition for evolutionary innovation, such as new RNA groups and increased complexity (Villarreal, 2009a, 2011, 2015).

If we look at some interactional motifs of RNA agents to form consortial biotic structures, (i.e., RNA groups which together share a functional identity), we must look at the group building of RNA stem-loop structures (Witzany, 2014a). As mentioned above, it has been found that single stem loops interact in a pure physical, chemical mode without selective forces, independent of whether they are derived randomly or are artificially constructed under *in vitro* conditions (Vaidya *et al.*, 2012). In contrast to this, if these single RNA stem loops build groups, they transcend a pure physical–chemical interaction pattern and biological selection emerges, causing biological identities capable of self/non-self-identification and preclusion, immune functions, and dynamically changing (adapting) membership roles.

A single alteration in a base-pairing RNA stem that leads to a new bulge may dynamically alter not only this single stem loop but may also change the whole group identity of which this stem loop is a part. This means that any RNA group transcribed out of the DNA storage medium, i.e., non-coding RNAs, mobile genetic elements, endogenous retroviruses and their defective and other persistent genetic parasites (not to forget their degraded parts), may play a new, modular, unexpected, and additionally non-predicable role in cooperating/competing RNAs, genetic regulations, counter-regulations, and diseases or infection events (Smit *et al.*, 2006; Manrubia and Briones, 2007; Geuking *et al.*, 2009; Stoddard, 2011; Villarreal and Witzany, 2015).

Simple self-ligating RNA stem loops can build much larger groups of RNA stem loops that serve to increase complexity (Salomon *et al.*, 2012). Significantly, RNA fragments that self-ligate into self-replicating ribozymes spontaneously form cooperative networks. For example, it was found that the three-membered networks showed highly cooperative growth dynamics. When such cooperative networks compete directly against selfish autocatalytic cycles, the former grow faster, indicating RNA populations evolve greater complexity through cooperation (Vaidya *et al.*, 2012; Higgs and Lehman, 2015). In this respect, cooperation outcompetes selfishness.

6. DNA Sequence Syntax with Several Different Meanings (Functions)

The genetic syntax, i.e., the nucleotide arrangement of genetic information is not unequivocal. This means that a given genetic sequence does not offer its final meaning. There are several processual steps in which the same sequence syntax may get several different meanings (functions). In natural languages and codes, this is a usual procedure and — additionally — it saves energy costs: A given sign sequence may represent various meanings/functions according to its contextual use. It is not necessary to generate a new sequence for every function. This indicates the primacy of pragmatics in determining the sequence structure, not its syntax (Witzany, 2014b; Cech, 2012; Doudna *et al.*, 1989). Several context-dependent modifications of the meaning (function) of a given DNA sequence are currently known and investigated. Let's have a look at the various techniques to modify meaning (function) of given genetic sequences.

6.1. *Epigenetics: Context-dependent markings*

Although all cell types of any known organism contain the organism-specific genetic information, they are expressed according to their spatiotemporal position and their contextual (pragmatic) needs, such as developmental stages, stress, damage repair, or changing environments. This means that depending on the context, for a cell located within an organ of an organism, the expression leads to a tissue- and site-specific new cell at the right time (Slotkin and Martienssen, 2007; Turner, 2002).

Chromatin marking enables a kind of identity programming (Barlow, 2011). This means that a specific cell within an organism is able to obtain or even change its identity through epigenetics, if developmental, environmental, nutritional, or stress-related conditions make it necessary. RNAs are mobile and can serve as signals throughout the tissues, organs, and whole organisms. In this respect, the imprinting of new experiences that leads to variable meanings of genetic information depends on the ability of non-coding RNAs. With epigenetic marking, life has an appropriate technique for the emergence of memory and learning processes for faster adaptation (Mattick, 2010; Blaze and Roth, 2012; Marshall and Bredy, 2016; Kim and Kaang, 2017).

Both the small RNAs and long non-coding RNAs are competent in terms of directing chromatin changes through histone modifications and DNA methylation. These non-coding RNAs are able to direct chromatin-modifying agents to specific targets. In small RNA-driven silencing pathways, the regulatory RNAs identify and mark potentially dangerous "non-self" elements for transcriptional silencing or elimination (Slotkin and Martienssen, 2007; Shapiro, 2014). In other networks, homology between the regulatory RNA and the target locus marks the region as "self" and protects it from silencing or elimination. Interestingly, epigenetic marking conceivably originally emerged to defend genomes against genetic invaders (Huda and Jordan, 2009; Huda *et al.*, 2010).

6.2. *RNA editing*

We can look at a shared behavioral motif of how to modify meaning out of a given DNA sequence syntax in RNA Editing (Göringer, 2008). RNA editing is a co- or post-transcriptional process, which alters the RNA sequence derived complementarily to the DNA from which it was transcribed. Before RNA editing, the editosome, i.e., small nuclear RNAs interconnected with a variety of proteins, must be assembled in a strictly coordinated process.

RNA editing changes gene sequences at the RNA level (Homann, 2008). The edited mRNA specifies an amino acid sequence that is different from the protein that could be expected and is encoded by the genomic DNA of the primary transcript. RNA editing alterations of such transcribed RNA sequences occur by modification, substitution, and insertion/deletion processes (Smith, 2008). Editing sites have to be identified individually to differentiate an A,T,G,C to be edited, from an A,T,G,C which should not be edited. The discriminating information can be found in the nucleotide sequence surrounding a given site. This means that context is relevant for recognition. Thus, each editing site carries its own recognition context (Gott, 2003).

6.3. *Alternative splicing*

RNA editing predates splicing and is timely and functionally interconnected. Editosome and spliceosome are important interacting agents (Matlin and Moore, 2007; House and Lynch, 2008; Yang, 2015). As in ribosome and editosome assembly, in spliceosome construction also, a ribonucleoprotein complex is assembled in various steps that cut out introns and splice exons together. The spliceosomal ribonucleoproteins are mainly small nuclear RNAs that are interconnected with at least 300 different proteins to five spliceosomal subunits.

Interestingly, a variety of steps in which the subunits of the final spliceosome are produced are counterbalanced within competing parts that regulate the stepwise processing of subcomplexes. After this final splicing procedure of the mature spliceosome, the remaining RNA products are actively discharged from the spliceosome and the remaining ribonucleoprotein particles are recycled for further catalytic processes as multi-use modules. Dependent on these regulations, the end product may vary with respect to the context dependency of the regulation process, which is highly sensitive to various needs and circumstances. In consequence, spliceosomal regulation differentiates the inclusion (splicing enhancers) or exclusion (splicing silencers) of exons in the final mRNA (Matlin and Moore, 2007). Splicing regulation occurs due to competing *cis*-acting elements that precisely balance regulatory proteins (Pyle and Lambowitz, 2006).

6.4. *tRNA-derived fragments*

Transfer-RNA (tRNA) is a key player between the RNA world and the protein world. tRNA-derived RNA fragments are small RNAs that are not random degradation products, but are specifically cleaved from mature tRNA transcripts. tRFs are microRNA, like small RNAs, involved in the regulation of genome stability. A variety of gene structures and RNA secondary structures of pre-tRNAs and mature RNAs have been identified in all domains of life. Coding sequences for these tRNAs are either separated within introns, fragmented, or permuted at the genome level (Keam and Hutvagner, 2015; Fujishima and Kanai, 2014; Gebetsberger *et al.*, 2016; Kanai, 2015).

Although evolutionary origins of the tRNA gene disruptions are unknown, a variety of tRNA structures seem to be co-evolved with RNA splicing endonuclease. Interestingly, tRNA-derived small RNAs are able to identify and target transcripts of transposable elements. This may indicate their relation to viruses and similar invading genetic parasites (Martinez *et al.*, 2017; Wang *et al.*, 2013).

6.5. *Pseudo-knotting*

Pseudo-knotting is a simple way in which a single strand of RNA can fold back on itself. This plays fundamental roles in structurally organizing complex RNAs, such as the assembly of ribonucleoprotein complexes, and additionally, in translational regulation and recoding when present within messenger RNAs (Staple and Butcher, 2005; Peselis and Serganov, 2014).

Because pseudo-knotting is extremely context sensitive, methods to predict pseudo-knotting based on stochastic context-free grammars were not very successful until recently (Staple and Butcher, 2005). The roles pseudo-knotting plays include generating the catalytic core of various ribozymes, self-splicing introns, and telomerase (Peselis and Serganov, 2014; Tholstrup *et al.*, 2012). Additionally, pseudo-knots play crucial roles in altering gene expression by inducing ribosomal frameshifting in many viruses. Pseudo-knots also induce ribosomes to slip into alternative reading frames. Because this behavioral motif is similar to that of self-splicing Group II introns, this may indicate their descent to genome-invading agents (Toor *et al.*, 2008).

6.6. *Frameshifting*

A ribosomal frameshift is a natural technique to process alternative translation of an mRNA sequence by changing the open-reading frame. It is usual in viruses, and enables the virus to encode multiple types of proteins from the same mRNA (Atkins *et al.*, 2016). Through frameshifting viruses are able to create many protein structures from small genomes. During frameshifting, the ribosome shifts forward or backward, in a way specific to the virus. Because three mRNA nucleotides always make a codon, which is translated into one amino acid, the frameshift makes every amino acid different, because it translates a different set of codons in the new reading frame (Giedroc and Cornish, 2009; Dinman, 2006). This means a different protein product is formed. Ribosomes typically translate mRNA without shifting the translational reading frame. Interestingly, a number of organisms have evolved mechanisms to cause site-specific or programmed frameshifting of the ribosome in either the +1 or −1 direction.

6.7. *Loop kissing*

An RNA group is built up of different members that represent a group identity (like gangs), as suggested by Villarreal (2015). One main behavioral motif here is the "kissing loop" motif. The single-stranded loop RNA is open for interactions with other (self or even non-self) single-stranded RNAs via base pairing, according to base-pairing combination rules (Brunel *et al.*, 2002; Cao *et al.*, 2014; Cao and Chen, 2011). This may have highly dynamic aspects such as the loop–loop interaction within an RNA stem-loop group (as represented by editosomes, spliceosomes, ribosomal subunits, and viruses) or between RNA stem loops with non-related RNA loops present within the cell, tissue, organ, or organism. Some of them may be remnants of formerly degraded RNAs. RNA recycling is an important resource for building and rebuilding RNA identity groups, i.e., formerly foreign (non-self) RNA group members may become relevant parts of a new (self) identity, according to the contextual need.

6.8. *Bypassing translation*

One of the most interesting explorations in the last decades was the multiple ways in which identical nucleotide sequences that serve as genetic storage mediums undergo a variety of dynamic changing modes during transcription as outlined above. The translational process from DNA via RNA and an abundance of regulatory processes of non-coding RNAs is object to a further process to dynamically modify the genetically stored meanings (functions).

During the translational process, ribosomes may bypass and, therefore, ignore parts of the mRNA sequence (Samatova *et al.*, 2014;

Agirrezabala, 2017). The translation is blocked at a certain codon and a different stem-loop RNA ensures the further translation process at a certain stage of the mRNA. For example, in the mitochondrial genome of a yeast species, 81 translational bypassing elements were recently found. The mitochondrial bypassing resembles a sequence of the t4 phage gene 60. Interestingly, some codons remain unused in mitochondrial translation and can be used as variable module-like parts in new ways, such as serving as translational barriers for bypassing undesired sequences, or they may be reassigned to new amino acid identities by codon capture. Such bypassing elements resemble dynamic modes of mobile genetic elements.

6.9. *Competing endogenous RNAs*

Competing endogenous RNAs (ceRNAs) designate a variety of RNA consortia that compete for the wide range of micro RNAs, which regulate messenger RNAs (Salmena *et al.*, 2011). Such RNA consortia are pseudogenes, long non-coding RNAs, and circular RNAs. Long non-coding RNA-derived microRNAs are very abundant and regulate nearly half of the genomes. ceRNAs possess microRNA-binding sites (miRNA response elements) and ceRNAs may also cooperate to repress targets (Denzler *et al.*, 2014). Such competing endogenous RNAs indicate competitive behavior as a biological phenomenon derived from the early RNA world of an abundance of RNA stem-loops competing for limited resources initiating biological selection processes. Competing endogenous RNAs are a further level of regulation processes that interact in a concerted manner with other RNA-determined regulation processes. If their competition gets counterbalanced, diseases such as cancer may be the result (Yang *et al.*, 2016).

7. Genetic Novelty and Genomic Variations by Viruses

After looking at the various techniques to modify the meaning/function of a given genetic sequence, we must be aware of genome-invading agents, such as viruses and RNA networks, which represent a very large and dynamic source of genetic novelty.

They can cooperate, build communities, generate nucleotide sequences de novo, and insert/delete them into host genetic content (Stedman, 2013, 2015). Viruses and RNA networks often remain as mobile genetic elements or similar "defectives" and determine host genetic identities throughout all kingdoms, including the virosphere. But, inclusion of a transmissive viral biology differs fundamentally from mainstream concepts of the last decades, in which it represents a vertical domain of life providing vast amounts of linked information not derived from direct ancestors (Villarreal, 2005).

Current knowledge about the virosphere and its role in evolution ("virolution") indicates interactions of RNA viruses, DNA viruses, viral swarms, and viral and RNA-based subviral networks that cooperate and coordinate (regulate) within cellular genomes either as replication-relevant coplayers or suppression-relevant silencers (Stedman, 2013, 2015; Ryan, 2009; Seligman and Raoult, 2016). Some represent infection-derived modular tools of non-coding RNAs which have built consortia of complementary agents that function together (Tycowski *et al.*, 2015).

Persistent viral lifestyles that we can identify as counterbalanced viral properties are found as addiction modules (see below), which transfer complete genetic datasets into host genomes and alter the DNA genetic identity of the host and, additionally, the formerly competing viral agents, without damage to the host genome content. As we saw earlier, RNA consortia may edit genetic code information without altering inherited DNA content. Transcribed from DNA cellular sequences, the RNA-activated inhabitants from former viral infection events act as modular tools for cellular needs in nearly all cellular processes (Witzany, 2009, 2016, 2017). They act not as lytic agents but as part of the host genetic identity because of domestication by counterbalanced competing genetic parasites (Villarreal, 2012a). Addiction modules not only change genetic identity but also enrich immune functions of host organisms to fight related parasites (Moelling and Broecker, 2015). We can easily identify these in the known complex adaptive immune systems (Villarreal, 2009b).

Endogenous retroviral competences in the persistent status are often characterized by features expressed only in the strict time window of a developmental process, such as axis formation, trophoblast formation (as in the case of endogenous retroviruses in placental mammals), or the S phase of the cell cycle (Sciamanna *et al.*, 2009; Villarreal and Witzany, 2010; Villarreal, 2016b). In these highly specialized contexts, they are replicated through signaling, which blocks the suppression of the replication process. After the function is fulfilled, a signal once again initiates the suppressor function. If the signal is missing, as in the case of regulation damage, the whole process can get out of control.

Diversity generating retroelements — with their (1) higher-order regulatory functions, (2) ability for genetic creativity and (3) capacity for innovation of new regulatory patterns and combinations — are most probably descended from retroviruses, which can easily be identified by their three essential parts: gag, pol, and env. Most endogenous retroviruses have been degraded into formerly connected domains, but they can still be recognized by retroposons or one of their three core genes (Boeke, 2003; Eickbush and Jamburuthugoda, 2008). This means that their formerly connected genomic content may be used by host organisms as single or networking modular tools for a variety of new regulatory functions (Belfort and Curcio, 2011; Witzany, 2012; Moelling *et al.*, 2017). Retroelements represent the most abundant genome invaders, especially in mammals.

8. How to Persist within an Infected Genome: Counterbalancing Addiction Modules

Addiction modules represent at least two competing genetic parasite consortia, which try to invade host genomes. The competition occurs not solely against each other, but primarily against immune functions of the host. In some cases, the competing genetic parasites and the third party, the host immune system, can together install persistent modus in which all remain as a counterbalancing module — more or less stable (Villarreal, 2009a, 2012b).

If these module-like agents are conserved into the genomic identity of the host, the former identity changes dramatically. New features are part of the host genomes which did not exist before. In such addiction module integration events, up until 100 new genes can be transferred into the host genome in a single event (Villarreal, 2005).

Addiction modules can be defined as features that consist, in general, of a stable, toxic component, which is counterbalanced by an unstable component inhibiting and suppressing the toxic component (Villarreal, 2012b). Both are necessary for transferring a feature to the host without harming the host. For example, in the case of a restriction/modification module, this means 52 restriction enzymes are counterbalanced by 52 modification enzymes. This indicates how complexly addiction modules are constructed and how difficult it can be to understand the evolution of such phenotypes (Kobayashi, 2001; Mruk and Kobayashi, 2014).

Several kinds of addiction modules are known. First of all, and most prominent, in bacterial life is the restriction/modification addiction module, which is a common feature with immune functions. One part consists of an antitoxic modification enzyme, which is an unstable beneficial (protective) agent. The counterpart consists of a toxic restriction enzyme component, which is a stable but harmful (destructive) agent. Another kind of addiction module consists of two related features. There is an antitoxic antipore toxin, which represents the unstable protective agent and a toxic component with a toxic pore, which represents the stable but destructive agent (Villarreal, 2012b; Gerdes and Wagner, 2007). A third kind of addiction module consists of the antitoxic viral immunity component and the toxic component of viral-mediated lysis. This third kind is the most obvious viral-derived immune function, because it necessarily consists of a persistent genetic parasite and an external lytic phage (Villarreal, 2012, 2016a).

Most importantly, this new information representing genetic novelty by a persistent integration of counterbalanced coded genetic elements is not the result of error replication, but a result of module-like linked genetic contents. This fundamental difference to error replication narratives proposes new nucleic acid sequence constructions by integration of larger content arrangements into a coherent syntax without destroying the already existing sequence content (Witzany, 2014).

Additionally, the basic motif of any counterregulated modules, such as known toxin/antitoxin modules, reminds us of similar structures with similar features such as the nuclease/ligase regulation as basic competence for self/non-self-differentiation within RNA stem-loop groups (Witzany, 2016), the kinein/dynein retroviral movement motif (Witzany, 2009), and the viral origin of some ribonucleoproteins (Villarreal, 2012b). Also, the telomere/telomerase-dependent genome maintenance (Witzany, 2008) and the function of centrosomes and spindles (Alliegro, 2011) most seemingly represent counterregulated addiction modules, although their origin still remains in the dark. Such counterbalancing interactions of former competing genetic parasites, now serving as a beneficial tool within host organisms, can be found generally in insertion/deletion modules and in adaptive immune systems such as VDJ of complex eukaryotes or in the bacterial CRIPR/Cas immune system (Villarreal, 2016a).

9. Genetic Novelty by Domesticated Infectious Agents

Transposable elements shape the structure and function of infected genomes (non-LTRs, LINEs, SINEs, Alus) and generate genomic variations (Slotkin and Martienssen, 2007). This means that sequence syntax may change and such changes may cause phenotypic variations. Also post-integration recombinations generate structural variations. Additionally, host cells may regulate the invading RNA consortia by restriction motifs, epigenetic methylation silencing, or by other ways to restrict retrotransposition (Richardson *et al.*, 2015). Depending on descent transposable elements may serve as regulatory elements for host cells, such as promotors, enhancers, or repressors.

Diversity-generating retroelements such as retrotransposons seem to derive from mobile self-splicing introns (a sister strand of retroviruses) together with retrons, retroplasmids, retrointrons, non-LTR and LTR retrotransposons, caulimoviruses, hepadnaviruses, and the DIRS and Bel family (Boeke 2003; Frost *et al.*, 2005). Their double role (as promotors and repressors) depends on the differences between retrotransposon families, even within the same family, whether they have co-evolved with host genomes, have been co-opted to serve in species-specific functions, and others that are not beneficial and may cause diseases (Robbez-Masson and Rowe, 2015). The evolutionarily original polymerase, the reverse transcriptase, with its dominant roles of retrotransposition and the transfer of RNA sequences into DNA sequences, plays important roles (Belfort and Curcio, 2011; Egan and Collins, 2012).

In this perspective, retrotransposons may change their function between repression and promotion of gene regulations in a highly species-specific manner without error replication, and act in a coordinated way to generate regulatory complexity for gene networks in developmental stages (Geuking *et al.*, 2009; Pathak and Nagy, 2009). If retrotransposons inactivate genes by inserting within them, or even more importantly, separating genes from a regulatory sequence, bringing new regulatory sequences to genes, or altering RNA splicing, or even RNA editing, they may generate genomic variations by incorporating themselves into the new sequences (Finnegan, 2012; Liscovitch *et al.*, 2017). According to the concerted temporal steps in the developmental processes, they may act in tissue-specific repression and activation complexity to coordinate the timely steps of tissue developments within organisms. As exapted and co-opted former infectious agents, they are essential tools of host genomes and they serve as drivers of genome plasticity and creation of novel genes (Villarreal and Witzany, 2015; Robbez-Masson and Rowe, 2015).

Additionally, all these interactional motifs play essential roles in warding off the invading genetic parasites. As they drive host evolution by genetic novelty and genomic variations, they may constantly serve as disease-causing enemies also. But some are co-opted and exapted, and they serve as essential parts in the dynamic adaptation of host immune systems (Goodier, 2016; Chuong *et al.*, 2016). Such sub-functionalization and neo-functionalization are essential characteristics in this network modularity (Kuniak and Makalowska, 2017).

Interestingly, all these invading RNA elements described in the previous sections may recombine and may even recombine with DNA viruses. This serves as the "ultimate modularity" on the genetic level (Koonin *et al.*, 2015; Koonin and Krupovic, 2017).

10. Masters of the RNA Universe: Reverse Transcriptase/RNAse H

A key agent in several aspects of biology, especially in genetics, is the diversity-generating retroelement reverse transcriptase (RT). The reverse transcriptase, derived from the ancient RNA world, and correctly termed, is the primary tool to transport information from RNA to DNA before other polymerases. RTs can be identified through the whole current RNA world activities in virions (naked RNA stem-loop consortia) and RNA viruses such as retroviruses, and are the essential tool to place RNA information within DNA-based genomes (Nakamura and Cech, 1998; Sciamanna *et al.*, 2009). In this respect, it is crucial to install persistent viral agents in host genomes.

The RTs care for virus replication cycles as well as the suppressed viral-derived inhabitants that serve as regulatory elements of the host genes. In special situations such as stress, the regulation of the host function may become weak and the original function may become virulent again. Especially if the RTs get into this modus, they may change the genetic identity of host genomes and fix it into the DNA storage medium as a part of an inheritable feature (Deng *et al.*, 2012; Spadafora, 2016). The telomerase, a subspecies of the reverse transcriptase family, cares for telomere replication at the ends of chromosomes and, therefore, is not only the key player of genome maintenance but is also a part of an immunity function that protects telomere ends from infection by genetic parasites (Nosek *et al.*, 2006; deLange, 2015; Podlevsky and Chen, 2016). This is a new role of a reverse transcriptase and it demonstrates that the RNA agents serve for both infection of the host genomes and — as coapted exaptations — immune functions against (related) infectious agents.

Reverse Transcriptase is the key player to transfer RNA sequences into DNA. This is a counterregulatory process that we can look at in embryogenesis and cell differentiation. In cancerogenesis, this gets out of control. We can look at such context-relevant loss of gene control in preimplantation of placenta, trophoblast differentiation, as well as tumor growth. In all these situations, the genetic identity producers that are silenced or conserved as regulatory tools may become virulent again, active for change, deregulation, and production of novel content arrangement. This may lead not only to genetic novelty, novel gene regulations, and changing dynamics but also to increased rates of disease, caused by deregulation of former counterbalanced networks of regulations (Spadafora, 2015). In this respect, carcinogenesis is the result of communication breakdown. Interestingly, invading cancer cells behave according to the retroviral-mediated embryological program of placenta invasion (Villarreal, 2016b). In such a case, the regulatory network may get into an unregulated, wrong direction. This means that, e.g., cell growth still remains the goal, but the stop signal does not function and the growth program goes on in an epigenetically induced deregulated way (Smith *et al.*, 2017), or invasion of the placenta into the uterus is the coordinated behavioral goal, but the invasive signaling network does not stop and spreads into other organs (metastasis) (Spadafora, 2015, 2016).

Retroviral infections are essential for evolutionary novelty, in that they serve as the main editors for new genetic identities to host genomes. The key natural engineers in this respect are RT and RNase H (Moelling and Broecker, 2015; Moelling *et al.*, 2017). They can transport new cellular functions and contribute to the host immune defense. Their high production rate of new sequences, formerly termed "error prone", renders RT–RNases H the drivers of evolutionary innovation. Integrated retroviruses and the related transposable elements (TEs) represent up to 80% of eukaryotic genomes and are also present in prokaryotes. Endogenous retroviruses not only regulate host genes but also introduce novel genes, such as the syncytins, that mediate maternal–fetal immune tolerance, and may become infectious again, under certain circumstances.

The RT and the RNase H represent the most ancient and abundant protein folds (Moelling *et al.*, 2017). RNases H most probably evolved from ribozymes in the ancient RNA world, forming ribosomes, RNA replicases, and polymerases. RT and RNases H are found in bacterial group II introns, the ancestors of transposable elements. They regulate viral replication, antiviral defence in eukaryotes and prokaryotes, splicing, DNA repair, and serve as various mediators of small regulatory RNAs. The evolution of DNA-based life, as we know it, would not have been possible without the RTs and the RNase H competencies.

11. Conclusions

In the decades where biology has been exclusively a subdiscipline of physics and chemistry, the only explanatory model to explain changes and novelties of replicated nucleotide sequences has been the "error replication" (mutations) narrative. This was the consequence of Erwin Schroedinger's answer "Life is physics and chemistry" to his question "What is Life". Within this mechanistic paradigm, natural genome editing, i.e., competent natural genetic content arrangement by cooperative consortia of RNA networks and viruses, is not part of theories and concepts in biology. The main problem of the error replication (mutations) narrative is that it cannot explain sufficiently these new and rich empirical data of the variety of RNA consortia and viruses as natural genome editors. No doubt they can insert and delete, can integrate themselves in complex genetic content without destroying relevant protein-coding content, arrange, rearrange, cut and paste, copy and paste, and serve as multiple reuseable modular tools in the genetic content architecture. In the old error replication narrative, all these consortia remain as randomly derived results of statistical events by genetic drift. While the old paradigm cannot integrate these new empirical data, the natural genome editing paradigm is essentially based on all these data. Current knowledge identified an abundance of processes with a variety of small interacting agents that actively edit genetic code sequences by different behavioral motifs; these can all be combined in a module-like manner. Natural genome editing is a technique that actively initiates "ultimate modularity" to generate genetic novelty and genomic variations, which can be fulfilled by cooperative interacting networks of viruses and subviral RNA consortia.

Funding

This research did not receive any specific grant from funding agencies in the public, commercial, or not-for-profit sectors.

References

Agirrezabala, X., Samatova, E., Klimova, M., Zamora, M., Gil-Carton, D., Rodnina, M. V. and Valle, M. (2017). Ribosome rearrangements at the onset of translational bypassing, *Sci. Adv.* 3(6), e1700147.

Alliegro, M. C. (2011). The centrosome and spindle as a ribonucleoprotein complex, *Chromosome Res.* 19, 367–376.

Atkins, J. F., Loughran, G., Bhatt, P. R., Firth, A. E. and Baranov, P. V. (2016). Ribosomal frameshifting and transcriptional slippage: From genetic steganography and cryptography to adventitious use, *Nucleic Acids Res.* 44, 7007–7078.

Barlow, D. P. (2011). Genomic imprinting: A mammalian epigenetic discovery model, *Annu. Rev. Genet.* 45, 379–403.

Belfort, M., Curcio, M. J. and Lue, M. J. (2011). Telomerase and retrotransposons: Reverse transcriptases that shaped genomes, *Proc. Natl. Acad. Sci. USA* 108, 20304–20310.

Blaze, J. and Roth, T. L. (2012). Epigenetic mechanisms in learning and memory, *Wiley Interdiscip. Rev. Cogn. Sci.* 4, 105–115.

Boeke, J. D. (2003). The unusual phylogenetic distribution of retrotransposons: A hypothesis, *Genome Res.* 13, 1975–1983.

Briones, C., Stich, M. and Manrubia, S. (2009). The dawn of the RNA World: Toward functional complexity through ligation of random RNA oligomers, *RNA* 15, 743–749.

Brunel, C., Marquet, R., Romby, P. and Ehresmann, C. (2002). RNA loop–loop interactions as dynamic functional motifs, *Biochimie* 84, 925–944.

Cao, S. and Chen, S. J. (2011). Structure and stability of RNA/RNA kissing complex: With application to HIV dimerization initiation signal, *RNA* 17, 2130–2143.

Cao, S., Xu, X. and Chen, S. J. (2014). Predicting structure and stability for RNA complexes with intermolecular loop-loop base-pairing, *RNA* 20, 835–845.

Carbonell, A., Flores, R. and Gago, S. (2012). Hammerhead ribozymes against virus and viroid RNAs, in *From Nucleic Acids Sequences to Molecular Medicine*, Erdmann V. A. and Barciszewski, J. (eds.), Springer, pp. 411–427.

Cech, T. R. (2012). The RNA worlds in context, *Cold Spring Harb. Perspect. Biol.* 4(7), a006742.

Cech, T. R. and Steitz, J. A. (2014). The noncoding RNA revolution — Trashing old rules to forge new ones, *Cell* 157, 77–94.

Chuong, E. B., Elde, N. C. and Feschotte, C. (2016). Regulatory evolution of innate immunity through co-option of endogenous retroviruses, *Science* 351, 1083–1087.

de Lange, T. (2015). A loopy view of telomere evolution, *Front Genet.* 6, 321.

Deng, Z., Wang, Z. and Lieberman, P. M. (2012). Telomeres and viruses: Common themes of genome maintenance, *Front. Oncol.* 2, 201.

Denzler, R., Agarwal, V., Stefano, J., Bartel, D. and Stoffel, M. (2014). Assessing the ceRNA hypothesis with quantitative measurements of miRNA and target abundance, *Mol. Cell* 54, 766–776.

Dinman, J. D. (2006). Programmed ribosomal frameshifting goes beyond viruses, *Microbe Wash. DC.* 1, 521–527.

Doudna, J. A., Cormack, B. P. and Szostak, J. W. (1989). RNA structure, not sequence, determines the 5′ splice-site specificity of a group I intron, *Proc. Natl. Acad. Sci. U.S.A.* 86, 7402–7406.

Egan, E. D. and Collins, K. (2012). Biogenesis of telomerase ribonucleoproteins, *RNA* 18, 1747–1759.

Eickbush, T. H. and Jamburuthugoda, V. K. (2008). The diversity of retrotransposons and the properties of their reverse transcriptases, *Virus Res.* 134, 221–234.

Eigen, M. (1971). Selforganization of matter and the evolution of biological macromolecules, *Naturwissenschaften* 58, 465–523.

Eigen, M. and Schuster, P. (1978). The hypercycle. A principle of natural self-organization. Part B: The abstract hypercycle, *Naturwissenschaften* 65, 7–41.

Finnegan, D. J. (2012). Retrotransposons, *Curr. Biol.* 22, 432–437.

Frost, L. S., Leplae, R., Summers, A. O. and Toussaint, A. (2005). Mobile genetic elements: The agents of open source evolution, *Nat. Rev. Microbiol.* 3, 722–732.

Fujishima, K. and Kanai, A. (2014). tRNA gene diversity in the three domains of life, *Front. Genet.* 5, 142.

Gebetsberger, J., Wyss, L., Mleczko, A. M., Reuther, J. and Polacek, N. (2016). A tRNA-derived fragment competes with mRNA for ribosome binding and regulates translation during stress, *RNA Biol.* 28, 1–10.

Gerdes, K. and Wagner, E. G. (2007). RNA antitoxins, *Curr. Opin. Microbiol.* 10, 117–124.

Geuking, M. B., Weber, J., Dewannieux, M., Gorelik, E., Heidmann, T., Hengartner, H., Zinkernagel, R. M. and Hangartner, L. (2009). Recombination of retrotransposon and exogenous RNA virus results in nonretroviral cDNA integration, *Science* 323, 393–396

Giedroc, D. P. and Cornish, P. V. (2009). Frameshifting RNA pseudoknots: Structure and mechanism, *Virus Res.* 139, 193–208.

Goodier, J. L. (2016). Restricting retrotransposons: A review, *Mob. DNA* 7, 16.

Göringer, H. U. (ed.) (2008). *RNA Editing*, Springer.

Gott, J. M. (2003). Expanding genome capacity via RNA editing, *C. R. Biol.* 326, 901–908.

Hayden, E. J. and Lehman, N. (2006). Self-assembly of a group I intron from inactive oligonucleotide fragments, *Chem. Biol.* 13, 909–918.

Higgs, P. G. and Lehman, N. (2015). The RNA world: Molecular cooperation at the origins of life, *Nat. Rev. Genet.* 16, 7–17.

Homann, M. (2008). Editing reactions from the perspective of RNA structure, in *RNA Editing*, Göringer, H. U. (ed.), Springer, pp. 1–32.

House, A. E. and Lynch, K. W. (2008). Regulation of alternative splicing: More than just the ABCs, *J. Biol. Chem.* 283, 1217–1221.

Huda, A. and Jordan, I. K. (2009). Epigenetic regulation of mammalian genomes by transposable elements, *Ann. N. Y. Acad. Sci.* 1178, 276–284.

Huda, A., Mariño-Ramírez, L. and Jordan, I. K. (2010). Epigenetic histone modifications of human transposable elements: Genome defense versus exaptation, *Mob. DNA* 1, 2.

Kanai, A. (2015). Disrupted tRNA genes and tRNA fragments: A perspective on tRNA gene evolution, *Life (Basel)* 5, 321–331.

Keam, S. P. and Hutvagner, G. (2015). tRNA-derived fragments (tRFs): Emerging new roles for an ancient RNA in the regulation of gene expression, *Life (Basel)* 5, 1638–1651.

Kim, S. and Kaang, B. K. (2017). Epigenetic regulation and chromatin remodeling in learning and memory, *Exp. Mol. Med.* 49, e281.

Kobayashi, L. (2001). Behavior of restriction-modification system as selfish mobile elements and their impact on genome evolution, *Nucleic Acid Res.* 29, 3742–3756.

Koonin, E. V. and Krupovic, M. (2017). Polintons, virophages and transpovirons: A tangled web linking viruses, transposons and immunity, *Curr. Opin. Virol.* 25, 7–15.

Koonin, E. V., Dolja, V. V. and Krupovic, M. (2015). Origins and evolution of viruses of eukaryotes: The ultimate modularity, *Virology* 479–480, 2–25.

Kubiak, M. R. and Makałowska, I. (2017). Protein-coding genes' retrocopies and their functions, *Viruses* 9(4), pii: E80.

Liscovitch-Brauer, N., Alon, S., Porath, H. T., Elstein, B., Unger, R., Ziv, T., Admon, A., Levanon, E. Y., Rosenthal, J. J. and Eisenberg, E. (2017). Trade-off between transcriptome plasticity and genome evolution in cephalopods, *Cell* 169, 191–202.

Manrubia, S. C. and Briones, C. (2007). Modular evolution and increase of functional complexity in replicating RNA molecules, *RNA* 13, 97–107.

Marshall, P. and Bredy, T. W. (2016). Cognitive neuroepigenetics: The next evolution in our understanding of the molecular mechanisms underlying learning and memory? *NPJ Sci. Learn.* 1, 16014.

Martinez, G., Choudury, S. G. and Slotkin, R. K. (2017). tRNA-derived small RNAs target transposable element transcripts, *Nucleic Acids Res.* 45, 5142–5152.

Matlin, A. J. and Moore, M. J. (2007). Spliceosome assembly and composition, *Adv. Exp. Med. Biol.* 623, 14–35.

Mattick, J. S. (2001). Non-coding RNAs: The architects of eukaryotic complexity, *EMBO Rep.* 2, 986–991.

Mattick, J. S. (2003). Challenging the dogma: The hidden layer of noncoding RNAs in complex organisms, *BioEssays* 25, 930–939.

Mattick, J. S. (2009). Deconstructing the dogma: A new view of the evolution and genetic programming of complex organisms, *Ann. Acad, N. Y. Sci.* 1178, 29–46.

Mattick, J. S. (2010). RNA as the substrate for epigenome–environment interactions: RNA guidance of epigenetic processes and the expansion of RNA editing in animals underpins development, phenotypic plasticity, learning, and cognition, *BioEssays* 32, 548–552.

Mercer, T. R. and Mattick, J. S. (2013). Structure and function of long noncoding RNAs in epigenetic regulation, *Nat. Struct. Mol. Biol.* 20, 300–307.

Moelling, K. and Broecker, F. (2015). The reverse transcriptase-RNase H: From viruses to antiviral defense, in *DNA Habitats and Their RNA Inhabitants*, Witzany, G. (ed.), Wiley & Sons, pp. 126–135.

Moelling, K., Broecker, F., Russo, G. and Sunagawa, S. (2017). RNase H as gene modifier, driver of evolution and antiviral defense, *Front. Microbiol.* doi: 10.3389/fmicb.2017.01745.

Mruk, I. and Kobayashi, I. (2014). To be or not to be: Regulation of restriction-modification systems and other toxin-antitoxin systems, *Nucleic Acids Res.* 42, 70–86.

Nakamura, T. M. and Cech, T. R. (1998). Reversing time: Origin of telomerase, *Cell* 92, 587–590.

Nosek, J., Kosa, P. and Tomaska, L. (2006). On the origin of telomeres: A glimpse at the pre-telomerase world, *BioEssays* 28, 182–190.

Pathak, K. B. and Nagy, P. D. (2009). Defective interfering RNAs: Foes of viruses and friends of virologists, *Viruses* 1, 895–919.

Peselis, A. and Serganov, A. (2014). Structure and function of pseudoknots involved in gene expression control, *Wiley Interdiscip. Rev. RNA* 5, 803–822.

Podlevsky, J. D. and Chen, J. J. L. (2016). Evolutionary perspectives of telomerase RNA structure and function, *RNA Biol.* 13, 720–732.

Pyle, A. M. and Lambowitz, A. M. (2006). Group II introns: Ribozymes that splice RNA and invade DNA, in *The RNA World*, Gesteland, R. F., Cech, T. R. and Atkin, J. F. (eds.), 3rd edn., Cold Spring Harbor Laboratory Press, pp. 469–506.

Qureshi, I. A. and Mehler, M. F. (2012). Emerging roles of non-coding RNAs in brain evolution, development, plasticity and disease, *Nat. Rev. Neurosci.* 13, 528–541.

Richardson, S. R., Doucet, A. J., Kopera, H. C., Moldovan, J. B., Garcia-Perez, J. L. and Moran, J. V. (2015). The influence of LINE-1 and SINE retrotransposons on mammalian genomes, *Microbiol. Spectr.* 3(2), MDNA3-0061-201499.

Robbez-Masson, L. and Rowe, H. M. (2015). Retrotransposons shape species-specific embryonic stem cell gene expression, *Retrovirol.* 12, 45.

Rohwer, F., Youle, M., Maughan, H. and Hisakawa, N. (2014). *Life in Our Phage World*, Wholon.

Ryan, F. (2009). *Virolution*, Collins.

Salmena, L., Poliseno, L., Tay, Y., Kats, L. and Pandolfi, P. P. (2011). A ceRNA hypothesis: The rosetta stone of a hidden RNA language? *Cell* 146, 353–358.

Salomon, K., Appel, B. and Müller, S. (2012). RNA self-ligation: From oligonucleotides to full length ribozymes, *Biochimie* 94, 1457–1463.

Samatova, E., Konevega, A. L., Wills, N. M., Atkins, J. F. and Rodnina, M. V. (2014). High-efficiency translational bypassing of non-coding nucleotides specified by mRNA structure and nascent peptide, *Nat. Commun.* 5, 4459.

Schuster, P. (2011). Mathematical modeling of evolution. Solved and open problems, *Theory Biosci.* 130, 71–89.

Sciamanna, I., Vitullo, P., Curatolo, A. and Spadafora, C. (2009). Retrotransposons, reverse transcriptase and the genesis of new genetic information, *Gene* 448, 180–186.

Seligman, H. and Raoult, D. (2016). Univying view of stem-loop hairpin RNA as origin of current and ancient parasitic and non-parasitic RNAs, including in giant viruses, *Curr. Opin. Microbiol.* 31, 1–8.

Shapiro, J. A. (2004). A 21st century view of evolution: Genome system architecture, repetitive DNA, and natural genetic engineering, *Gene* 345, 91–100.

Shapiro, J. A. (2009). Revisiting the central dogma in the 21st century, *Ann. N.Y. Acad. Sci.* 1178, 6–28.

Shapiro, J. A. (2014). Epigenetic control of mobile DNA as an interface between experience and genome change, *Frontiers Genet.* 5, 1–16.

Slotkin, R. K. and Martienssen, R. (2007). Transposable elements and the epigenetic regulation of the genome, *Nat. Rev. Genet.* 8, 272–285.

Smit, S., Yarus, M. and Knight, R. (2006). Natural selection is not required to explain universal compositional patterns in rRNA secondary structure categories, *RNA* 12, 1–14.

Smith, H. C. (2008). Editing informational content of expressed DNA sequences and their transcripts, in *RNA Editing*, Göringer, H. U. (ed.), Springer, pp. 249–265.

Smith, Z. D., Shi, S., Gu, H., Donaghey, J., Clement, K., Cacchiarelli, D., Gnirke, A., Michor, F. and Meissner, A. (2017). Epigenetic restriction of extraembryonic lineages mirrors the somatic transition to cancer, *Nature* 549, 543–547.

Spadafora, C. (2015). A LINE-1–encoded reverse transcriptase–dependent regulatory mechanism is active in embryogenesis and tumorigenesis, *Ann. N.Y. Acad. Sci.* 1341, 164–171.

Spadafora, C. (2016). Soma to germline inheritance of extrachromosomal genetic information via a LINE-1 reverse transcriptase-based mechanism, *BioEssays* 38, 726–733.

Spadaro, P. A. and Bredy, T. W. (2012). Emerging role of non-coding RNA in neural plasticity, cognitive function, and neuropsychiatric disorders, *Front. Genet.* 3, 132.

Staple, D. W. and Butcher, S. E. (2005). Pseudoknots: RNA structures with diverse functions, *PLoS Biol.* 3(6), e213.

Stedman, K. M. (2013). Mechanisms for RNA capture by ssDNA viruses: Grand theft RNA, *J. Mol. Evol.* 76, 59–364.

Stedman, K. M. (2015). Deep recombination: RNA and ssDNA virus genes in DNA virus and host genomes, *Annu. Rev. Virol.* 2, 203–217.

Stoddard, B. L. (2011). Homing endonucleases: From microbial genetic invaders to reagents for targeted DNA modification, *Structure* 19, 7–15.

Tholstrup, J., Oddershede, L. B. and Sørensen, M. A. (2012). mRNA pseudoknot structures can act as ribosomal roadblocks, *Nucleic Acids Res.* 40, 303–313.

Toor, N., Keating, K. S., Taylor, S. D. and Pyle, A. M. (2008). Crystal structure of a self-spliced group II intron, *Science* 320, 77–82.

Turner, B. M. (2002). Cellular memory and the histone code, *Cell* 111, 285–291.

Tycowski, K. T., Guo, Y. E., Lee, N., Moss, W. N., Vallery, T. K., Xie, M. and Steitz, J. A. (2015). Viral noncoding RNAs: More surprises, *Genes Dev.* 29, 567–584.

Vaidya, N., Manapat, M. L., Chen, I. A., Xulvi-Brunet, R., Hayden, E. J. and Lehman, N. (2012). Spontaneous network formation among cooperative RNA replicators, *Nature* 491, 72–77.

Villarreal, L. P. (2004). Can viruses make us human? *Proc. Am. Phil. Soc.* 148, 296–323.

Villarreal, L. P. (2005). *Viruses and the Evolution of Life*, ASM Press.

Villarreal, L. P. (2009). *Origin of Group Identity: Viruses, Addiction and Cooperation*, Springer.

Villarreal, L. P. (2009). The source of self: Genetic parasites and the origin of adaptive immunity, *Ann. N.Y. Acad. Sci.* 1178, 194–232.

Villarreal, L. P. (2011). Viral ancestors of antiviral systems, *Viruses* 3, 1933–1958.

Villarreal, L. P. (2012a). Viruses and host evolution: Virus-mediated self identity, in *Self and Non-Self*, Lopez-Larrea C. (ed.), Landes Bioscience and Springer Science Business Media, pp. 185–217.

Villarreal, L. P. (2012b). The Addiction Module as a Social Force, in *Viruses: Essential Agents of Life*, Witzany, G. (ed.), Springer, pp. 107–145.

Villarreal, L. P. (2015). Force for ancient and recent life: Viral and stem-loop RNA consortia promote life, *Ann. N.Y. Acad. Sci.* 1341, 25–34.

Villarreal, L. P. (2016a). Persistent virus and addiction modules: An engine of symbiosis, *Curr. Opin. Microbiol.* 31, 70–79.

Villarreal, L. P. (2016b). Viruses and the placenta: The essential virus first view, *APMIS* 124, 20–30.

Villarreal, L. P. and Witzany, G. (2010). Viruses are essential agents within the roots and stem of the tree of life, *J. Theor. Biol.* 262, 698–710.

Villarreal, L. P. and Witzany, G. (2013a). The DNA habitat and its RNA inhabitants: At the dawn of RNA sociology, *Genom. Ins.* 6, 1–12.

Villarreal, L. P. and Witzany, G. (2013b). Rethinking quasispecies theory: From fittest type to cooperative consortia, *World J. Biol. Chem.* 4, 79–90.

Villarreal, L. P. and Witzany, G. (2015). When competing viruses unify: Evolution, conservation, and plasticity of genetic identities, *J. Mol. Evol.* 80, 305–318.

Villarreal, L. P. and Witzany, G. (2018). Editorial: Genome invading RNA networks, *Frontiers Microbiol.* 9, 581, doi: 10.3389/fmicb.2018.00581.

Wang, Q., Lee, I., Ren, J., Ajay, S. S., Lee, Y. S. and Bao, X. (2013). Identification and functional characterization of tRNA-derived RNA fragments (tRFs) in respiratory syncytial virus infection, *Mol. Ther.* 21, 368–379.

Witzany, G. (1995). From the "logic of the molecular syntax" to molecular pragmatism. Explanatory deficits in Manfred Eigen's concept of language and communication, *Evol. Cog.* 1, 148–168.

Witzany, G. (2008). The viral origins of telomeres, telomerases and their important role in eukaryogenesis and genome maintenance, *Biosemiotics* 1, 191–206.

Witzany, G. (2009). Noncoding RNAs: Persistent viral agents as modular tools for cellular needs, *Ann. N.Y. Acad. Sci.* 1178, 244–267.

Witzany, G. (2012). From molecular entities to competent agents: Viral infection-derived consortia act as natural genetic engineers, in Witzany, G. (ed.) *Viruses: Essential Agents of Life*, Springer, pp. 407–419.

Witzany, G. (2014a). RNA sociology: Group behavioral motifs of RNA consortia, *Life(Basel)* 4, 800–818.

Witzany, G. (2014b). Language and communication as universal requirements for life, in *Astrobiology: An Evolutionary Approach*, Kolb, V. (ed.), CRC Press, pp. 349–369.

Witzany, G. (2015). Life is physics and chemistry and communication, *Ann. N.Y. Acad. Sci.* 1341, 1–9.

Witzany, G., (ed.) (2015). *DNA Habitats and Their RNA Inhabitants*, Wiley & Sons.

Witzany, G. (2016). Crucial steps to life: From chemical reactions to code using agents, *BioSystems* 140, 49–57.

Witzany, G. (2017). Two genetic codes: Repetitive syntax for active non-coding RNAs; non-repetitive syntax for the DNA archives, *Comm. Integr. Biol.* 10(2), e1297352.

Witzany, G. and Baluska, F. (2012). Life's code script does not code itself. The machine metaphor for living organisms is outdated, *EMBO Rep.* 13, 1054–1056.

Yablonovitch, A. L., Deng, P., Jacobson, D. and Li, J. B. (2017). The evolution and adaptation of A-to-I RNA editing, *PLoS Genet.* 13(11), e1007064.

Yang, L. (2015). Splicing noncoding RNAs from the inside out, *Wiley Interdiscip. Rev. RNA* 6, 651–660.

Yang, C., Wu, D., Gao, L., Liu, X., Jin, Y., Wang, D., Wang, T. and Li, X. (2016). Competing endogenous RNA networks in human cancer: Hypothesis, validation, and perspectives, *Oncotarget.* 7, 13479–13490.

Zinad, H. S., Natasya, I. and Werner, A. (2017). Natural antisense transcripts at the interface between host genome and mobile genetic elements, *Front Microbiol.* 8, 2292.

PART III

Cognition and Intelligence in Natural and Artificial Systems

Chapter 14

Information Processing by Structural Machines

Mark Burgin

University of California, Los Angeles, CA, USA

mburgin@math.ucla.edu

In this chapter, we explore information processing by the model of computation called structural machine. Structures contain information as matter contains energy and structural machines work with structures of arbitrary order transforming not only elements of these structures or the content of these elements, as conventional models of computation do, but also relations of different orders in the processed structures. This allows achieving higher flexibility and efficiency in comparison with regular models of computation, including both conventional and unconventional computing systems. We pay special attention to information processes such as observation, information transmission and information reception. To show superior flexibility of structural machines, we demonstrate that they can simulate such advanced computational models as Kolmogorov algorithms (Theorem 1) and simple limit Turing machines (Theorem 2). In essence, many important and popular models of computation are special cases of structural machines. To illustrate higher efficiency of structural machines, we demonstrate that they can solve the ordering problem with much lower time complexity than conventional algorithms (Theorem 3). Being structurally universal abstract automata, structural machines can directly work with molecular structures, with structures studied and utilized in the topological quantum field theory and with structures of quantum information such as qubits.

1. Introduction: From Data Processing to Information Processing

Computers and other computing devices transform input data into output data. This situation is adequately reflected by mathematical models of automata, algorithms and computations, when input strings of symbols are transformed into output strings of symbols. Why then is it assumed that computers and other computing devices process information?

The answer to this question is given by the quantum theory of knowledge (Burgin, 1997, 2004a, 2016b), namely, data contain two types of information — information for material and abstract computing devices such as computers and Turing machines, and information for people. Information in data for computing devices together with computing hardware and software (programs) defines the process of computation when input data and data that are already in the device bring about an operation of the device, e.g., state transition. Information for people is explicated (observed and comprehended) by means of interpretation when raw data are interpreted and attributed becoming knowledge as it is explained in Burgin (2016b, Chapters 4 and 8).

For instance, when a Turing machine U or a simple inductive Turing machine W taking an input string x gives the resulting symbol 1, that is data transformation. However, if x contains a description of a Turing machine T and a word z, while the machine U (W) is programmed for solving the Halting Problem (the Definability Problem), then the result (output) 1 means that given the input z, the Turing machine T halts giving the result of computation (Burgin, 2010).

To extend our understanding and further develop information processing, researchers constructed various mathematical models of automata, algorithms and computations building the foundations of theoretical computer science. The most advanced of these models is *structural machine* because a structural machine can process arbitrary structures as data, while any other model of automata, algorithms and computations processes only a limited type of data structures (Burgin and Adamatzky, 2017a,b). For instance, Turing machines with one tape work only with linear structures in the form of words or strings. In addition, the hardware of structural machines is more flexible and adaptable than the hardware of other theoretical models of automata, algorithms and computations. As a result, structural machines allow adequate modeling not only of conventional computers, networks and other automata, algorithms and computations but also of different devices, such as Physarum machines (Adamatzky, 2010), performing unconventional computations as it is demonstrated in Burgin and Adamatzky (2017a,b).

An important peculiarity of structural machines is that they work not with data stored in and placed into the memory of a computing device as the majority of theoretical models of automata, algorithms and computations do. In contrast to this, structural machines process structures in the functional space. Note that the memory of the computing device can be a part of the functional space. This feature of structural machines provides additional possibilities for modeling different processes. On the one hand, it is possible to represent the memory of a computing device as one structure as it is done in Kolmogorov algorithms (cf. Kolmogorov, 1953; Uspensky and Semenov, 1993) or as a functional space, which contains many structures and in a particular case can coincide with a structure used by Kolmogorov algorithms. The memory of contemporary computers can be treated as such a functional space. On the other hand, we can interpret the functional space as the physical space and explore its computing abilities in the context of info-computationalism (Dodig-Crnković, 2006, 2011, 2012; Dodig-Crnković and Mueller, 2011) and pancomputationalism (cf., e.g., Zuse, 1969; Fredkin, 1990; Wolfram, 2002; Lloyd, 2006; Piccinini, 2007) when the universe is treated as a huge computational structure or a network of computational elements and processes, which perform computations following fundamental physical laws.

Here, we further develop the theory of structural machines exploring capabilities of structural machines in information processing, knowledge acquisition and data transformation. The chapter is organized as follows. In Section 2, after Section 1, an informal description of structural machines and their functioning is given. It is explained that while a universal Turing machine is universal in the class of all Turing machines with a given alphabet and tapes, e.g., three two-dimensional tapes, a structural machine is universal in the class of all computing automata. As a result, all known abstract automata, prototype algorithms and other models of computation become special cases of structural machines. In Section 3, we give a rigorous mathematical description of structural machines and study their properties delineating characteristics and parameters of processors. We pay special attention to information processes such as observation, information transmission and information reception. In Section 4, we study relations between structural machines and other basic computational models. In particular, we show how structural machines model the Kolmogorov algorithm (Theorem 1) and limit Turing machines (Theorem 2) demonstrating (Theorem 3) that structural machines are rather efficient allowing data ordering with logarithmic complexity $O(\log n)$, while the complexity of conventional algorithms is $O(n \cdot \log n)$. In Section 5, we discuss the obtained results and suggest future directions for research.

2. What Structural Machines are and What their Purpose is

The introduction to structural machines has three objectives:

(1) Elaboration of a more adequate model of contemporary computers, which work with high-level programming languages.
(2) Formation of mathematical tools for transition from computing with data to knowledge processing in the framework of artificial intelligence (AI).
(3) Construction of a theoretical model for natural computation.

Considering the first objective, we see that high-level programming languages allow working with diverse data types representing a variety of data structures. For instance, Java has such data types as stacks, queues, lists, linked lists, double-ended queues, vectors, trees and priority queues (Goodrich and Tamassia, 2001). At the same time, theoretical models of computation, such as finite automata, Turing machines or neural networks, work with a fixed data type. For instance, finite automata and Turing machines with one one-dimensional tape each work with words, while Turing machines with three two-dimensional tapes work with such data type as three two-dimensional arrays. The most advanced, with respect to data structures, model of automata — Kolmogorov algorithm — works with graphs.

Thus, there is a necessity to have a model of computing that can operate with a variety of data structures of different types. Structural machines solve this problem because a structural machine is exactly such a model as it works with arbitrary structures. This feature makes structural machines a structurally universal model of automata, algorithms and computations. As a result, any data and knowledge structures processed by some abstract automata can also be processed by a relevant class of structural machines.

Considering the second objective, it is necessary to take into account that knowledge is a kind of structures (Burgin, 2016b) and knowledge processing is structure transformation (Burgin *et al.*, 2017). Thus, to work directly with knowledge, it is necessary to be able to process structures. Only in such a way would it be possible to develop high-level AI. As a result, there is a necessity to have a model of computing that can operate with knowledge structures. Structural machines solve this problem because they work with arbitrary structures.

Besides, actual AI has to be able to efficiently carry out research in different areas, including mathematics. According to the prevailing opinion, mathematics is a science of structures (Bourbaki, 1993; Burgin,

1998; Resnick, 1997). Contemporary computers can efficiently operate with material symbols and, thus, do mathematics in the symbolic form. However, mathematical systems are much more complex structures and to efficiently operate with them, it is necessary to have much more advanced computing systems, for which structural machines provide a guiding blueprint.

As a rule, pictures can better represent complex structures than symbolic systems. This explains why pictures appear throughout mathematical history, being central for visualizing insights and for motivating proofs in many mathematical areas, especially in geometry, topology, algebra and combinatorics. They extend from ancient work in the schools of Euclid and Pythagoras to modern ideas in particle physics, topological quantum field theory (TQFT) and category theory (Jaffe and Liu, 2018).

That is why mathematicians started developing picture languages, such as the three-dimensional quon language, with the goal to support image reasoning in mathematics and unify ideas from different subjects (Liu *et al.*, 2017). Structural machines allow efficient realization of picture languages as they directly work with various structures.

What is also important for information processing in general and artificial intelligence in particular is efficiency and complexity of computation. As it is demonstrated in this work and in the previous publications (Burgin and Adamatzky (2017a,b), structural machines can simulate the main computing automata, at least with the same efficiency and the same or even smaller complexity, while in many cases outperforming traditional models of computation.

Finally, coming to the third objective we see that natural computation involves even more diverse structures in comparison with computers and programming languages (Dodig-Crnković, 2011). To correctly model natural computation, it is necessary to provide essential flexibility for processed data structures and corresponding data types. Due to their features, structural machines also solve this problem because they work accurately with arbitrary structures in general, being able to directly operate geometrical and topological objects, such as molecular structures, structures from the TQFT and structures of quantum information (Jaffe and Liu, 2018; Liu *et al.*, 2017). It was demonstrated that structural machines provide efficient and adequate models for biological and chemical computers (Burgin and Adamatzky, 2017a,b).

By its structure, a structural machine M has three basic components: the *control device*, which organizes functioning of the machine; the *processor*, which performs computations; and *functional space* where computations with the prescribed types of structures are taking place. In turn, the functional space consists of three components: the *input space*, the *output space* and the *processing space.* The simplest control device changes the states of M, while a more advanced control device can perform information exchange between different components of M. The processor of M processes information by structure transformation in the functional space, which is a structure itself and contains other structures. For instance, in a Turing machine T, its head h is the processor and its tape is the functional space in which the input, output and processing spaces coincide (Burgin, 2005). It is important to understand that basic components of a structural machine can have rather sophisticated structures. For instance, the input, output and processing spaces can be arbitrary graphs or labeled graphs.

The structural machine M functions in the following way. It is assumed that at the beginning of functioning, the input of the machine M is in the input space while two other spaces are empty. As the first step, the machine M rewrites input from the input space to the processing space and continues working with structures there. In the recursive mode, the machine M obtains the final result, rewrites it into the output tape and informs the user, e.g., by halting, that the result is obtained. In the inductive mode, while working with a structure in the processing space, the machine M sends some results to the output tape from time to time and these outputs determine the final result (Burgin, 2015).

In this chapter, we also formalize the concept of locality in computation, which means that computing operations are always performed with data represented only in computer memory. To do this, we assume that each processor p is situated in the structure R processed by its structural machine. The place occupied by p is called the topos T_p of p. Locality of operation means that only elements from some neighborhood of T_p, which is called the observation zone, can impact functioning of p and p can transform (change) only the part of R that belongs to the operation zone of the processor p. For instance, in a Turing machine T, its head h is the processor for which its topos, observation zone and operation zone coincide with the cell of the machine's tape where the head h is situated.

A natural condition for these zones is that both observation and operation zones of the processor p contain the topos T_p of p. However, this is not always the case because the processor p can be at one place, say, Q, of the functional space transforming a very different part W of the functional space, though in such situations it is always possible to include the topos T_p in both observation and operation zones.

An important characteristic/parameter of a unit processor p is its transition zone, which includes all topoi where p can move in one step from its present topos T_p. For instance, the transition zone of the head h of a classical Turing machine T consists of the cell where h is situated and two of its neighbors — one cell from the left and one cell from the right. However, there are representations of Turing machines in which the head h has to move in each step (Burgin, 2005). In this case, the transition zone of the head h consists only of two neighbors of the cell where h is situated.

Sometimes it is useful to consider also the *prohibited operation zone* and *inaccessible observation zone*. In this case, the processor p is prohibited to make any changes in its prohibited operation zone while any parts (elements) from the inaccessible observation zone of p cannot influence functioning of p. Definitions imply the following relations between zones:

- The *prohibited operation zone* of p is the complement of the operation zone of p.
- The *inaccessible observation zone* of p is the complement of the observation zone of p.

At the same time, the processor p of some structural machines called multiprocessor structural machines can include an arrangement of unit processors, which can be distributed through the whole structure R while each of them operates locally. A cellular automaton C gives an example of such a structural machine. In C, unit processors are finite automata, which cover the whole structure processed by the cellular automaton C (cf., e.g., Burgin, 2005).

Another example of a multiprocessor structural machine is an artificial neural network. In it, artificial neurons are unit processors. Moreover, unit processors of a structural machine can function not only as first-order (linear) neurons but also as high-order (nonlinear) neurons allowing the machine to model high-order and heterogeneous neural networks (Grossberg, 1988; Siegelman, 1999).

Note that some unit processors in structural machines can play the role of oracles with respect to other unit processors, as it is explained in the mathematical theory of oracles (Burgin, 2017). Consequently, structural machines can naturally model different automata with oracles, such as Turing machines with oracles or evolutionary Turing machines with oracles, as well as various network structures with oracles, such as DIME (Mikkilineni *et al.*, 2015). In a similar way, some unit processors can be agents of other unit processors. This allows utilization of structural machines as mathematical models of agent technology, which has become very popular in the software industry.

In essence, many important and popular models of computation are special cases of structural machines. Examples are all kinds of Turing machines, including Turing machines with multidimensional tapes, inductive Turing machines, limit Turing machines (Burgin, 2005), alternating Turing machines (Chandra *et al.*, 1981), evolutionary Turing machines (Burgin and Eberbach, 2009) as well as register machines, random access machines, cellular automata and relational machines (Abiteboul *et al.*, 1994, 1997, 1998).

After introducing a formal mathematical description of structural machines and related concepts, we study their properties, delineating characteristics and parameters of processors in order to construct an adequate model of physical computing devices, such as networks, computers or calculators. Some of these properties are described by axioms, while others are deduced from definitions and axioms. In addition, we introduce and explore various forms of processors as basic mechanisms of computation in structural machines. To show computing power of structural machines, we prove that they can model such advanced computational models as Kolmogorov algorithms and limit Turing machines.

3. Organization and Functioning of Structural Machines

There is a diversity of types, kinds and forms of structural machines, which depend on different parameters. One of these parameters is the type of structures (data) processed by structural machines. Data structures determine orders of structural machines. Here we study structural machines of the first order, which work with first-order structures, and structural machines of the second order, which work with first-order and second-order structures.

Definition 1 (Burgin, 2012). A *first-order structure* is a triad of the form

$$\boldsymbol{A} = (A, r, \mathsf{R}),$$

where A is a set, which is called the *substance* of the structure $\boldsymbol{A}$ and consists of elements of the structure $\boldsymbol{A}$, which are called *structure elements* of the $\boldsymbol{A}$; r is the *incidence relation*, which connects groups of elements from A with the names of relations from R; and R is a set, which is called the *arrangement* of the structure $\boldsymbol{A}$ and consists of relations between elements from A in the structure $\boldsymbol{A}$. These relations have the first order and are called *structure relations* of the structure $\boldsymbol{A}$.

For instance, if $\boldsymbol{R}$ is an n-ary relation from R and $a_1, a_2, a_3, \ldots, a_n$ are elements from A, then the expression $r(R; a_1, a_2, a_3, \ldots, a_n)$ means that the elements $a_1, a_2, a_3, \ldots, a_n$ belong to the relation $\boldsymbol{R}$ with the name R, i.e., the incidence relation r connects the elements $a_1, a_2, a_3, \ldots, a_n$ with R if they belong to $\boldsymbol{R}$.

Describing structures, it is important to distinguish relations and their names because when a structural machine functions, relations, as a rule, are changing, while their names can remain the same, indicating that this relation is dynamically the same but its statics is changing. For instance, when a structure $\boldsymbol{A}$ on a set A has a ternary relation $\boldsymbol{R}$ with the name R as its component, and in the process of computation, the machine connects three elements that were not related by $\boldsymbol{R}$ by a link including this triple into the relation $\boldsymbol{R}$. As a result, $\boldsymbol{R}$ becomes larger but stays dynamically the same preserving the same name R. This is similar to the situation when a human being is changing, remaining, at the same time, the same person.

However, a structural machine M can also change names of the structure relations, for example, by splitting one relation into two new relations of the same arity and giving them new names. In this case, the structural machine M works with relations of higher order because the relation between a relation $\boldsymbol{R}$ and its name R has higher order than the relation $\boldsymbol{R}$ (cf. Definition 2).

Lists, queues, arrays, words, texts, graphs, multigraphs, directed graphs, mathematical and chemical formulas, tapes of Turing machines and Kolmogorov complexes are particular cases of structures of the first order that have only unary and binary relations. Note that labels, names, types and properties are unary relations.

In the case when the arrangement R of relations consists of one binary and several unary relations, the first-order structure is a labeled (named) directed graph. Transition diagrams of finite automata are examples of labeled directed graphs. When the arrangement R contains only binary and unary relations, the first-order structure is a labeled (named) directed multigraph (Berge, 1973).

Example 1. Let us consider the first-order structure $\boldsymbol{A} = (A, r, \mathsf{R})$ where $A = \{a, b, c, d, e, f, g, h, k\}$ and R consists of one ternary relation $\boldsymbol{Q}$ and three binary relations $\boldsymbol{P}$, $\boldsymbol{H}$ and $\boldsymbol{F}$, the graphical representation of which is given in Figure 1.

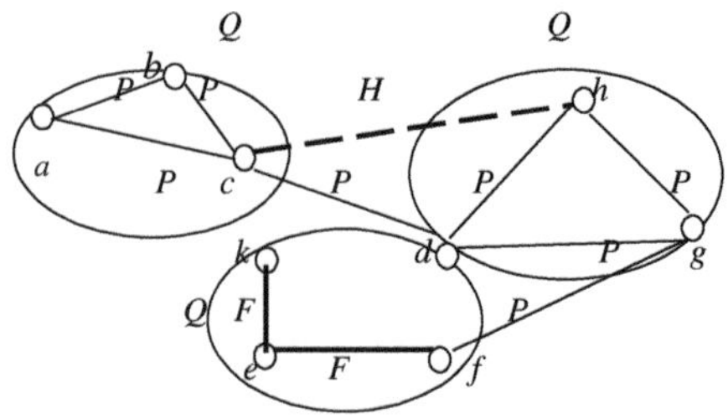

Figure 1. The graphical representation of a first-order structure.

However, in many cases, it is not enough to have first-order structures and we need second-order structures. For instance, we need them to model Physarum machines (Burgin and Adamatzky, 2017a) or to represent labeled graphs used in the models of finite automata (Hopcroft *et al.*, 2001).

Definition 2 (Burgin, 2012). A *second-order structure* is a triad of the form

$$\boldsymbol{A} = (A, r, \mathsf{R}),$$

where the set A, which is called the *substance* of the structure $\boldsymbol{A}$, consists of elements of the structure $\boldsymbol{A}$, which are called *structure elements* of the structure $\boldsymbol{A}$; the set R, which is called the *arrangement* of the structure $\boldsymbol{A}$, consists of relations in the structure $\boldsymbol{A}$, which are called *structure relations* of the structure $\boldsymbol{A}$; r is the incidence relation that connects groups of elements from A and/or relations from R with names of relations from R; $\mathsf{R} = \mathsf{R}_1 \cup \mathsf{R}_2 \cup \mathsf{R}_3$; R_1 is the set of relations between the elements from the set A; R_2 is the set of relations in the set R_1, i.e., elements from R_2 are relations between relations from R_1; and R_3 is the set of relations between elements from A and relations from R_1.

Relations from R_2 and R_3 are called relations of the *second level* in $\boldsymbol{A}$.

A second-order structure $\boldsymbol{A} = (A, r, \mathsf{R})$ is called *austere* if the set R_3 is empty. In many cases, it is sufficient to use only austere second-order structures. For instance, in papers by Burgin and Adamatzky (2017a,b), only austere second-order structures are considered.

Second-order structures have many applications. However, here they are only used to represent data processed by structural machines of the second order.

In a similar way, we determine relations and structures of higher orders, which are useful for building and using metadata.

Definition 3 (Burgin, 2012). Relations of the *level* n in a structure $\boldsymbol{A}$ are relations between relations of level $n-1$ and relations of orders less than n as well as relations between elements of A and relations of level $n-1$ in $\boldsymbol{A}$.

Let us consider some examples.

(a) A categorical example of relations of the second level and structures of the second order

Taking an abstract category C, it is possible to consider each morphism from C, e.g., f, as a relation between two objects, e.g., objects A and B.

$$A \xrightarrow{\quad f \quad} B. \tag{1}$$

Then a category is a structure of the first order while morphisms are relations of the first level.

Now let us consider the category $M(C)$ of morphisms from C. Its morphisms have the form

$$\begin{array}{ccc} K & \longrightarrow & H \\ \downarrow & & \downarrow \\ A & \longrightarrow & B \end{array} \tag{2}$$

where the square (2) is commutative. Morphisms from the category $M(C)$ are relations between morphisms from C and thus are at the second level. Consequently, the category $M(C)$ is an austere structure of the second order.

(b) A set-theoretical example of relations of the second level and structures of the second order

Let us take a collection P of sets and relations between sets from P denoting the set of all these relations by $R(P)$. It makes P a structure of the first order. Then let us consider operations with (compositions of) relations from $R(P)$ denoting this set by $C(R(P))$. As we know, it is possible to treat each operation (composition) as a relation. Consequently, elements from $C(R(P))$ have the second level and, consequently, the set $C(R(P))$ is a structure of the second order.

Continuing this process of taking operations with (compositions of) relations, we can build a structure of an arbitrary big order with relations at the corresponding level.

Definition 4 (Burgin, 2012). A *structure of order n* is a triad of the form

$$\boldsymbol{A} = (A, r, \mathsf{R}),$$

where the set A, which is called the *substance* of the structure $\boldsymbol{A}$, consists of elements of the structure $\boldsymbol{A}$, which are called *structure elements* of the structure $\boldsymbol{A}$; the set R, which is called the *arrangement* of the structure $\boldsymbol{A}$, consists of relations in the structure $\boldsymbol{A}$, which are called *structure relations* of the structure $\boldsymbol{A}$ and the order of which is not larger than n; and r is the incidence relation that connects groups of elements from A and/or relations from R with names of relations from R.

Relations from the arrangement R determine the *intrinsic structure* of the system $\boldsymbol{A} = (A, r, \mathsf{R})$. However, modeling, efficient operation with and utilization of first-order and higher-order structures demands additional constructions, which define extrinsic structures (Burgin, 2012). One of these *extrinsic structures* is pretopology determined by neighborhoods in the set A (Čech, 1966). There are different types of such neighborhoods.

Definition 5. Let $\boldsymbol{R}$ be an n-ary relation of the first order from the arrangement R. Then,

(a) the *substantial R-neighborhood* of a structure element a in a structure $\boldsymbol{A} = (A, r, \mathsf{R})$ is a set of the form

$$\begin{aligned} \mathrm{O}_{RS}a = \{a\} \cup \{d \in A; \exists i, k \exists a_2, \ldots, a_n \in A\,((1 \leq k \leq n) \\ \&(a_1, a_2, \ldots, a_i = a, \ldots, a_k = d, \ldots, a_n) \in \boldsymbol{R})\}; \end{aligned}$$

(b) the *link R-neighborhood* of a structure element a in a structure $\boldsymbol{A} = (A, r, \mathsf{R})$ is a set

$$\mathrm{O}_{RL}a = \{(a_1, a_2, \ldots, a_n) \in \boldsymbol{R}; a_1, a_2, \ldots, a_n \in \mathrm{O}_{R}a\};$$

(c) the *full R-neighborhood* of a structure element a in a structure $\boldsymbol{A} = (A, r, \mathsf{R})$ is the set

$$\mathrm{O}_{RF}a = \mathrm{O}_{RS}a \cup \mathrm{O}_{RL}a.$$

Informally, the substantial R-neighborhood of a structure element a consists of all structure elements connected to a by the relation $\boldsymbol{R}$.

The link R-neighborhood of a structure element a consists of all constituents from the relation $\boldsymbol{R}$ that contain a.

The full R-neighborhood of a structure element a is the union of the substantial R-neighborhood and link R-neighborhood of a.

Note that all such neighborhoods are defined only for relations of the first order and any element belongs to any of its substantial neighborhoods.

Examples.

1. In the structure from Figure 1, the substantial F-neighborhood of the structure element e is equal to the set $\{k, e, f\}$.
2. In the structure from Figure 1, the set $\{a, b, c\}$ is the substantial P-neighborhood of the structure element a determined by the relation P.
3. In the structure from Figure 1, the set $\{a, b, c\}$ is the substantial Q-neighborhood of the structure element a determined by the relation Q.
4. In the structure from Figure 1, the set $\{c, d, h, g\}$ is the substantial P-neighborhood of the structure element d determined by the relation P.
5. In the structure from Figure 1, the set $\{f, k, e\}$ is the substantial Q-neighborhood of the structure element f determined by the relation Q.
6. In the structure from Figure 1, the set $\{(a, b), (a, c)\}$ is the link P-neighborhood of the structure element a determined by the relation Q.
7. In the structure from Figure 1, the set $\{(a, b, c)\}$ is the link Q-neighborhood of the structure element a and of the structure element b where both neighborhoods are determined by the relation Q.

Note that it is possible to treat all neighborhoods as relations of a special type in the set A.

Definition 5 implies the following result.

Lemma 1. *For any relation* $\boldsymbol{R}$ *of the first order from the arrangement* R, *each structure element* a *in a structure* $\boldsymbol{A} = (A, r, \mathsf{R})$ *has uniquely defined substantial* R*-neighborhood, link* R*-neighborhood and full* R*-neighborhood.*

Relations between and operations with relations from the arrangement R imply relations between and operations with neighborhoods.

Proposition 1. *For any relation $\boldsymbol{R}$ of the first order from the arrangement R and any structure element a, we have the following properties:*

(a) *The inclusion $\boldsymbol{R} \subseteq \boldsymbol{Q}$ of two relations $\boldsymbol{R}$, $\boldsymbol{Q} \in \mathsf{R}$ implies inclusions of neighborhoods*

$$\mathrm{O}_{RS}a \subseteq \mathrm{O}_{QS}a, \quad \mathrm{O}_{RL}a \subseteq \mathrm{O}_{QL}a, \quad \mathrm{O}_{RF}a \subseteq \mathrm{O}_{QF}a.$$

(b) *If $\boldsymbol{R} = \boldsymbol{Q} \cup \boldsymbol{P}$ for relations $\boldsymbol{R}$, $\boldsymbol{P}$, $\boldsymbol{Q} \in \mathsf{R}$, then*

$$\mathrm{O}_{QS}a \cup \mathrm{O}_{PS}a = \mathrm{O}_{RS}a,$$

$$\mathrm{O}_{PL}a \cup \mathrm{O}_{QL}a = \mathrm{O}_{RL}a,$$

$$\mathrm{O}_{PF}a \cup \mathrm{O}_{QF}a = \mathrm{O}_{RF}a.$$

(c) *If $\boldsymbol{R} = \boldsymbol{Q} \cap \boldsymbol{P}$ for relations $\boldsymbol{R}$, $\boldsymbol{P}$, $\boldsymbol{Q} \in \mathsf{R}$, then*

$$\mathrm{O}_{QS}a \cap \mathrm{O}_{PS}a = \mathrm{O}_{RS}a,$$

$$\mathrm{O}_{PL}a \cap \mathrm{O}_{QL}a = \mathrm{O}_{RL}a,$$

$$\mathrm{O}_{PF}a \cap \mathrm{O}_{QF}a = \mathrm{O}_{RF}a.$$

Neighborhoods of elements allow us to build neighborhoods of sets of elements.

Definition 6. Let $\boldsymbol{R}$ be an n-ary relation of the first order from the arrangement R and $Z \subseteq A$. Then,

(a) the *substantial R-neighborhood* of the set Z in a structure $\boldsymbol{A} = (A, r, \mathsf{R})$ is the set

$$\mathrm{O}_{RS}Z = \bigcup_{a \in Z} \mathrm{O}_{RS}a,$$

(b) the *link R-neighborhood* of the set Z in a structure $\boldsymbol{A} = (A, r, \mathsf{R})$ is a set

$$\mathrm{O}_{RL}Z = \bigcup_{a \in Z} \mathrm{O}_{RL}a,$$

(c) the *full R-neighborhood* of the set Z in a structure $\boldsymbol{A} = (A, r, \mathsf{R})$ is the set

$$\mathrm{O}_{RF}Z = \mathrm{O}_{RS}Z \cup \mathrm{O}_{RL}Z.$$

Informally, the substantial R-neighborhood of a set Z of structure elements consists of all structure elements connected to elements from Z by the relation $\boldsymbol{R}$. For instance, in the structure from Figure 1, the substantial P-neighborhood of the set $\{c, d\}$ is equal to the set $\{a, b, c, d, h, g\}$.

Lemma 1 implies the following result.

Lemma 2. *The substantial (link or full) R-neighborhood of any set Z of structure elements is uniquely defined.*

Proposition 1 implies the following result.

Proposition 2. *For any set Z of structure elements, we have the following properties:*

(a) *The inclusion $\boldsymbol{R} \subseteq \boldsymbol{Q}$ of two relations $\boldsymbol{R}$, $\boldsymbol{Q} \in \mathsf{R}$ implies inclusions of neighborhoods*

$$\mathrm{O}_{RS}Z \subseteq \mathrm{O}_{QS}Z, \quad \mathrm{O}_{RL}Z \subseteq \mathrm{O}_{QL}Z, \quad \mathrm{O}_{RF}Z \subseteq \mathrm{O}_{QF}Z.$$

(b) *If $\boldsymbol{R} = \boldsymbol{Q} \cup \boldsymbol{P}$ for relations $\boldsymbol{R}$, $\boldsymbol{P}$, $\boldsymbol{Q} \in \mathsf{R}$, then*

$$\mathrm{O}_{QS}Z \cup \mathrm{O}_{PS}Z = \mathrm{O}_{RS}Z,$$

$$\mathrm{O}_{PL}Z \cup \mathrm{O}_{QL}Z = \mathrm{O}_{RL}Z,$$

$$\mathrm{O}_{PF}Z \cup \mathrm{O}_{QF}Z = \mathrm{O}_{RF}Z.$$

(c) *If $\boldsymbol{R} = \boldsymbol{Q} \cap \boldsymbol{P}$ for relations $\boldsymbol{R}$, $\boldsymbol{P}$, $\boldsymbol{Q} \in \mathsf{R}$, then*

$$\mathrm{O}_{QS}Z \cap \mathrm{O}_{PS}Z = \mathrm{O}_{RS}Z,$$

$$\mathrm{O}_{PL}Z \cap \mathrm{O}_{QL}Z = \mathrm{O}_{RL}Z,$$

$$\mathrm{O}_{PF}Z \cap \mathrm{O}_{QF}Z = \mathrm{O}_{RF}Z.$$

(d) *The inclusion $Z \subseteq Y \subseteq X$ implies inclusions of neighborhoods*

$$\mathrm{O}_{RS}Z \subseteq \mathrm{O}_{RS}Y,$$

$$\mathrm{O}_{RL}Z \subseteq \mathrm{O}_{RL}Y,$$

$$\mathrm{O}_{RF}Z \subseteq \mathrm{O}_{RF}Y.$$

(e) *If $V = Z \cup Y$, then*

$$\mathrm{O}_{RS}Z \cup \mathrm{O}_{RS}Y = \mathrm{O}_{RS}V,$$

$$\mathrm{O}_{RL}Z \cup \mathrm{O}_{RL}Y = \mathrm{O}_{RL}V,$$

$$\mathrm{O}_{RF}Z \cup \mathrm{O}_{RF}Y = \mathrm{O}_{RF}V.$$

(f) *If* $W = Z \cap Y$, *then*

$$\mathrm{O}_{RS}Z \cap \mathrm{O}_{RS}Y = \mathrm{O}_{RS}W,$$
$$\mathrm{O}_{RL}Z \cap \mathrm{O}_{RL}Y = \mathrm{O}_{RL}W$$
$$\mathrm{O}_{RF}Z \cap \mathrm{O}_{RF}Y = \mathrm{O}_{RF}W.$$

The constructed neighborhoods have the first level. Using composition of binary relations, it is possible to define neighborhoods of higher levels.

We remind (cf., e.g., Burgin, 2011) that for binary relations $T \subseteq X \times Y$ and $R \subseteq Y \times Z$, it is possible to build a new binary relation $RT \subseteq X \times Z$ that is called the (*sequential*) *composition* or *superposition* of binary relations T and R and is defined as

$$R \circ T = \{(x, z); x \in X, z \in Z; \text{ where } (x, y) \in T$$
$$\text{and } (y, z) \in R \text{ for some } y \in Y\}.$$

We denote the composition $R \circ R$ by R^2 and in a general case, the composition $R^n = R^{n-1} \circ R$ for all $n = 2, 3, \ldots$.

It is also possible to define composition of relations of any arity. Let us consider a k-ary relation T in the set X and an h-ary relation T in the set X.

Definition 7. The (*sequential*) *composition* or *superposition* $W = R \circ T$ of relations T and R is an $(h+k-2)$-ary relation W in the set X defined as the set of all tuples $(x_1, x_2, \ldots, x_{j-1}, x_{j+1}, \ldots, x_k, y_1, y_2, \ldots, y_{t-1}, y_{t+1}, \ldots, y_h)$ with $(x_1, x_2, \ldots, x_{j-1}, x, x_{j+1}, \ldots, x_k) \in T$ and $(y_1, y_2, \ldots, y_{t-1}, x, y_{t+1}, \ldots, y_h) \in R$ for some $x \in X$, $1 \leq j \leq k$ and $1 \leq t \leq h$.

Note that when T and R are binary relations, we have the conventional composition of relations. At the same time, if $T \subseteq X^k$ and $R \subseteq X^h$, then $R \circ T \subseteq X^{k+h-2}$.

As before, taking an h-ary relation R in the set X, we define the composition $R^n = R^{n-1} \circ R$ for all $n = 2, 3, \ldots$.

Definition 8. Let $\boldsymbol{R}$ be an n-ary relation of the first order from the arrangement R. Then,

(a) the *substantial R-neighborhood* $\mathrm{O}^n_{RS}a$ of level n of a structure element a in a structure $\boldsymbol{A} = (A, r, \mathsf{R})$ is the substantial R^n-neighborhood of the first level of the structure element a;

(b) the *link R-neighborhood* $O^n_{RL}a$ of level n of a structure element a in a structure $\boldsymbol{A} = (A, r, \mathsf{R})$ is the link R^n-neighborhood of the first level of the structure element a;
(c) the *full R-neighborhood* $O^n_{RF}a$ of level n of a structure element a in a structure $\boldsymbol{A} = (A, r, \mathsf{R})$ is the full R^n-neighborhood of the first level of the structure element a.

Informally, the substantial R-neighborhood of the level n of a structure element a consists of all structure elements connected to a by applying n times links from the relation $\boldsymbol{R}$. For instance, in the structure from Figure 1, the substantial P-neighborhood of the second level of the element c is equal to the set $\{a, b, c, d, h, g\}$.

The link R-neighborhood of the level n of a structure element a consists of all links from the relation $\boldsymbol{R}^n$ connected to a.

The full R-neighborhood of the level n of a structure element a is the union of the substantial R-neighborhood of the level n and link R-neighborhood of the level n of the element a.

Note that here all such neighborhoods are defined only for relations of the first order, although it is possible to extend these concepts for relations of any order.

Definition 8 implies the following result.

Lemma 3. *For any binary relation $\boldsymbol{R}$ of the first order from the arrangement R, each structure element a in a structure $\boldsymbol{A} = (A, r, \mathsf{R})$ has a uniquely defined substantial R-neighborhood of level n, link R-neighborhood of level n and full R-neighborhood of level n.*

Relations between and operations with relations from the arrangement R imply relations between and operations with neighborhoods on all levels.

Proposition 3. *For any relation $\boldsymbol{R}$ of the first order from the arrangement R and any structure element a, we have the following properties:*

(a) *The inclusion $\boldsymbol{R} \subseteq \boldsymbol{Q}$ of two relations $\boldsymbol{R}$, $\boldsymbol{Q} \in \mathsf{R}$ implies inclusions of neighborhoods of level n,*

$$O^n_{RS}a \subseteq O^n_{QS}a,\ O^n_{RL}a \subseteq O^n_{QL}a \text{ and } O^n_{RF}a \subseteq O^n_{QF}a.$$

(b) *If $\boldsymbol{R} = \boldsymbol{Q} \cup \boldsymbol{P}$ of relations $\boldsymbol{R}$, $\boldsymbol{P}$, $\boldsymbol{Q} \in \mathsf{R}$ implies equalities of neighborhoods of level n,*

$$O_{QS}a \cup O_{PS}a = O^n_{RS}a,\ O^n_{PL}a \cup O^n_{QL}a = O^n_{RL}a \quad \text{and}$$

$$O^n_{PF}a \cup O^n_{QF}a = O^n_{RF}a.$$

(c) *If $\boldsymbol{R} = \boldsymbol{Q} \cap \boldsymbol{P}$ of relations $\boldsymbol{R}$, $\boldsymbol{P}$, $\boldsymbol{Q} \in \mathsf{R}$ implies equalities of neighborhoods of level n,*

$$\mathrm{O}^n_{QS}a \cap \mathrm{O}^n_{PS}a = \mathrm{O}^n_{RS}a, \ \mathrm{O}^n_{PL}a \cap \mathrm{O}^n_{QL}a = \mathrm{O}^n_{RL}a \quad \textit{and}$$

$$\mathrm{O}^n_{PF}a \cap \mathrm{O}^n_{QF}a = \mathrm{O}^n_{RF}a.$$

Neighborhoods of level n of elements allow us to build neighborhoods of level n of sets of elements.

Definition 9. Let $\boldsymbol{R}$ be an h-ary relation of the first order from the arrangement R and $Z \subseteq A$. Then,

(a) the *substantial R-neighborhood* of level n of the set Z in a structure $\boldsymbol{A} = (A, r, \mathsf{R})$ is the set

$$\mathrm{O}^n_{RS}Z = \bigcup_{a \in Z} \mathrm{O}^n_{RS}a,$$

(b) the *link R-neighborhood* of level n of the set Z in a structure $\boldsymbol{A} = (A, r, \mathsf{R})$ is a set

$$\mathrm{O}^n_{RL}Z = \bigcup_{a \in Z} \mathrm{O}^n_{RL}a,$$

(c) the *full R-neighborhood* of level n of the set Z in a structure $\boldsymbol{A} = (A, r, \mathsf{R})$ is the set

$$\mathrm{O}^n_{RF}Z = \mathrm{O}^n_{RS}Z \cup \mathrm{O}^n_{RL}Z.$$

Lemma 3 implies the following result.

Lemma 4. *If $\boldsymbol{R}$ is an h-ary relation of the first order from R, then the substantial (link or full) R-neighborhood of level n of any set Z of structure elements is uniquely defined.*

Proposition 3 implies the following result.

Proposition 4. *If $\boldsymbol{R}$ is an h-ary relation of the first order from R, then for any set Z of structure elements, we have the following properties:*

(a) *The inclusion $\boldsymbol{R} \subseteq \boldsymbol{Q}$ of two relations $\boldsymbol{R}$, $\boldsymbol{Q} \in \mathsf{R}$ implies inclusions of neighborhoods of level n*

$$\mathrm{O}^n_{RS}Z \subseteq \mathrm{O}^n_{QS}Z, \quad \mathrm{O}^n_{RL}Z \subseteq \mathrm{O}^n_{QL}Z, \quad \mathrm{O}^n_{RF}Z \subseteq \mathrm{O}^n_{QF}Z.$$

(b) *If* $\boldsymbol{R} = \boldsymbol{Q} \cup \boldsymbol{P}$ *for relations* $\boldsymbol{R}$, $\boldsymbol{P}$, $\boldsymbol{Q} \in \mathsf{R}$, *then*

$$\mathrm{O}^n_{QS}Z \cup \mathrm{O}^n_{PS}Z = \mathrm{O}^n_{RS}Z,$$

$$\mathrm{O}^n_{PL}Z \cup \mathrm{O}^n_{QL}Z = \mathrm{O}^n_{RL}Z,$$

$$\mathrm{O}^n_{PF}Z \cup \mathrm{O}^n_{QF}Z = \mathrm{O}^n_{RF}Z.$$

(c) *If* $\boldsymbol{R} = \boldsymbol{Q} \cap \boldsymbol{P}$ *for relations* $\boldsymbol{R}$, $\boldsymbol{P}$, $\boldsymbol{Q} \in \mathsf{R}$, *then*

$$\mathrm{O}^n_{QS}Z \cap \mathrm{O}^n_{PS}Z = \mathrm{O}^n_{RS}Z,$$

$$\mathrm{O}^n_{PL}Z \cap \mathrm{O}^n_{QL}Z = \mathrm{O}^n_{RL}Z,$$

$$\mathrm{O}^n_{PF}Z \cap \mathrm{O}^n_{QF}Z = \mathrm{O}_{RF}Z.$$

(d) *The inclusion* $Z \subseteq Y \subseteq X$ *implies inclusions of neighborhoods of level* n

$$\mathrm{O}^n_{RS}Z \subseteq \mathrm{O}^n_{RS}Y, \quad \mathrm{O}^n_{RL}Z \subseteq \mathrm{O}^n_{RL}Y, \quad \mathrm{O}^n_{RF}Z \subseteq \mathrm{O}^n_{RF}Y.$$

(e) *If* $V = Z \cup Y$, *then*

$$\mathrm{O}^n_{RS}Z \cup \mathrm{O}^n_{RS}Y = \mathrm{O}^n_{RS}V,$$

$$\mathrm{O}^n_{RL}Z \cup \mathrm{O}^n_{RL}Y = \mathrm{O}^n_{RL}V,$$

$$\mathrm{O}^n_{RF}Z \cup \mathrm{O}^n_{RF}Y = \mathrm{O}^n_{RF}V.$$

(f) *If* $W = Z \cap Y$, *then*

$$\mathrm{O}^n_{RS}Z \cap \mathrm{O}^n_{RS}Y = \mathrm{O}^n_{RS}W,$$

$$\mathrm{O}^n_{RL}Z \cap \mathrm{O}^n_{RL}Y = \mathrm{O}^n_{RL}W,$$

$$\mathrm{O}^n_{RF}Z \cap \mathrm{O}^n_{RF}Y = \mathrm{O}^n_{RF}W.$$

Substantial neighborhoods of sets of structure elements determine pretopology in the set A of all structure elements.

We remind that a *pretopological space* is defined as a set X with a preclosure operator, also called Čech closure operator, *cl*. Let 2^X be the power set of X.

A *preclosure operator* on a set X is a mapping *cl*: $2^X \to 2^X$ that satisfies the following axioms (Čech, 1966):

1. $cl(\oslash) = \oslash$;
2. $Z \subseteq cl(Z)$ for any $Z \subseteq X$;
3. $cl(Z \cup Y) \subseteq cl(Z) \cup cl(Y)$ for any $Z, Y \subseteq X$;
4. $Y \subseteq Z$ implies $cl(Y) \subseteq cl(Z)$ for any $Z, Y \subseteq X$.

Let $\boldsymbol{R}$ be a relation from R. Properties of substantial R-neighborhoods imply the following result.

Proposition 5. *Substantial R-neighborhoods of level n define a pretopology in the set A.*

Proof. Let us define the closure $cl(Z)$ of a set $Z \subseteq A$ equal to the substantial R-neighborhood $\mathrm{O}^n_{RS}Z$ of level n of Z and check the axioms of pretopological spaces.

Axioms 1 and 2 are true by definition as $cl(\oslash) = \oslash$ and $Z \subseteq cl(Z)$ for any $Z \subseteq A$.

In addition,

$$cl(Z \cup Y) = \mathrm{O}^n_{RS}(Z \cup Y) = \bigcup_{a \in Z \cup Y} \mathrm{O}^n_{RS}a = \left(\bigcup_{a \in Z} \mathrm{O}^n_{RS}a\right) \cup \left(\bigcup_{a \in Y} \mathrm{O}^n_{RS}a\right)$$
$$= \mathrm{O}^n_{RS}Z \cup \mathrm{O}^n_{RS}Y = cl(Z) \cup cl(Y).$$

This gives us Axiom 3.

Axiom 4 is implied by Proposition 4.

The Proposition 5 is proved. □

Corollary 1 (Burgin and Adamatzky, 2017b). *Substantial R-neighborhoods of the first level define a pretopology in the set A.*

Let us consider a subset P of the set R.

Definition 10. (a) The *substantial* P-*neighborhood* of a structure element a in a structure $\boldsymbol{A} = (A, r, \mathsf{R})$ is the set

$$\mathrm{O}_{\mathbf{PS}}a = \bigcup_{\boldsymbol{R} \in \mathsf{P}} \mathrm{O}_{RS}a.$$

(b) The *link* P-*neighborhood* of a structure element a in a structure $\boldsymbol{A} = (A, r, \mathsf{R})$ is a set of the form

$$\mathrm{O}_{\mathbf{PL}}a = \bigcup_{\boldsymbol{R} \in \mathsf{P}} \mathrm{O}_{RL}a.$$

(c) The *full* P-*neighborhood* of a structure element a in a structure $\boldsymbol{A} = (A, r, \mathsf{R})$ is the set

$$\mathrm{O}_{\mathbf{PF}}a = \bigcup_{\boldsymbol{R} \in \mathsf{P}} \mathrm{O}_{RF}a.$$

Definition 10 implies the following result.

Lemma 5. *If* P *is a subset of the set* R, *then the substantial (link or full)* P-*neighborhood of a structure element* a *in the structure* $\boldsymbol{A} = (A, r, \mathsf{R})$ *is uniquely defined.*

Corollary 2 (Burgin and Adamatzky, 2017a). *The substantial (link or full)* R-*neighborhood of a structure element* a *in the structure* $\boldsymbol{A} = (A, r, \mathsf{R})$ *is uniquely defined.*

Properties of union of sets and Definitions 5 and 9 imply the following result.

Lemma 6. $\mathrm{O}_{\boldsymbol{PF}}a = \mathrm{O}_{\boldsymbol{PS}}a \cup \mathrm{O}_{\boldsymbol{PL}}a$.

Neighborhoods of elements allow us to build neighborhoods of sets of elements.

Definition 11. If P is a subset of the set R and $Z \subseteq A$, then

(α) the *substantial* P-*neighborhood* of the set Z in a structure $\boldsymbol{A} = (A, r, \mathsf{R})$ is the set

$$\mathrm{O}_{\boldsymbol{PS}}Z = \bigcup_{a \in Z} \mathrm{O}_{\boldsymbol{PS}}a;$$

(β) the *link* P-*neighborhood* of the set Z in a structure $\boldsymbol{A} = (A, r, \mathsf{R})$ is a set of the form

$$\mathrm{O}_{\boldsymbol{PL}}Z = \bigcup_{a \in Z} \mathrm{O}_{\boldsymbol{PL}}a;$$

(γ) the *full* P-*neighborhood* of the set Z in a structure $\boldsymbol{A} = (A, r, \mathsf{R})$ is the set

$$\mathrm{O}_{\boldsymbol{PF}}Z = \mathrm{O}_{\boldsymbol{PS}}Z \cup \mathrm{O}_{\boldsymbol{PL}}Z.$$

Lemma 5 implies the following result.

Lemma 7. *If* P *is a subset of the set* R, *then the substantial (link or full)* P-*neighborhood of any set* Z *of structure elements is uniquely defined.*

Proposition 5 and Definition 11 imply the following results.

Proposition 6. *If* P *is a subset of the set* R, *then the substantial* P-*neighborhoods define a pretopology in the set* A.

Corollary 3 (Burgin and Adamatzky, 2017b). *Substantial* R*-neighborhoods define a pretopology in the set A.*

Let us study properties of neighborhoods in structures.

Definition 12. If $\boldsymbol{R}$ is an n-ary relation from R, then the substantial R-neighborhood of a structure element a in a structure $\boldsymbol{A} = (A, r, \mathsf{R})$ is *symmetric* if for any elements $a_1, a_2, a_3, \ldots, a_n$ from A and any permutation $i_1, i_2, i_3, \ldots, i_n$ of the numbers $1, 2, 3, \ldots, n$, we have

$$(a_1, a_2, a_3, \ldots, a_n) \in \boldsymbol{R} \text{ if and only if } (a_{i1}, a_{i2}, a_{i3}, \ldots, a_{in}) \in \boldsymbol{R}.$$

Lemma 8. *If the substantial* R*-neighborhood* $\mathrm{O}_{RS}a$ *of a structure element* a *is symmetric, a structure element* b *belongs to the* R*-neighborhood* $\mathrm{O}_{RS}a$ *if and only if* a *belongs to the substantial* R*-neighborhood* $\mathrm{O}_{RS}b$ *of* b.

Definition 13. If $\boldsymbol{R}$ is an n-ary relation from R, then the R-neighborhood of a structure element a in a structure $\boldsymbol{A} = (A, r, \mathsf{R})$ is *transitive* if for any elements $a_1, a_2, a_3, \ldots, a_n, d_1, d_2, d_3, \ldots, d_n$ from A and any numbers i and j from the set $\{1, 2, 3, \ldots, n\}$, we have

If $(a_1, a_2, a_3, \ldots, a_n) \in \boldsymbol{R}$ and $(d_1, \ldots, a_i, \ldots, d_n) \in \boldsymbol{R}$, then there are elements $b_1, b_2, b_3, \ldots, b_n$ from A such that $(b_1, \ldots, a_i, \ldots, d_j, \ldots, b_n) \in \boldsymbol{R}$.

Proposition 7. *If a relation* $\boldsymbol{R} \in \mathsf{R}$ *is symmetric and transitive, then substantial* R*-neighborhoods of two structure elements either coincide or do not intersect.*

Corollary 4. *If* P *is a subset of the set* R *and all relations in* P *are symmetric and transitive, then substantial* P*-neighborhoods of two structure elements either coincide or do not intersect.*

Corollary 5. *If all relations in* R *are symmetric and transitive, then substantial* R*-neighborhoods of two structure elements either coincide or do not intersect.*

Corollary 6. *If* R *consists of one symmetric binary relation* $\boldsymbol{R}$*, i.e.,* $\boldsymbol{A}$ *is a graph, and* R *is also transitive, then any substantial* R*-neighborhood (*R*-neighborhood) is a complete graph.*

Proposition 8. *If a relation* $\boldsymbol{R} \in \mathsf{R}$ *is symmetric and transitive, then the system of substantial* R*-neighborhoods forms a base of a topology in* A.

Corollary 7. *If* P *is a subset of the set* R *and all relations in* R *are symmetric and transitive, then the system of substantial* P*-neighborhoods forms a base of a topology in* A.

Corollary 8. *If all relations in* R *are symmetric and transitive, then the system of substantial* R*-neighborhoods forms a base of a topology in* A.

Definition 14. The *type* $\mathrm{T}(\boldsymbol{A})$ of a first-order structure $\boldsymbol{A} = (A, r, \mathsf{R})$ is the set $\{(R, \alpha(R)); \boldsymbol{R} \in \mathsf{R}\}$ of pairs $(R, \alpha(R))$ where $\alpha(R)$ is the arity of the relation $\boldsymbol{R}$ with the name R.

We assume that two first-order structures $\boldsymbol{A} = (A, r, \mathsf{R})$ and $\boldsymbol{B} = (B, p, \mathsf{P})$ have the *same type* if there is a one-to-one mapping $f\colon \mathrm{T}(\boldsymbol{A}) \to \mathrm{T}(\boldsymbol{B})$ such that if $f(R, \alpha(R)) = (P, \alpha(P))$, then $\alpha(R) = \alpha(P)$.

For instance, all binary relations are of the same type.

A structural machine M works with structures of a given type and has three components:

(1) The *control device* C_M regulates the state of the machine M.
(2) The *processor* P_M performs transformation of the processed structures, and its actions (operations) depend on the state of the machine M and the state of the processed structures.
(3) The *functional space* Sp_M consists of three components:
    (i) the *input space* In_M, which contains the input structure;
    (ii) the *output space* Out_M, which contains the output structure;
    (iii) the *processing space* PS_M, in which the input structure(s) is transformed into the output structure(s).

We assume that all structures — the input structure, the output structure and the processed structures — are of the same type.

Computation of a structural machine M determines the *trajectory of computation*, which is a tree in a general case and a sequence when the computation is a deterministic case and is performed by a single processor unit.

There are two forms of functional spaces, Sp_M and USp_M:

(1) Sp_M is the set of all structures that can be processed by the structural machine M and is called a *categorical functional space.*
(2) USp_M is a structure for which all structures that can be processed by the structural machine M are substructures and is called a *universal functional space.*

For instance, if the machine M can process only letters of some alphabet Σ, then $Sp_M = \Sigma$ while USp_M is any word that contains all letters from Σ.

There are three basic types of control devices:

(1) A *central control device* controls all processors of the structural machine.
(2) A *cluster control device* controls a cluster of processors in the structural machine.
(3) An *individual control device* controls a single processor in the structural machine.

There are two basic types of processors:

(1) A *localized processor* is a single abstract device (processor unit or unit processor).
(2) A *distributed processor*, which is also called a *total processor*, consists of a system of *unit processors* or *processor units.*

It is possible to treat a localized processor as a singular unit processor although it can be constructed of several processor units.

Note that a structural machine can have several distributed processors.

In turn, there are three basic types of distributed processors:

(1) A *homogeneous distributed processor* consists of a system of identical unit processors, i.e., all these unit processors are copies of one processor.
(2) An *almost homogeneous distributed processor* consists of a system in which almost all unit processors are identical.
(3) A *heterogeneous distributed processor* consists of a system of different unit processors.

As a result, we have three *structural types* of processors.

There are also three basic *temporal classes* of distributed processors:

(1) In a *synchronous distributed processor*, all unit processors perform each operation at the same time.
(2) In an *almost synchronous distributed processor*, almost all unit processors perform each operation at the same time.
(3) In an *asynchronous distributed processor*, unit processors function independently.

A cellular automaton is a synchronous distributed processor, while a Petri net is an asynchronous distributed processor (Petri, 1962).

Organization of control in distributed processors determines three *control categories* of distributed processors:

(1) A distributed processor with the *centralized control* has one control device, which controls all its unit processors.

(2) A distributed processor with the *dispersed control* has several control devices, each of which controls a group of its unit processors.
(3) A distributed processor with the *individualized control* has a control device for each of its unit processors.

It is natural to suppose that each unit processor performs only local operations. In a general case, each unit processor moves from one structure element to another, performing operations in their neighborhoods (Figure 2). This makes it possible to consider a localized processor as a special type of a distributed processor with one unit processor.

A standard example of a localized processor is the head of a Turing machine with one head or a finite automaton. One head of Turing machine corresponds to one growth zone of the slime mold.

A standard example of a homogeneous distributed processor is the system of all heads of a Turing machine with several heads. It is also possible to perceive a Physarum machine as a multiprocessor structural machine (Burgin and Adamatzky, 2017a,b).

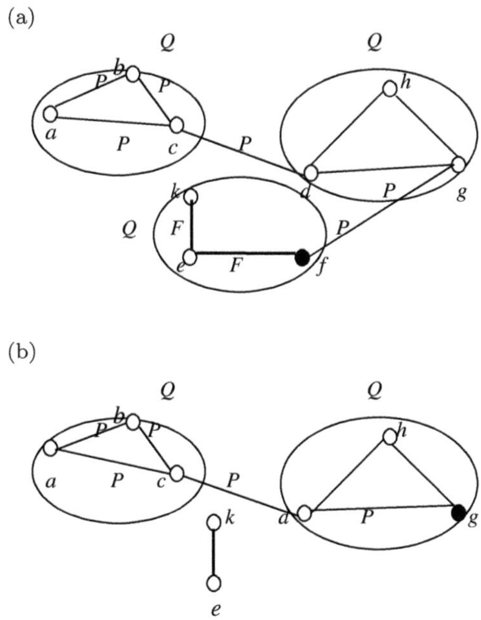

Figure 2. The graphical representation of an operation on first-order structures.

Examples of heterogeneous distributed processors are processing devices in evolutionary automata such as evolutionary finite automata, evolutionary Turing machines or evolutionary inductive Turing machines (Burgin and Eberbach, 2009).

Note that not all heterogeneous distributed processors are the same and to discern their properties, it is possible to use measures of homogeneity constructed by Burgin and Bratalskii (1986). These measures can also be useful for exploring and utilizing structures processed by computers and their models, such as structural machines.

There are three types of localized processors:

(1) a processor localized to one structure element (e.g., a node);
(2) a processor localized to an R-neighborhood of one structure element (e.g., a node) where $\boldsymbol{R}$ is a relation from R;
(3) a processor localized to an R-neighborhood of one structure element (e.g., a node).

In what follows, localization of a processor is formalized by the concept of the processor topos.

There are three sorts of distributed processors:

(1) A *constant distributed processor* has a fixed number of localized unit processors.
(2) A *variable distributed processor* can change the number of localized unit processors.
(3) A *growing distributed processor* can increase the number of localized unit processors.

Growing distributed processors are special kinds of variable distributed processors.

There are three types of variable (growing) distributed processors:

(1) In a *bounded variable* (*bounded growing*) *distributed processor*, the quantity of localized unit processors is always between two numbers, e.g., between 3 and 10 (is bounded by some number, e.g., bounded by 100).
(2) An *unbounded variable distributed processor* can use any finite number of localized unit processors in its functioning.
(3) An *infinite distributed processor* can use any (even infinite) number of localized unit processors.

Cellular automata give examples of structural machines with infinite distributed processors, in which unit processors are identical finite automata.

One-dimensional cellular automata work with such structures as words. Two-dimensional cellular automata work with such structures as two-dimensional arrays.

Now let us consider characteristics of unit processors in structural machines. It is possible to discern positional characteristics and information characteristics. Positional characteristics describe position of a processor and its movements, while information characteristics depict information functioning of the processor.

Each unit processor p of a structural machine M has its *spatial triad* TR_p, which consists of the topos, observation zone and operation zone, i.e., $\mathrm{TR}_p = (T_p, Ob_p, Op_p)$. In it, the first positional characteristic is the topos of a processor.

Definition 15. The *topos* T_p of a unit processor p is the part of the structure occupied by this processor.

When we take into account time of processing, the topos T_p of the processor p is denoted by $T_p(t)$.

This definition allows defining topoi for total processors of structural machines.

Definition 16. The *topos* T_A of a total (distributed) processor A is a combination of the topoi of all its unit processors, i.e., $T_A = \{T_{p1}, T_{p2}, T_{p3}, \ldots\}$ where all p_j are unit processors from A.

When we take into account time of processing, the topos T_A of the processor A is denoted by $T_A(t)$, i.e., $T_A(t)$ is the topos of the processor A at time t.

Definition 17. If $T_A(t_1) = T_A(t_2)$ for any two moments t_1 and t_2 of time, then the processor A is called *stationary*. Otherwise, the processor A is called *moving*.

For instance, in a cellular automaton, all processors are stationary, while the head of a Turing machine is a moving processor.

Note that in a general case, the topos of a processor can be infinite although the constructive conditions on algorithms prohibit infinite topoi. Cellular automata give an example of a processor with an infinite topos if we treat all finite automata of a cellular automaton as one distributed processor. When each of these finite automata is treated as a separate processor, each of them has a finite topos, which consists of one point (vertex) in the grid.

It is natural to accept the following axiom.

Axiom T. Topoi of different unit processors do not intersect.

Localization of unit processors implies restrictions on their topoi. Namely, the topos of a unit processor localized to one structure element consists of this structure element, the topos of a unit processor localized to an R-neighborhood of one structure element is a part of this neighborhood and the topos of a unit processor localized to an R-neighborhood of one structure element is a part of that neighborhood.

Each processor of a structural machine performs information operations of the following types:

(1) obtains information about structures processed by the machine;
(2) receives information from other processors;
(3) sends information to other processors;
(4) transforms structures, processing information.

Each of these processes goes in some vicinity, e.g., a neighborhood, of the processor, and is described by spatial and temporal characteristics.

The key characteristic of a processor for information processes of the type (1) is its observation zone. Observation of a processor means a possibility to obtain information from all elements and relations that belong to the observation zone of this processor situated in its topos.

Definition 18. The *observation zone* $Ob_p(T_p)$ of a unit processor p is the part of the structure Sp_M observed by this processor from its topos T_p.

We see that the operation zone $Ob_p(T_p)$ of p is a function with the argument T_p. This function can also be dependent on other parameters. For instance, when we take into account time of processing, the operation zone of the processor p is a function $Ob_p(T_p, t)$ of two arguments where t is time of observation. Note that in this context, time can be the point when the operation starts or the whole interval when the observation of the processor p continues. The operation zone can also depend on the state of the processor or/and its control device. In this case, we have a function $Ob_p(T_p, t, q)$ of three arguments where q is a state of the unit processor p or/and of its control device.

As in the case of topoi, we define observation zones for distributed processors.

Definition 19. The *observation zone* $Ob_A(T_A)$ of a distributed (compound) processor A is the union of the observation zones of all its unit processors where T_A is the set that contains one topos for each of its unit processors.

We see that the observation zone $Ob_A(T_A)$ is a function with the argument $T_A = \{T_{p1}, T_{p2}, T_{p3}, \ldots\}$ where all p_j are unit processors from A. This function can be also dependent on other parameters. For instance, when we take into account time of processing, the observation zone of the processor A is a function $Ob_A(T_A, t)$ of two arguments where t is time of observation. Note that in this context, time can be the point when the observation starts or the whole interval when the observation of all unit processors in A continues. The observation zone can also depend on the state of the processor. In this case, we have a function $Ob_A(T_A, t, Q)$ of three arguments where q is a state of the processor A. Usually, a state Q of the processor A has the form $\{q_{p1}, q_{p2}, q_{p3}, \ldots\}$ where q_{pj} is a state of a unit processor p_j from A.

It is natural to suppose that observation zones of processors affect their functioning.

Axiom ZI$_p$. Operations performed by a unit processor p depend only on what is inside its observation zone.

Note that this axiom is weaker than Axiom Z from Burgin and Adamatzky (2017a,b).

The key characteristic of a processor for information processes of the type (4) is its operation zone.

Definition 20. The *operation zone* $Op_p(T_p)$ of a unit processor p is the part of the structure Sp_M that can be changed by this processor from its topos.

We see that the operation zone $Op_p(T_p)$ of p is a function with the argument T_p. This function can also be dependent on other parameters. For instance, when we take into account time of processing, the operation zone of the processor p is a function $Op_p(T_p, t)$ of two arguments where t is the time of observation. Note that in this context, time can be the point when the operation starts or the whole interval when the operation of the processor p continues. The operation zone can also depend on the state of the processor or/and its control device. In this case, we have a function $Op_p(T_p, t, q)$ of three arguments where q is a state of the unit processor p or/and of its control device.

For instance, the head of a Turing machine with one linear tape is a unit processor. The topos of the head is one cell in the tape. Its observation zone is the same cell and the symbol written in it. Its operation zone is the symbol written in the cell occupied by the head.

Usually, these parts of the functional space PS_M satisfy the following conditions:

$$T_p \subseteq Op_p(T_p) \subseteq Ob_p(T_p)$$

and

$$T_p(T_p, t) \subseteq Op_p(T_p, t) \subseteq Ob_p(T_p, t).$$

Informally, it means that the topos of a processor is inside its operation zone, while it is possible to perform operations only inside the observation zone.

These conditions are true, for example, for Turing machines because the head (processor) of a Turing machine changes symbols only in the memory cell it observes. At the same time, these conditions are not satisfied for pushdown automata because the head (processor) of a pushdown automaton always observes only one memory cell but can change information (symbols) in several memory cells when it pushes a word into its stack (Hopcroft *et al.*, 2001).

Often we have $Op_p = Ob_p$, while the topos T_p consists of a single node (an element of the structure $\boldsymbol{A}$).

Proposition 9. *Axiom ZI_p implies that the operation zone of the unit processor p is a subset of its observation zone.*

As in the case of observation zones, we define operation zones for distributed processors.

Definition 21. The *operation zone* $Op_A(T_A)$ of a distributed (compound) processor A is the union of the operation zones of all its unit processors.

Similar to the observation zone, the operation zone $Op_A(T_A)$ of A is a function with the argument $T_A = \{T_{p1}, T_{p2}, T_{p3}, \ldots\}$ where all p_j are unit processors from A. This function can also be dependent on other parameters. For instance, when we take into account time of processing, the operation zone of the processor A is a function $Op_A(T_A, t)$ of two arguments where t is time of observation. In this setting, time can be the point when the operation starts or the whole interval when the operation of the processor A continues. The operation zone can also depend on the state of the processor or/and its control device. In this case, we have a function $Op_A(T_A, t, Q)$ of three arguments where Q is a state of the processor A.

Usually, a state Q of the processor A has the form $\{q_{p1}, q_{p2}, q_{p3}, \ldots\}$ where q_{pj} is a state of a unit processor p_j from A or of its control device.

Axiom ZI. Operations performed by unit processors from a distributed processor A depend only on what is inside the observation zones of each of them.

Proposition 10. *Axiom ZI implies that the operation zone of each unit processor p from a distributed processor A is a subset of its observation zone.*

Corollary 9. *Axiom ZI implies that the operation zone of a distributed processor A is a subset of its observation zone.*

Functioning of processors in structural machines includes not only structure transformations but also transitions from one topos to another. These transitions are determined by the second positional characteristic of a processor, which is called the transition zone.

Definition 22. The *transition zone* $Tr_p(T_p)$ of a unit processor p consists of all topoi where p can move in one step from its present topos T_p.

For instance, the transition zone $Tr_h(T_p)$ of such a processor as the head h of a Turing machine with one linear tape consists of three adjacent cells when h is situated in the middle cell.

Note that a stationary processor has a void transition zone.

In some cases, it is useful to assume that the transition zone of a unit processor is included in its observation zone. However, this is not always true. For instance, the observation zone of the head h of a Turing machine consists of the cell where h is situated while the transition zone consists of three cells.

Functioning of processors in structural machines also has other parameters. The key characteristic of a processor for information processes of the type (3) is its transmission zone.

Definition 23. The *transmission zone* $Tm_p(T_p)$ of a unit processor p specifies where p can transmit information.

The transmission zone of a unit processor p can be fixed (constant) for p or it can depend on such parameters as the topos T_p, time t and state q of the processor or/and its control device.

There are three kinds of transmission zones — spatial, functional and combined.

Definition 24. The *spatial transmission zone* $STm_p(T_p)$ of a unit processor p consists of all topoi where another unit processor q can be placed to receive information from p.

For instance, in a two-dimensional cellular automaton, the spatial transmission zone $STm_p(T_p)$ of a unit processor p consists of all four neighbors of p.

Definition 25. The *functional transmission zone* $FTm_p(T_p)$ of a unit processor p consists of all other unit processors, which can receive information from p.

In a cellular automaton, the functional spatial transmission zone of a unit processor p coincides with the spatial transmission zone of p because all unit processors are stationary.

Definition 26. The *combined transmission zone* $SFTm_p(T_p)$ of a unit processor p consists of all other unit processors, which can receive information from p when they are in the specified area (specified topoi).

The key characteristic of a processor for information processes of the type (2) is its reception zone.

Definition 27. The *reception zone* $Rm_p(T_p)$ of a unit processor p specifies from where p can receive information.

Note that in a general case, reception and observation zones do not coincide because a unit processor p can receive information from its observation zone whenever it is necessary, while the processor p can receive information from its reception zone when this information is sent to p.

The reception zone of a unit processor p can be fixed (constant) for p or it can depend on such parameters as the topos T_p, time t and state q of the processor or/and its control device.

There are three kinds of reception zones — spatial, functional and combined.

Definition 28. The *spatial reception zone* $SRm_p(T_p)$ of a unit processor p consists of all topoi such that p can receive information from another unit processor q when q is in one of these topoi.

For instance, in a two-dimensional cellular automaton, the spatial reception zone $SRm_p(T_p)$ of a unit processor p consists of all four neighbors of p.

Definition 29. The *functional reception zone* $FRm_p(T_p)$ of a unit processor p consists of all other unit processors, from which p can receive information.

In a cellular automaton, the functional spatial reception zone of a unit processor p coincides with the spatial reception zone of p because all unit processors are stationary.

Definition 30. The *combined reception zone* $SFRm_p(T_p)$ of a unit processor p consists of all other unit processors, from which p can receive information when they are in the specified area (specified topoi).

The intersection of the reception zone and transmission zone is called the *exchange zone* or *communication zone* of a unit processor p, i.e.,

$$Tm_p(T_p) \cap Rm_p(T_p) = Cm_p(T_p).$$

All these characteristics are external and form two triads — the *spatial triad* $\mathrm{TR}_p = (T_p, Ob_p, Op_p)$ and the *functional triad* $\mathrm{FR}_p = (Op_p, Tm_p, Rm_p)$ of a unit processor p.

It is also possible to define reception, transmission and transition zones for distributed (compound) processors.

These characteristics of a unit processor p or a compound processor A can belong to one of the following three categories:

(1) constant characteristics, e.g., a constant topos;
(2) variable characteristics, e.g., a variable operation zone;
(3) growing/diminishing characteristics, e.g., a growing transmission zone.

Having six external characteristics — topos and observation, operation, reception, transmission and transition zones — and three categories of each of them, we obtain 3^6 classes of unit processors. However, each category of external characteristics contains potentially infinite number of forms because there are infinitely many structures that can be used as topoi, observation zones, operation zones, transmission zones, transition zones and reception zones. Thus, even modifications of external characteristics provide researchers with infinite resources for building various types and kinds of structural machines and using them for construction of computers, networks, cell phones and other information devices. Internal characteristics increase these resources even more. On the one hand, this feature shows computing power of structural machines. On the other hand, the same feature reveals high flexibility and adaptability of structural machines.

It is necessary to understand that by themselves external characteristics do not make unit processors more powerful or more efficient because these processor parameters also depend on internal characteristics. For instance, a unit processor with a big operation zone can have very limited operations for changing its functional space.

Let us consider additional classes of unit processors.

Definition 31. A processor unit p is called

- *topologically uniform*: if all its topoi are isomorphic;
- *operationally uniform*: if all its operation zones are isomorphic;
- *transitionally uniform*: if all its transition zones are isomorphic;
- *observationally uniform*: if all its observation zones are isomorphic;
- *transmissionally uniform*: if all its transmission zones are isomorphic;
- *receptionally uniform*: if all its reception zones are isomorphic;
- *topologically standardized*: if all its topoi have the same type, e.g., are R-neighborhoods;
- *operationally standardized*: if all its operation zones have the same type, e.g., are R-neighborhoods;
- *transitionally standardized*: if all its transition zones have the same type, e.g., are R-neighborhoods;
- *observationally standardized*: if all its observation zones have the same type, e.g., are R-neighborhoods;
- *transmissionally standardized*: if all its transmission zones have the same type, e.g., are R-neighborhoods;
- *receptionally standardized*: if all its reception zones have the same type, e.g., are R-neighborhoods.

For instance, a processor unit p is topologically uniform if all its topoi consist of a single node, e.g., the head of a Turing machine is topologically uniform.

Lemma 9. *Any topologically (operationally, transitionally or observationally) uniform processor unit p is topologically (correspondingly, operationally, transitionally or observationally) standardized.*

Usually, processors of abstract and physical automata (machines) are topologically, operationally, transitionally and observationally uniform. At the same time, processors in chemical and biological automata (machines) can be non-uniform. An example of a non-uniform processor is a processor that can read from or write up to five cells in the memory. Thus, when this processor writes to one cell, its topos consists of one cell. At the same time, when this processor writes to three cells in the memory, its topos consists of these three cells.

Topoi, observation zones and operation zones of unit processors allow us to define the topoi, the observation zones and operation zones of distributed processors.

There are different types of processor units. A processor unit can be

- *controlled* (by the central control device of the structural machine).
- *autonomous*, when it has its own control device.
- *cooperative*, when it has its own control device but the functioning of this processor unit depends on the states both of its own control device and of the central control device of the structural machine.

For instance, in a multihead Turing machine T, all heads are controlled processor units. The control device of T controls them. At the same time, all finite automata in a cellular automaton are autonomous processor units.

We remind that a *finite state machine* also called a *finite state automaton* is an abstract system that can be in a finite number of different finite states and the functioning of which is described as changes of its states.

Proposition 11. *A structural machine M is a finite state machine if and only if*

- *its functional space Sp_M is finite, i.e., in the case of universal functional space, it is a finite structure, or in the case of categorical functional space, it consists of a finite number of finite structures;*
- *the number of unit processors is finite and each of them can be in a finite number of different finite states.*

For instance, a finite automaton is a finite state machine, while a Turing machine is not a finite state machine.

Definition 32. A *temporally finite state machine* is an abstract system that can be in a finite number of different finite states at any moment of time and the functioning of which is described as changes of its states.

Proposition 12. *A structural machine M is a temporally finite state machine if and only if*

- *at any moment of time, its functional space Sp_M is finite, i.e., in the case of universal functional space, it is a finite structure, or in the case of categorical functional space, it consists of a finite number of finite structures;*

- *at any moment of time, the number of unit processors is finite and each of them can be in a finite number of different finite states;*
- *any operation of each unit processor involves only a finite number of structure elements and relations.*

For instance, a Turing machine is a temporally finite state machine, while finite dimensional and general machines of Blum, Shub and Smale (1989) are not temporally finite state machines.

Definition 33. An operation of a processor is *local* or, more exactly, *unilocal* if it is performed with one structural element (e.g., node) and some (all) of its relations (a *pointed operation*), e.g., deleting a structural element (e.g., a node) and all its binary connections (links or edges), adding a link to a structural element or changing a label of a structural element.

For instance, the head h of a Turing machine performs only local operations, while the head of a pushdown automaton can perform non-local operations (Hopcroft *et al.*, 2001). Processors of automata that perform operations of unrestricted formal grammars are mostly non-local (Hopcroft *et al.*, 2001).

Definition 34.

(a) An operation of a processor P is R-*local* if it is performed with elements (e.g., nodes) from the R-neighborhood of a definite element (e.g., node) and with some (all) of their relations (a *singularly local operation*).
(b) An operation of a processor P is topologically R-*local* if it is performed with elements (e.g., nodes) from the R-neighborhood of a topos of P (e.g., node) and with some (all) of their relations (a *topologically local operation*).

Proposition 13. *An operation O of a topologically uniform processor P of a structural machine M is singularly local if the following conditions are satisfied:*

(1) *the structural machine M operates with structures defined by an arrangement* R, *which contains only one binary relation;*
(2) *a topos of the processor P of M is one structural element;*
(3) *an operation O of P is topologically local.*

Definition 35.

(a) An operation of a processor is R-*local* or *totally local* if it is performed with elements (e.g., nodes) from the R-neighborhood of a definite element (e.g., node) and with some (all) of their relations.

(b) An operation of a processor P is topologically R-*local* if it is performed with elements (e.g., nodes) from the R-neighborhood of a topos of P (e.g., node) and with some (all) of their relations (a *wholly local operation*).

Proposition 14. *If a topos of a topologically uniform processor P is one structural element and an operation O of P is wholly local, then O is totally local.*

Definitions imply the following result.

Proposition 15. *If $\boldsymbol{R}$ belongs to* R, *then any R-local operation is* R-*local.*

Let us consider operations performed by processors of structural machines.

The first group of operations consists of the *transition operations*:

(1) *moving* the processor from one topos, e.g., a structure element, to another topos: This operation is local when both elements belong to some relation from R;
(2) *changing* the operation zone of the processor;
(3) *changing* the observation zone of the processor.

The second group of operations consists of the *substantial transforming operations*:

(1) *adding* a structure element, e.g., a node;
(2) *deleting* (*removing*) a structure element, e.g., a node, from a neighborhood of the element where the processor is situated and all relations that include this element;
(3) *deleting* (*removing*) a link from a relation $\boldsymbol{R}$ that connects some structure elements with the element where the processor is situated;
(4) *adding* a link to a relation **R** that connects some structure elements with the element where the processor is situated;
(5) *deleting* (*removing*) a relation $\boldsymbol{R}$ from R;
(6) *adding* a new relation to R.

The third group of operations consists of the *symbolic transforming operations*:

(1) *renaming* a node;
(2) *naming* a node;

(3) *denaming* a node, i.e., deleting the name of a node;
(4) *renaming* a link;
(5) *naming* a link;
(6) *denaming* a link, i.e., deleting the name of a link.

Example 2. Operation of deleting the element f from the first-order structure $\boldsymbol{A} = (A, r, \mathsf{R})$ in Figure 2 where (A) shows the structure before operation and (B) shows the structure after operation. Note that such operations often change relations in the processed structure. Besides, the processor (processor unit) moves from the place (position) f to the place (position) g (see Figure 2). This operation is performed according to the instruction $(q, f, f) \rightarrow (q, g, \sim f)$, in which q is the state of the processor (processor unit), f is the place (position) of the processor (processor unit) before the operation, g is the place (position) of the processor (processor unit) after the operation and $\sim f$ means elimination of f.

Example 3. Operation of adding the relation P for elements g and f in the first-order structure $\boldsymbol{A} = (A, r, \mathsf{R})$ where (A) shows the structure before operation and (B) shows the structure after operation (see Figure 3). Besides, the processor (processor unit) moves from the place (position) g to the place (position) e. This operation is performed according to the instruction $(p, g, f) \rightarrow (p, e, P(g, f))$, in which p is the state of the processor (processor unit), g is the place (position) of the processor (processor unit) before the operation, e is the place (position) of the processor (processor unit) after the operation and f is an observed element and means addition of the pair (g, f) to the relation P.

Structural machines can simulate Turing machines, Kolmogorov algorithms (machines), storage modification machines and cellular automata. We prove some of these results in Section 4.

Structural machines can also simulate processes generated by logical calculi (Schumann and Adamatzky, 2011), as well as λ-calculus and formal grammars because structural machines are able to perform operations used in logic, linguistic and various databases. Structural machines can also compute partial recursive functions and limit partial recursive functions.

It is possible to build structural machines that can work not only with discrete but also with continuous data because structures can be continuous and there are no restrictions on relations in processed structures. This

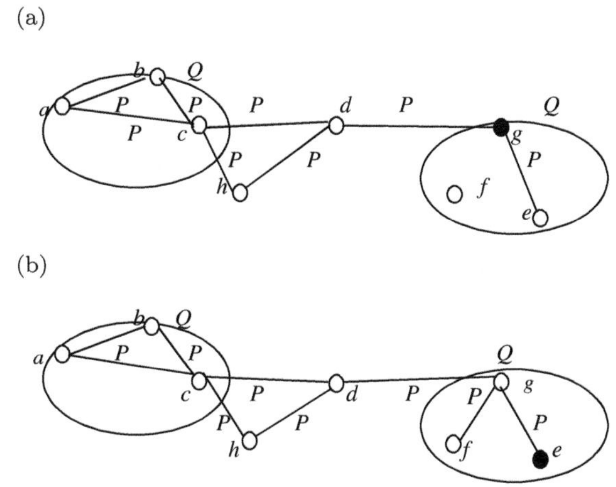

Figure 3. The graphical representation of an operation on first-order structures.

possibility turns artificial neural networks (Haykin, 1994), Shannon's differential analyzer (Shannon, 1941), finite dimensional and general machines of Blum *et al.* (1989) and Type 2 Turing machines (Weihrauch, 2000) into special cases of structural machines.

This shows that it is practical to discern discrete structural machines, which work with discrete structures, have discrete systems of states and operations, and continuous structural machines. In continuous structural machines, one, two or all three of the following components can be continuous: continuous processed structures, continuous system of states and/or continuous operations.

4. Structural Simulation of Kolmogorov Algorithms and Limit Turing Machines

In this section, we study relations between structural machines and other popular computational models. At first, we consider Kolmogorov algorithms.

Theorem 1. *A structural machine with a centralized processor can simulate any Kolmogorov algorithm with the linear time complexity.*

Proof. A Kolmogorov algorithm works with Kolmogorov complexes, performs operations but, similar to finite automata, does not have the inner structure (Kolmogorov, 1953; Uspensky and Semenov, 1993).

A Kolmogorov complex over alphabet B, or Kolmogorov B-complex, is a connected undirected graph with the initial vertex, all vertices of which are labeled (named) by letters from B and have the Kolmogorov property, which demands that for any vertex u, all vertices connected to u by an edge have different labels (names). These complexes play the role of states of Kolmogorov algorithms because on each step of computation, the algorithm obtains a total Kolmogorov complex.

Functioning of a Kolmogorov algorithm A is determined by the operator of direct transformations Ω_A, which consists of a finite number of local rules. Each rule is a triad of the form

$$(U, W, f),$$

where U and W are Kolmogorov complexes and f is a one-to-one mapping of the set $\mathrm{v}(U)$ of all vertices of the complex U into the set $\mathrm{v}(W)$ of all vertices of the complex W, which satisfies additional conditions of locality. This rule determines substitution of the complex U by the complex W in the total Kolmogorov complex. Locality means that for a given Kolmogorov algorithm A, there is a fixed number r such that all vertices in both U and W are reachable from the initial vertex by a path with the length not larger than r. In this case, the operation performed according to the rule (U, W, f) is called r-local.

In the generalization from Uspensky and Semenov (1993) of Kolmogorov complexes, edges are also labeled (named) by natural numbers.

It is easy to see that Kolmogorov complexes and their generalizations are first-order structures with one binary and several unary relations. Consequently, all local operations can be performed by a processor operation zone which contains all neighborhoods of the level r. It means that a structural machine M with this processor can simulate any r-local Kolmogorov algorithm A. Because M can perform each r-local operation of the algorithm A in one step, this simulation will have linear complexity.

The theorem is proved.

Now let us look at how structural machines can simulate super-recursive algorithms. □

Theorem 2. *A structural machine with a centralized processor can simulate any simple limit Turing machine.*

Proof. First, let us describe the structure of a simple limit Turing machine (Burgin, 2005).

Any limit Turing machine has three components, which exactly replicate the structure of a contemporary computer. Namely, a limit Turing machine M has hardware, software and infoware.

The *hardware* of a limit Turing machine M consists of three abstract devices:

(1) the *control device* (*controller*) A, which is a finite automaton and controls performance of M;
(2) the *operating device* (*processor*) H, which includes three processing units;
(3) the *memory* E.

The *software* of a limit Turing machine M consists of rules according to which the operating device H functions.

The *infoware* of a limit Turing machine M consists of those symbols, words and languages which are processes by M. In addition, there is a topology τ on the set of all words processed by M.

In the constructive hierarchy of limit Turing machines, the simplest devices are simple limit Turing machines.

The *memory* E of a simple limit Turing machine is exactly the same as the memory of a Turing machine with three tapes: one of them is the input tape, another is the working tape and the third is the output tape. In contrast to Turing machines, which are usually considered with one tape, this structure properly models the structure of computers representing not only the processing device but also input and output devices of a computer.

The *operating device* (*processor*) H of a simple limit Turing machine consists of three heads, which are similar to those that are used in Turing machines, and each head works in its separate tape. The head h_1 works in the input tape, the head h_2 works in the working tape and the head h_3 works in the output tape.

The *software* R of a simple limit Turing machine M is also a program that consists of three groups of simple rules:

$$q_h a_i \to a_j q_k X, \tag{3}$$

where q_h and q_k are states of A, a_i and a_j are symbols of the alphabet of M, and X is one of the following symbols R, L and S. Each group of rules determines the functioning of one head of H. The rule (1) means that if the state of the control device A of M is q_h and the corresponding head h_t $(t = 1, 2, 3)$ of the processor H observes the symbol a_i in the cell where this head is situated, then the state of A becomes q_k, while the head h_t writes the symbol a_j in the cell where it is situated and moves right to the next cell from its tape when $X = \text{R}$, moves left to the next cell from its tape when $X = \text{L}$ and stays in the same cell when $X = \text{S}$. Each rule directs one step of computation of the simple limit Turing machine M. Rules of the limit Turing machine M define the transition function of M and describe changes of A, H and E. Consequently, these rules also determine the transition functions of A, H and E.

We see that the rules according to which limit Turing machines function are the same as the rules of Turing machines.

The machine M functions in the following way. At the beginning, the head h_1 of the processor H works in agreement with the head h_2, rewriting the input word from the input tape into the working tape. Then the head h_2 starts working independently. It observes some cell with a symbol a_i (it may be the symbol Λ, which denotes an empty cell) and the control device A is in some state q_h. Then the control device A (and/or the head h_2) chooses from the system R of rules a rule r with the left part equal to $q_h a_i$ and performs the operation prescribed by this rule. In some cases, head h_3 writes some symbol in the output tape. If there is no rule in R with such a left part, the machine M stops functioning. If there are several rules with the same left part, then M works as a non-deterministic Turing machine, performing all possible operations from the program of the machine. When A comes to one of the final states from F, the machine M also stops functioning. In all other cases, it continues operation without stopping.

Thus, on each step, operation of a simple limit Turing machine is the same as operation of some Turing machine. The difference is in the definition of the result. The result of a Turing machine is achieved when it comes to a final state. Usually, after obtaining the result, the Turing machine halts because for a Turing machine, it is unreasonable to continue to work after this. A simple limit Turing machine works in a different way. In some cases, it stops without the result. In other cases, it gives the result coming to a final state just as a Turing machine. However, there are situations when a simple limit Turing machine does not halt but gives the result, which is defined by the behavior of the output tape. Namely, the result is the limit of the sequence of words that appear in the output tape during the whole computational process.

Limit Turing machines form an algorithmic base for numerical analysis (Burgin, 2007).

When topology τ is discrete, limit Turing machines coincide with inductive Turing machines.

In Theorem 1 from Burgin and Adamatzky (2017b), it is demonstrated how structural machines with a centralized processor can simulate any Turing machine with one tape. In the theory of algorithms and computations, it is proved that a Turing machine with one tape can simulate a Turing machine with any number of tapes (cf., e.g., Burgin (2005); Hopcroft *et al.* (2001)).

At the same time, a simple limit Turing machine has exactly the same structure and operations (instructions) as a Turing machine with three tapes (Burgin, 2005). Consequently, working in the limit mode (Burgin, 2015), a structural machine with a centralized processor can simulate any simple limit Turing machine.

The theorem is proved.

It is also possible to simulate membrane computations (Păun, 2002) with structural machines, while some classes of structural machines can work as neural Turing machines (Graves *et al.*, 2014). However, these results are presented elsewhere.

Now we show that structural machines can achieve high efficiency in solving some important problems. To do this, we consider such a practical problem as ordering arrays of data. The most popular and efficient algorithm for doing this is called *MergeSort* (cf., e.g., Kleinberg and Tardos (2006)). It was invented by John von Neumann in 1945. MergeSort has an average and worst-case complexity of $O(n \log n)$. Structural machines can achieve essentially better complexity of performance. □

Theorem 3. *Multiprocessor structural machines can perform MergeSort with logarithmic complexity, i.e., with complexity* $O(\log n)$.

For the sake of compactness, we give an outline of a proof.

To prove the theorem, we show that a relevant structural machine can merge two ordered arrays creating one ordered array in four steps. Performing these steps in a parallel mode gives logarithmic complexity.

MergeSort sorts a given set of numbers or, in general, elements of an ordered set by first dividing them into two equal parts, sorting each half separately by recursion and then combining the results of these recursive calls — in the form of two sorted halves — using the linear-time algorithm for merging sorted lists (cf. Kleinberg and Tardos, 2006). A relevant structural machine M using enough processors can perform merging of sorted lists (arrays) in four steps, reducing time to $O(\log n)$.

To achieve this goal, the following relations are utilized:

- Binary relations
  - $1NP$ determines the first basic linear array (sequence or list) of data;
  - $2NP$ determines the second basic linear array (sequence or list) of data;
  - NP determines the merged array (sequence or list) of data;
  - relation $\leq$ determines the order of data;
  - $l_j NP$ determines the linear array (sequence or list) with number j.
- Unary relations
  - $l_j b$ determines the beginning of the linear array (sequence) with number j;
  - $l_j e$ determines the end of the linear array (sequence) with number j.

An array (list) is ordered when $NP(a, b)$ implies $a \leq b$.

To perform merging of two ordered arrays (lists) with n elements each, the structural machine M uses $n+1$ processors, which perform the following operations, which include simple operations, such as a comparison of two elements (numbers), executed in the parallel mode:

(1) adding the necessary structure, which consists of $n + 1$ lists, and each of these lists contains n empty cells;
(2) comparing data from the two merged arrays (lists) in the parallel mode;
(3) filling some of the empty cells from the added structure;
(4) cleaning the added structure.

The structural machine M has enough processors to do any finite number of merges at the same time. As the number of steps is $\mathrm{O}(\log n)$, the complexity of the whole algorithm is $4 \cdot \mathrm{O}(\log n) = \mathrm{O}(\log n)$.

The theorem is proved.

5. Conclusion

We determined and studied information processing by structural machines demonstrating that they can simulate Kolmogorov algorithms (Theorem 1) and limit Turing machines (Theorem 2). We also established high efficiency of structural machines demonstrating (Theorem 3) that structural machines are rather efficient allowing data ordering with logarithmic complexity while complexity of conventional algorithms is $n \cdot \log n$. Further work can go in two directions: implementation of practical algorithms on structural machines and development of structural machine models for different kinds of computations, such as quantum, biological and chemical computations.

For the better development of practical algorithms and more efficient computing devices, it is necessary to allow structural machines to "enter the real world" being realized by physical devices and not just remain one of the many formal accomplishments of theoretical computer science.

It is necessary to remark that due to their flexibility, definite classes of structural machines allow much better modeling of quantum computations than conventional models. Now there are various theoretical models of quantum computation: quantum Turing machines (Deutsch, 1985) and quantum circuits (Feynman, 1986) correspond to discrete computing, while modular functors describe topological quantum computation (Freedman *et al.*, 2002), to mention but a few. It is interesting that while some theoretical models are recursive algorithms, which are not more powerful than Turing machines (Deutsch, 1985), other theoretical models are more powerful than Turing machines (Kieu, 2003). Structural machines provide a theoretical framework for unification of different models of quantum computation. This opportunity brings us to the open problem of modeling quantum computation by structural machines.

It would also be interesting to show how structural machines can simulate and in some cases outperform other abstract automata and models of computation. For instance, it is possible to suggest the following conjecture:

Structural machines can simulate Gandy–Paun–Rozenberg machines (Obtułowicz, 2010) *and Gandy machines* (Gandy, 1980) *without increasing computational complexity.*

One more possible direction of research in this area is related to databases and knowledge bases. To model processes in relational databases, Abiteboul, Papadimitriou and Vianu introduced important classes of relational machines and reflective relational machines (Abiteboul *et al.*, 1994, 1997, 1998). A relational machine M is a Turing machine with a standard Turing tape, a query tape, and a relational store, capable of storing an unbounded but finite number of arbitrary conventional relations $R_0, R_1, R_2, \ldots$, i.e., relations of the first order. All these machines are notable special cases of structural machines because conventional set-theoretical relations are structures of the first order. That is why it is possible to use the theory of structural machines to study properties of relational machines and reflective relational machines. Moreover, application of this theory allows the introduction of relational machines and reflective relational machines of higher orders with the goal to improve the database performance.

One more possible direction of research in this area is related to optimization of computation. For instance, it is possible to decrease computational complexity of many problems by using alternating Turing machines (Immerman, 1988; Kozen, 2006). Using expanding systems of unit processors, structural machines can simulate alternating Turing machines with the same or even lesser complexity, achieving in such a way more efficient information processing.

Acknowledgment

The author would like to express gratitude to Gordana Dodig-Crnković and Robert Logan for their useful remarks.

References

Abiteboul, S., Papadimitriou, C. H. and Vianu, V. (1997). Fixpoint logics, relational machines and computational complexity, *J. ACM* 44(1), 30–56.

Abiteboul, S., Papadimitriou, C. H. and Vianu, V. (1998). Reflective relational machines, *Inform. Comput.* 143(2), 110–136.

Abiteboul, S., Papadimitriou, C. H. and Vianu, V. (1994). The power of reflective relational machines, in *Proceedings of the Symposium on Logic in Computer Science 1994. LICS '94*, pp. 230–240.

Adamatzky, A. (2010). *Physarum Machines: Computers from Slime Mould*, World Scientific, New York.

Berge, C. (1973). *Graphs and Hypergraphs*, North Holland P.C., Amsterdam/ New York.

Blum, L., Shub, M. and Smale, S. (1989). On a theory of computation and complexity over the real numbers: NP-completeness, recursive functions and universal machines, *Bull. Amer. Math. Soc.* 21(1), 1–46.

Bourbaki, N. (1993). *Elements of the History of Mathematics*, Springer-Verlag, New York.

Burgin, M. (1997). *Fundamental Structures of Knowledge and Information*, Ukrainian Academy of Information Sciences, Kiev (in Russian).

Burgin, M. (1998). *On the Nature and Essence of Mathematics*, Ukrainian Academy of Information Sciences, Kiev (in Russian).

Burgin, M. (2003). Cluster Computers and Grid Automata, in *Proceedings of the ISCA 17th International Conference Computers and their Applications*, International Society for Computers and their Applications, Honolulu, Hawaii, pp. 106–109.

Burgin, M. (2004a). Data, information, and knowledge, *Information* 7(1), 47–57.

Burgin, M. (2004b). Algorithmic complexity of recursive and inductive algorithms, *Theoret. Comput. Sci.* 317(1/3), 31–60.

Burgin, M. (2005). *Super-recursive Algorithms*, Springer, New York.
Burgin, M. (2007). Numerical analysis, limit computations, and neoclassical analysis, in *Abstracts of Lectures: 4th International Conference of Applied Mathematics and Computing*, Plovdiv, Bulgaria, August 12–18, 2007, p. 66. Available online: http://math.uctm.edu/conference2008/content/abstracts.pdf.
Burgin, M. (2009). Grid automata as a tool for network design, in *Proceedings of the 8th WSEAS International Conference on Data Networks, Communications, Computers* (DNCOCO'09), Baltimore, Maryland, USA, November, pp. 147–151.
Burgin, M. (2010). *Measuring Power of Algorithms, Computer Programs, and Information Automata*, Nova Science Publishers, New York.
Burgin, M. (2012). *Structural Reality*, Nova Science Publishers, New York.
Burgin, M. (2015). Super-recursive algorithms and modes of computation, in *Proceedings of the 2015 European Conference on Software Architecture Workshops*, Dubrovnik/Cavtat, Croatia, pp. 10:1–10:5.
Burgin, M. (2016a). Inductively computable hierarchies and inductive algorithmic complexity, *Global Journal of Computer Science and Technology*, Ser. E: Network, Web & Security, 16(1), pp. 35–45.
Burgin, M. (2016b). *Theory of Knowledge: Structures and Processes*, World Scientific, New York.
Burgin, M. (2017). Inaccessible information and the mathematical theory of oracles, in *Information Studies and the Quest for Transdisciplinarity: Unity through Diversity*, World Scientific, New York, pp. 59–114.
Burgin, M. and Adamatzky, A. (2017a). Structural machines and slime mold computation, *Int. J. General Syst.* 45(3), 201–224.
Burgin, M. and Adamatzky, A. (2017b). Structural machines as a mathematical model of biological and chemical computers, *Theory Appl. Math. Comput. Sci.* 7(2), 1–30.
Burgin, M. and Bratalskii, E. (1986). The principle of asymptotic uniformity in complex system modeling, in *Operation Research and Automated Control Systems*, Institute of Cybernetics, Kiev, pp. 115–122 (in Russian).
Burgin, M. and Eberbach, E. (2009). On foundations of evolutionary computation: an evolutionary automata approach, in *Handbook of Research on Artificial Immune Systems and Natural Computing: Applying Complex Adaptive Technologies* (Hongwei Mo, Ed.), IGI Global, Hershey, Pennsylvania, pp. 342–360.
Burgin, M., Mikkilineni, R. and Mittal, S. (2017). Knowledge processing as structure transformation, *Proceedings*, 1(3), 58; doi:10.3390/IS4SI-2017-03988.
Čech, E. (1966). *Topological Spaces*, John Wiley & Sons, New York.
Chandra, A. K., Kozen, D. C. and Stockmeyer, L. J. (1981). Alternation, *J. ACM* 28, 114–133.
Deutsch, D. (1985). Quantum theory, the Church-Turing principle, and the universal quantum Turing machine, *Proc. Roy. Soc. Ser. A* 400, 97–117.
Dodig-Crnković, G. (2006). *Investigations into Information Semantics and Ethics of Computing*, Mälardalen University Press, Västerås.

Dodig-Crnković, G. (2012). Alan Turing's legacy: Info-computational philosophy of nature, In *Proceedings of Symposium on Natural Computing/ Unconventional Computing and its Philosophical Significance*, AISB/IACAP World Congress 2012, Birmingham, UK, July 2–6, pp. 47–50.

Dodig-Crnković, G. (2011). Dynamics of information as natural computation, *Information* 2(3), 460–477.

Dodig-Crnković, G. and Müller, V. (2009). A dialogue concerning two world systems: info-computational vs. mechanistic, in *Information and Computation*, Dodig Crnković, G., Burgin, M. (eds.) World Scientific Publishing Co., Inc., Singapore, pp. 149–184.

Feynman, R. P. (1986). Quantum mechanical computers, *Found. Phys.* 16(6), 507–531.

Fredkin, E. (1992). *Finite Nature*, XXVIIth Rencontre de Moriond.

Freedman, M. H., Kitaev, A., Larsen, M. J., and Wang, Z. (2002). Topological quantum computation, *Bull. Amer. Math. Soc.* 40(1), 31–38.

Gandy, R. (1980). Church's thesis and principles for mechanisms, *The Kleene Symposium*, North-Holland, Amsterdam, pp. 123–148.

Goodrich, M. T. and Tamassia, R. (2004). *Data Structures and Algorithms in Java*, Wiley Higher Education, John Wiley & Sons, Inc., New York.

Graves, A., Wayne, G. and Danihelka, I. (2014). *Neural Turing machines*, preprint, arXiv:1410.5401.

Grossberg, S. (1988). Nonlinear neural networks: Principles, mechanisms, and architectures, *Neural Networks* 1, 17–61.

Haykin, S. (1994). *Neural Networks: A Comprehensive Foundation*, Macmillan, New York.

Hopcroft, J. E., Motwani, R. and Ullman, J. D. (2001). *Introduction to Automata Theory, Languages, and Computation*, Addison-Wesley, New York.

Immerman, N. (1988). Nondeterministic space is closed under complementation, *SIAM J. Comput.* 17, 935–938.

Jaffe, A. M. and Liu, Z. (2018). Mathematical picture language program, *Proc. Natl. Acad. Sci. USA* 115(1), 81–86.

Kieu, T. D. (2003). Computing the noncomputable, *Contemp. Phys.* 44, 51–77.

Kleinberg, J. and Tardos, E. (2006). *Algorithm Design*, Pearson–Addison-Wesley.

Kolmogorov, A. N. On the concept of algorithm, *Uspekhi Mat. Nauk.* 8(4), 175–176.

Kozen, D. (2006). *Theory of Computation.* Springer-Verlag.

Liu, Z., Wozniakowski, A. and Jaffe, A. M. (2017). Quon 3D language for quantum information, *PNAS* 114(10), 2497–2502.

Lloyd, S. (2006). *A theory of quantum gravity based on quantum computation*, preprint, arXiv:quant-ph/0501135.

Mikkilineni, R., Morana, G. and Burgin, M. (2015). Oracles in software networks: A new scientific and technological approach to designing self-managing distributed computing processes, in *Proceedings of the 2015 European Conference on Software Architecture Workshops*, Dubrovnik/Cavtat, Croatia, September 7–11, 2015, ACM, pp. 11:1–11:8.

Obtułowicz, A. (2010). Randomized Gandy–Păun–Rozenberg machines, *Rom. J. Inform. Sci. Technol.* 13(2), 181–196.

Paun, Gh. (2002). *Computing with Membranes: An Introduction*, Springer, Berlin.
Petri, C. (1962). *Kommunikation mit Automaten*, Ph.D. Dissertation, University of Bonn, Bonn, Germany.
Piccinini, G. (2007). Computational modeling vs. computational explanation: Is everything a turing machine, and does it matter to the philosophy of mind? *Austr. J. Philos.* 85(1), 93–115.
Resnick, M. (1997). *Mathematics as a Science of Structures*, Oxford University Press, Oxford.
Schumann, A. and Adamatzky, A. (2011). Physarum spatial logic, *New Math. Nat. Comput.* 7(3), 483–498.
Shannon, C. (1941). Mathematical theory of the differential analyzer, *J. Math. Physics* 20, 337–354.
Siegelman, H. T. (1999). *Neural Networks and Analog Computation: Beyond the Turing Limit*, Birkhauser, Berlin.
Uspensky, V. and Semenov, A. L. (1993). *Algorithms: Main Ideas and Applications*, Springer, Dordrecht.
Weihrauch, K. (2000). *Introduction to Computable Analysis*, Springer, Berlin.
Wolfram, S. (2002). A *New Kind of Science*, Champaign, Il., Wolfram Media.
Zuse, K. (1969). *Rechnender Raum*, Friedrich Vieweg & Sohn, Braunschweig, Germany.

Chapter 15

Consciousness and Information: Before or After the Event?

Ron Cottam* and Roger Vounckx†

Department of Electronics and Informatics,
Vrije Universiteit Brussel,
Brussels, Belgium
**ricottam@etrovub.be*
†*rvounckx@etrovub.be*

At first sight, consciousness appears to be the context within which action or though, is planned. The question we now raise is whether this is really the case, or whether consciousness of an event is simply an "irrelevant" by-product, and whether information is the interface between unconscious and conscious processing. We take as our starting point the output of one or other of the many current models for the generation of awareness or consciousness, and develop the idea that high-level consciousness is an amplified form of this which results from interscalar interactions in the network hierarchy of the brain or of cellular structure. Our question visualizes action-planning as an unconscious precursor of consciousness, and places consciousness as the "tip of the iceberg" of neural processing through the medium of information. If this point of view is valid, then information is naturally ephemeral, and only a valid description in the context of a multiscalar conscious interpretant.

1. Introduction

Conventionally, we look on consciousness as a precursor of action or thought, as a control environment within which action- or thought-planning takes place. This places consciousness temporally in advance of conscious action, as opposed to action's many manifestations which take place outside the realm of consciousness, for example, that of breathing, or of the often-experienced driving of a car along a familiar route while being totally occupied with unconnected thoughts.

Suggested models for the generation of consciousness are many and varied (Crick and Koch, 1990; Dennett, 1991; Coaen *et al.*, 1994; Köhler and Moscovitch, 1997; Kihlstrom, 1987; Leahey, 1997; Solso, 1999; Tononi, 2004; Lefton *et al.*, 2005; Damasio, 2010; Taylor, 2011; Wang, 2012; Schroeder, 2012). We will begin from the assumption of an initially almost insignificant degree of awareness, as derived from one or other of the currently considered viable models, and as relevant to organisms lacking an extensive neural structure. We do not address this initial appearance of awareness, but only its further amplification through a number of hierarchically organized levels of neural processing, ultimately up to the high degree of consciousness which we as humans experience. In doing so, we intentionally and provisionally keep out of this hotly debated area of initial generation of consciousness, thus avoiding giving implicit support to one or other of the above cited propositions. However, recent articles (Tononi, 2004; Schroeder, 2012) have attributed the generation of consciousness to the integration of information. In this chapter, we take a "devil's advocate" position with respect to consciousness, and therefore also with respect to information, in that we address the question of primacy of consciousness and information with respect to *all* facets of action.

The English language as it stands is insufficient for our purposes, as it attaches the word "information" to a wide range of different concepts, and we will need a different expression to refer to this wide panoply without diluting the specific concept of *information* which we will derive. The word we have chosen for this task is *info*. Merriam-Webster provides as its primary meaning for information "the communication or reception of knowledge or intelligence", adding also "intelligence, news", "facts, data" and "a numerical quantity that measures the uncertainty in the outcome of an experiment to be performed", among other definitions. "As Luciano Floridi (2010, 2011) stresses, it is a polysemantic concept associated with different phenomena, such as communication, knowledge, reference, meaning, truth, etc." (Lombardi *et al.*, 2014). Most particularly, even in the scientific research domain, there is much disagreement over how the concept of information should be represented. The most well-known version of this derives from the work of Shannon (1948), where average information is portrayed in the context of message transmission through a restricted channel as entropy. Although entropy is often used in this way as a characterization of the information content of a data source, the information content is not absolutely defined, as it depends directly on the probabilistic model being used. Consequently, for example, the definition of information in a thermodynamic context — where entropy is defined somewhat differently — is to a degree at odds with the content of Shannon information. In our current context, where we will derive a very specific and unique meaning for *information*, it is in our interest to group together the various other uses of the word under our umbrella term of *info*, which as a consequence of its multiple uses is unsurprisingly impossible to define exactly, either linguistically or mathematically. Conversely, our use of the word *information* is extremely well definable, although not directly in linguistic terms, as it represents the unique result of the hierarchical processing style we will present. A mathematical definition will also be absent from this chapter, as the hierarchical processing style is presented in conceptual and not mathematical form. Such a mathematical definition is, however, currently under investigation, in terms of quantum logical considerations (Schroeder, 2009) and generic quantum error correction.[1]

The basic hypothesis we will present is that consciousness is an irrelevant by-product of the integration of current sensory and historically stored contents of {*info* and/or *information*} which is essential to support the survival of an organism in a possibly hazardous environment, and that the principal result of such a living integration is *information* itself. This makes *information* uniquely an ever-changing feature of any living system, and this ephemeral character is at odds with its more usually assumed "factual" nature. From this point of view, more habitual expressions of "information" — Shannon "information" (Shannon, 1948), for example — are particular instantiations of *info* and not of *information* as we would ourselves describe it. As such, they are indeed related to *information*, but are drastically reduced in completeness by the process of becoming identifiably inert "facts". In general, we would maintain that *at any given moment in time*, *information* is unquantifiable, but has maximal temporal precision, as opposed to *info*, which will often present a formally expressed aspect, but whose temporal precision is usually far reduced, if not virtually absent. This difficult concept makes the relationship between quantifiability and temporal precision for *info* and *information* in some ways resemble a Heisenberg uncertainty between position and momentum (Busch *et al.*, 2007), along with the establishment of a maximal mathematical product of the two, which is related to the "livingness" of the system under consideration. It also flies in the face of a conventionally assumable relationship between

[1]An idea due to Walter Schempp: private communication.

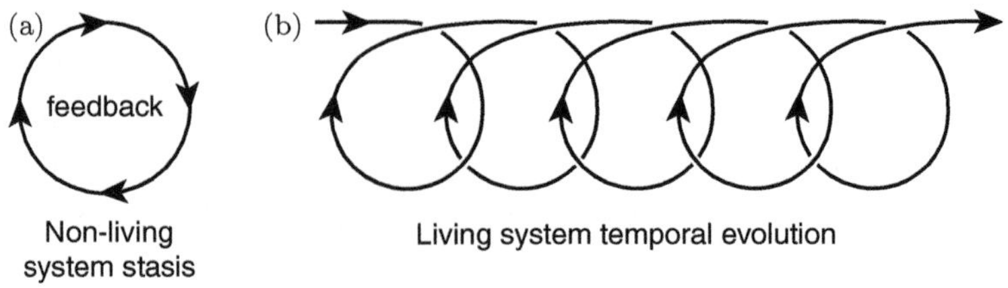

Figure 1. (a) The stable feedback characteristic of a non-living system's relationship between process and structure, and (b) the temporally evolving relationship of a living system.

quantifiability and precision *per se*: namely that the more quantifiable a property is, the greater can automatically be the precision of its characterization.

Our distinction between "non-living" and "living" systems is illustrated in Figure 1 in terms of a system-stabilizing feedback of the relationship between process and structure (Cottam and Ranson, 2018). Non-living systems exhibit a temporally static feedback characteristic (see Figure 1(a)), while living systems exhibit a temporally progressive feedback (see Figure 1(b)). This distinction is the reason for the apparently contradictory relationship between quantifiability and temporal precision that we have cited above.

At this point in the chapter, the stated important distinction between *information* and *info* will probably appear to be rather abstract, and we will need to see the following implications of a hierarchical approach to living systems to be any clearer. Suffice it to say that we provisionally propose that all previous scientific and societal references to "information" fall into our category of *info*, and that *information* itself corresponds to a very particular result of the internal hierarchical processing[2] of living systems to which we will refer.

In defending our hypothesis, we will of course need to elaborate as far as possible the(se) mechanism(s) by which {*info* and/or *information*} integration results in a further generation of *information*. We will relate this integration directly to the hierarchical nature of living systems, both in

[2]Notably, here, we do not use the expression "information processing" in living systems, as this would immediately invalidate our thesis! The processing which is apparent in a hierarchical structure is neither uniquely of info, nor uniquely of information, as we will see later — thus our references in the following paragraphs to {*info* and/or *information*} processing.

their physical makeup and their consequent {*info* and/or *information*} processing. We should note in passing that the approach to hierarchy we will take — that of *model* hierarchy (Cottam *et al.*, 2003, 2004a) — inevitably also encompasses heterarchical aspects of life.

2. Model Hierarchy

Much of both general and particular work on hierarchical structures has been published. A large part of this output has specifically concerned business contexts (see, e.g., Anderson and Brown, 2010; Widhiastuti, 2012), and only a part is directly related to natural biological systems (see, e.g., Salthe, 1985; Pattee, 1973; Chauvet, 1993). However, the only work on *model* hierarchy of which we are aware has originated from this laboratory (the Department of Electronics and Informatics of the Vrije Universiteit Brussel) (see, e.g., Cottam *et al.*, 2008a, 2008b).

We take as an important example of prior effort the publications of Stan Salthe (see, e.g., Salthe, 1993, 2012), which define an exclusive pair of hierarchical structures which he lately refers to as *compositional* and *subsumption* hierarchies. A compositional hierarchy consists of a nested set of levels of physical composition in terms of scale, for example, atoms, molecules, cells, organisms, etc. A subsumption hierarchy consists of a set of related descriptions of a single context, for example, physics, chemistry, biology, etc. A model hierarchy, however, is one organizational level above these two (Cottam *et al.*, 2013): it is the parent of Salthe's two hierarchies. Salthe[3] himself has commented that a model hierarchy resembles a subsumption hierarchy constructed in terms of scale. It is important also to note that we believe hierarchy in nature to be an ontological aspect, rather than purely an epistemological one. This point of view is supported by the stability in nature of different organizational levels of atoms, molecules, cells, organisms, etc., which is confirmed by their successful temporally independent treatment as such in the various domains of science.

The simplest form of this arrangement is found in crystals, where the (micro) atomic level of organization is directly and stably related to the (macro) whole-crystal level (Cottam and Saunders, 1973), albeit without exhibiting large-scale *info* differences between the levels (a configuration often referred to as a "flat" hierarchy).

[3]Stan Salthe: private communication.

A model hierarchy consists of a scaled set of models of a single system. An example could be {a tree as atoms}, {a tree as molecules}, {a tree as cells}, ... {a tree as itself}. Each level of the hierarchy constitutes a self-consistent model of the tree, and adjacent levels are related through what we will refer to as a Rosennean complex interface.

Rosen (1991) has defined a style of complexity for which an infinite number of formal models would be necessary to completely describe the context; description in terms of a single formalism then always leaves a possibly large degree of uncertainty: Rosennean complexity implies a sense of incompleteness of definition. In our example of a tree, the number of molecules far outweighs the number of cells, so their mutual interface must take account of this reduction in going from {a tree as molecules} to {a tree as cells}, or in going in the other direction. These processes are impossible to describe within a single formalism; thus, the interface has the characteristic of Rosennean complexity (Rosen, 1991). Figure 2 indicates the general form of a model hierarchy. Unconventionally, we represent hierarchy lying on its side, to avoid the presupposition that the apex is in overall control of the structure: this is counter to belief in a business environment, for example, where the CEO is nominally the controlling head of the organization. In our context, however, it is not necessarily the case that the apex — which will always be an organism "as itself" — is in overall control of the entire structure: even below the cellular level, DNA errors can ultimately cause organism demise, let alone viral infections.

Toward the left-hand side of this structure, the model levels become progressively more complicated. We should first be clear as to what we mean by the word "complicated". We distinguish simplistically 3 degrees

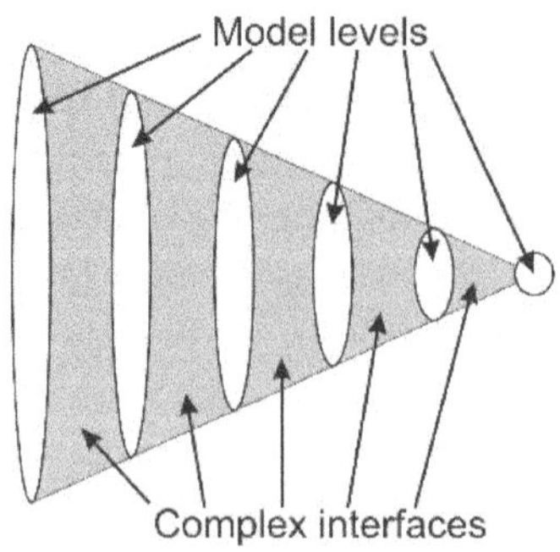

Figure 2. The general form of a model hierarchy, with the apex on the right-hand side. The gray regions indicate the interlevel complex regions.

of representation: simple, complicated and complex. We have already addressed "complex" above. "Simple" implies ease and precision of understanding or computation, and "complicated" implies more elaborate but still easy to compute structure, with an almost complete absence of complexity. We say "almost" complete, keeping in mind the restriction on system completeness implied by Gödel's work (Berto, 2010). We have earlier proposed (Cottam *et al.*, 2005) that a model hierarchy is a principal characteristic of life. In a living physical model hierarchy, the different levels each represent an organizational style of the organism, from {an organism as atoms} up to {an organism as itself}, through levels associated with biomolecules, organs, etc. Each model level maintains its completeness by expelling complexity into its surrounding (complex) interfaces (Cottam *et al.*, 2004b). The effect of the surrounding complexity is to partially isolate model-levels from each other, supporting a locally scaled autonomy. The result is very different from that of an artificial hierarchy — a digital computer, for example — in that higher model levels (to the right-hand side of Figure 2) can run faster than those lower down, permitting the organism "as itself" to rapidly react to external threats without being encumbered by the complicated slowness of lower organizational levels, such as that of cells.

So, individual model levels are partially isolated, partially intercommunicating. But by what means does this intercommunication take place? Walter Schempp[4] has suggested that interlevel communication in a model hierarchy takes place through a generic form of quantum error correction (Gottesman, 2009). This would be consistent with a description of the complex interfaces as repositories of quantum-style logic (Cottam *et al.*, 2018). Schroeder (2009) has pointed out that quantum-style coherence is feasible in the absence of quantum mechanics.

Somewhat unsurprisingly, considering the close relation between a complex interface and its adjacent model level, the assembly of complex interfaces forms a second partial hierarchical scheme, counter-oriented to the model-level partial hierarchy, as shown in Figure 3 (Cottam *et al.*, 2000). There is a particularly close relationship between a model level and the immediate precursor complex region to its left. The model level itself is established from "containing" *info*, which constitutes detail of the model at that scale and which subsumes all the lower-level systemic detail; the

[4]Walter Schempp: private communication.

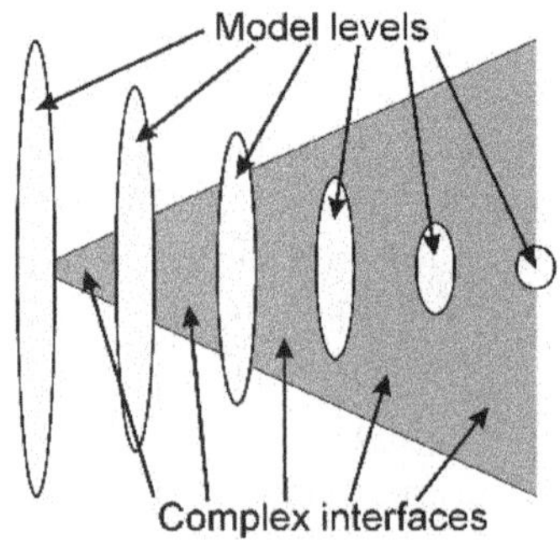

Figure 3. The appearance of a second partial hierarchy, now of complex interfaces, counter-oriented to the original hierarchy of models.

complex region is characterized by "contained" *info* which constitutes the lower-level systemic detail subsumed by the model. Consequently, the complex regions are more extensive in *info* to the right-hand side of the assembly than to the left: this is why the complex partial hierarchy is more correctly laterally inverted in Figure 3 from its appearance in Figure 2. As might be expected, the two partial hierarchies are not completely independent of each other, being coupled together in a similarly partial manner. Thus, the first hierarchy consists of a scaled set of organism models, while it is notable that the second constitutes a historically assembled scaled set of internal representations of the organism's environment (Cottam *et al.*, 2017a).

3. Hierarchical {*Info* and/or *Information*} Processing

We have so far related model hierarchy to the physical structure of an organism, but there is more. We propose that model hierarchy can successfully represent the internal processing of {*info* and/or *information*} which is required for organism survival (Cottam *et al.*, 1999). Incoming results of biosensor reactions to the organism's environment must be fused together to enable the organism to draw overall conclusions leading to action. This process itself leads to the creation of an internal representation of the environment, which is used in defining the action to be taken: the first "port of call" for such a reactive processor will be this internal hierarchical representation of the hierarchical environment, and not uniquely the immediate incoming results of biosensor activity.

We consider it justified to attribute all {*info* and/or *information*} processing to model hierarchy, encompassing as it does heterarchy. Does such processing ever take place outside the confines of an organism? We would say not. Consequently, *info* in a lifeless environment would automatically have the character of fragmentation, and never that of integration.

{*Info* and/or *information*} processing in the mammal brain provides a prime example of hierarchical processing (Striedter *et al.*, 2013), although hierarchical character is incomplete, presumably due to the history-based scavenging nature of evolution — we again emphasize that model hierarchy subsumes heterarchy. However, we now come to a problem. If the different model levels are partially autonomous and partially isolated, how is it that we can form a mental model of the entire structure to a good degree of approximation? We have proposed (Cottam *et al.*, 2012) that a vital function of a model hierarchical processor is the generation of such an approximate representation, which we refer to as "hyperscale" (Cottam *et al.*, 2006). Remembering that we have two partial hierarchies, we will also have two hyperscalar representations, one associated with the model levels, the other with their complex interfaces, as indicated in Figure 4.

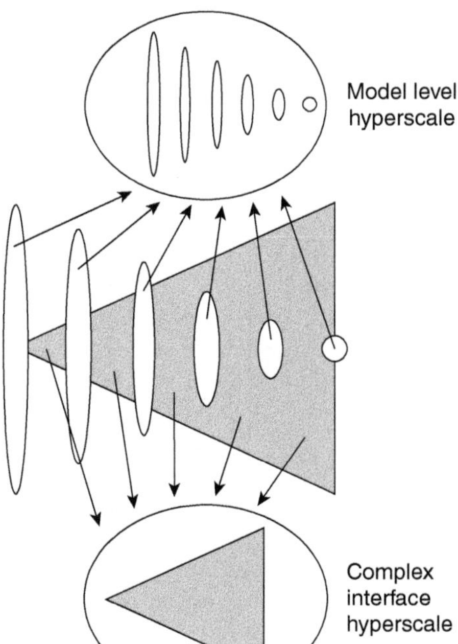

Figure 4. The appearance of dual hyperscales from the model partial hierarchy and the complex interface partial hierarchy.

4. {*Info* and/or *Information*} Reunification to Information

Although often forgotten, the prime characteristic of a system is its unification. But we now appear to have fragmented our model hierarchy into two parts. These two must ultimately reunify — to give a singular entity which we refer to as "metascale" (Cottam *et al.*, 2017b) (see Figure 5).

Although we have described this entire process as a simple sequence of events, it is likely that these are to some extent concurrent, to some extent sequential. This distinction has far-reaching effects in the definition of information and consciousness: full concurrence would imply perfect synchronization of all the resultant system properties. We consider it most probable that a degree of sequential nature remains: this is a delicate point. Notice that the reason for the original splitting up of the hierarchy was that otherwise the individual model levels would remain partially isolated and unification would be impossible.

The finally unified representation of {*info* and/or *information*} integration is now our target — it is *information* in its most complete guise. This

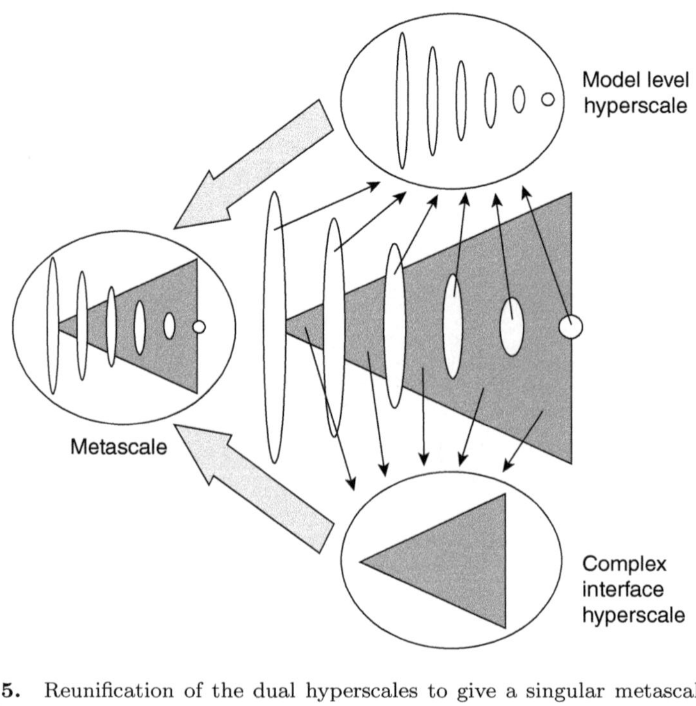

Figure 5. Reunification of the dual hyperscales to give a singular metascalar entity description.

is the result of {*info* and/or *information*} processing by the organism, and it is consequently fully subjective to that organism. This makes *information* very different from *info*, which remains objective or quasi-objective in character.

Remembering that the development of hyperscale immediately permits reactive processing of the {*info* and/or *information*}, and that hyperscale quasi-predates reunification, *information* can be qualified as existing after reactive processing, and not before, this depending on the degree of processing concurrence involved in the transitions from hierarchy to hyperscale and metascale.

5. The Question of Consciousness

The brain appears to be layered in composition from (1) the brainstem and the cerebellum..., through (2) the limbic system, amygdala, hypothalamus, cingulate cortex..., to (3) the cerebral neocortex, which is found in higher mammals and humans. These different layers presumably operate as a unified "whole" (not forgetting the hemispheric split and the *corpus calossum*'s coupling of the two hemispheres[5]), but the "top" (3) does not directly control the "bottom" (1) [Grandin and Johnson, 2005], in concert with our maintenance for a model hierarchy. However, we hypothesize provisionally that high-level consciousness as we experience it appeared with the evolution of the "top" neocortical complex.

An important aspect of {*info* and/or *information*} processing is that of stasis neglect (Cottam *et al.*, 2013). Between model levels there will be a combination of communicated {*info* and/or *information*} which consists of both static detail and the temporally novel. Stasis neglect causes long-term static {*info* and/or *information*} to be neglected in updating the levels, as a means of reducing the imposed computational load. This provides an important aspect of the appearance of high-level consciousness, as well as in the associated generation of information through integration of {*info* and/or *information*}.

[5]Notably, in the case of operations to separate the two hemispheres, to cure particularly extreme forms of epilepsy (see, for example, Akelaitis, 1941, Akelaitis *et al.*, 1942), the two hemispheres appear to operate quite independently, even to the extent of exhibiting dual consciousness (see, for example, Sperry *et al.*, 1969).

We are left with the question of how a high level of consciousness can develop in what we are assuming is a multilevel neural-processing structure (Cottam and Ranson, 2012). Our hypothesis is that an initially low level of awareness can be amplified by events taking place in an {*info* and/or *information*}-processing network. This would present us with a progressively increasing degree of integrated awareness of {*info* and/or *information*} across the organism model hierarchy from the left-hand side toward the right, but still not give us the high level of consciousness we can associate with information integration between the partial model hierarchy and its contextual complex region partial hierarchy. For that, we need to look at the relationship between these two — between the partial model hierarchy and its contextual complex region partial hierarchy.

We call upon the ideas of Koichiro Matsuno at this point (Matsuno, 2000), and his description of interrelationships as mutual observations. In the interaction between the two hyperscales to create metascalar information, we suggest that they mutually observe, leading to the "stasis neglect" truncation of an otherwise endless process, creating the high level of consciousness which we as mammals experience — see Figure 6. This mirrors Rosen's (1991) description of truncation of the otherwise endless sequence of interactions which leads to the stabilization of Newton's laws of motion.

Our maintenance of the remaining partial sequential nature is clearly vital here, but it seems reasonable: mutual observation between the two hyperscales is only possible if they already exist. However, neural action-planning can take place immediately on generation of the approximating hyperscales. This suggests a sequence of events which can be simply modeled as follows. First the system is nominally unified by the creation of the assembly of model levels and the initial complexity in their interfacings.

This is, of course, a very primitive viewpoint, as we are here talking about many aeons of evolution. Progressively with environmental interactions, sensor inputs update the system to mature the internal representation of the encountered environment, by developing the complex interfaces and ejection of complexity from the model levels. Ultimately, this process continues in real time as the organism continues to live. Again, in real time, the two hyperscales are continuously updated by, and for, survival maintenance, which again is only possible if the model levels and complex interfaces have already been developed. Next, the two hyperscales integrate to reunify and stabilize the system, producing a stable environment within which mutual observation between them generates — lastly — the phenomenon of high-level consciousness.

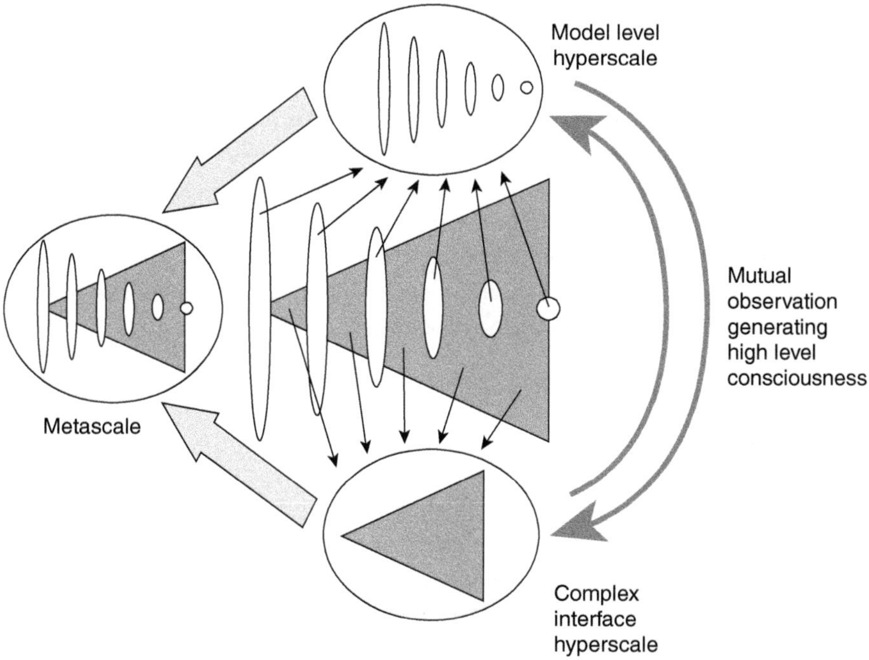

Figure 6. Mutual observation between the two hyperscales generates high-level consciousness.

The character of high-level consciousness must be somewhere between a purely internal phenomenon and an external one, being associated as it is with the internal representation of the encountered environment. This is counter to the conventional view that high-level consciousness is a purely internal matter. We would expect, prior to this last event of high-level generation, that mutual observation between individual lower model levels and their associated complex interfaces will produce a scaled set of purely internal awarenesses which, through their being at subsystemic levels, are partway between our assumed initially low level of awareness and high-level consciousness. These most probably appear as low-level "noise" at the higher level of consciousness, and consequently they can possibly be associated with the "unconscious mind".

However, the origin of these purely internal awarenesses lies in the same {*info* and/or *information*} processing as for high-level consciousness. Consequently, each is *also* associated with an internal integration to *information* in the overall scheme of things. This is why we have systematically referred to {*info* and/or *information*} processing, rather than simply *info* processing, in our development.

If this overall hypothesis is correct, then high-level consciousness is real-time predated by the capacity to react to external stimuli. This makes high-level-conscious organisms consistent with the wide range of cognitive entities found in nature, from the presumably feebly conscious capability of amoebas to hunting in packs (Geisen *et al.*, 2015) and aligning themselves with respect to infrared light sources (Albrecht-Buehler, 2018), through the insect world and up to mammals, in that within this representation, reaction to stimuli is always pre-conscious.

6. Conclusion

Our development indicates the way that {*info* and/or *information*} is integrated to *information* in the model hierarchy representation of an organism. We do not identify *info* specifically as the combination of data and context, although this will often be the case. *Information*, however, is *always* a combination of model constituent and environmental context, as we have shown.

We have presented a "devil's advocate" rejection of the controlling function of high-level consciousness, which depends on a specific interpretation of the relationship between sequentiality and concurrence in the generations of system hyperscales, metascale and high-level consciousness. We do not pretend that this approach settles the matter — we merely propose that careful examination of these systemic phenomena should follow to evaluate the position of consciousness in the operation of organisms. It is worth noting that even if all of the processes we have described are entirely concurrent, it is still not obvious that high-level consciousness predates in real time the planning of actions and reactions.

It may appear that this conclusion does not hold up in the case of anesthetic loss of consciousness, but it is unclear whether anesthetics act on a neural substrate of consciousness itself or on the means by which models and their complex interfaces interact, which would also remove consciousness.

From a more philosophical point of view, if consciousness is indeed an irrelevant by-product of {*info* and/or *information*} processing, then what are the implications for moral responsibility in the performance of apparently conscious actions?

Information appears in this scenario as the integration of system models and internal environmental modeling, as the immediate precursor of high-level consciousness, thus supporting current viewpoints as to the relevance of {*info* and/or *information*} integration to the appearance of high-level consciousness, while not agreeing with the identification of *info* integration with consciousness itself. The distinction that we have made between *info* and *information* appears to be an important one, which supports the view that *information* in its complete state is purely subjective (Cottam *et al.*, 2017c), as opposed to *info* in our description, which can be varyingly objective or subjective in character. It remains to be seen how this viewpoint will play out in the future.

References

Akelaitis, A. J. (1941). Psychobiological studies following section of the corpus callosum: A preliminary report, *Am. J. Psychiat.* 97, 1147–1157.

Akelaitis, A. J., Risteen, W. A., Herren, R. Y. and van Wagenen, W. P. (1942). Studies on the corpus callosum. III. A contribution to the study of dyspraxia in epileptics following partial and complete section of the corpus callosum, *Arch. Neuro. Psychiatr.* 47, 971–1008.

Albrecht-Buehler, G. (2018). Cell intelligence. Available online: http://www.basic.northwestern.edu/g-buehler/FRAME.HTM (accessed on 30/01/2018).

Anderson, C. and Brown, C. E. (2010). The functions and disfunctions of hierarchy, *Res. Organ. Behav.* 27, 1–35.

Berto, F. (2010). *There's Something About Gödel: The Complete Guide to the Incompleteness Theorem*, John Wiley & Sons, Hoboken, NJ.

Busch, P., Heinonen, T. and Lahti, P. (2007). Heisenberg's uncertainty principle, *Phys. Rep.* 452, 155–176.

Chauvet, G. A. (1993). Hierarchical functional organization of formal biological systems: A dynamical approach. I. The increase of complexity by self-association increases the domain of stability of a biological system, *Philos. Trans. R. Soc. Lond. B. Biol. Sci.* 339, 425–444.

Coaen, S., Ward, L. M. and Enns, J. T. (1994). *Sensation and Perception*, Harcourt Brace, New York.

Cottam, R. and Ranson, W. (2012). The unity of mind, brain and world: Current perspectives on a science of consciousness, in *A Biosemiotic View on Consciousness Derived from System Hierarchy*, Pereira Jr., A. and Lehmann D. (eds.), Cambridge University Press, Cambridge, pp. 77–112.

Cottam, R. and Ranson, W. (2018). *Bridging the Gap between Life and Physics*, Springer, Dordrechy.

Cottam, R. and Saunders, G. A. (1973). The elastic constants of GaAs from 2K to 320K, *J. Phys. C Solid State* 6, 2015–2118.

Cottam, R., Ranson, W. and Vounckx, R. (1999). A biologically consistent hierarchical framework for self-referencing survivalist computation, in *Computing Anticipatory Systems, AIP Conference Proceedings*, Dubois D. M. (ed.), 517, pp. 252–262.

Cottam, R., Ranson, W. and Vounckx, R. (2000). A diffuse biosemiotic model for cell-to-tissue computational closure, *BioSystems* 55, 159–171.

Cottam, R., Ranson, W. and Vounckx, R. (2003). Autocreative hierarchy II: Dynamics — Self-organization, emergence and level-changing, in *International Conference on Integration of Knowledge Intensive Multi-Agent Systems*, Hexmoor H. (ed.), IEEE, Piscataway, NJ, pp. 766–773.

Cottam, R., Ranson, W. and Vounckx, R. (2004a). Autocreative hierarchy I: Structure — Ecosystemic dependence and autonomy, *SEED J.* 4, 24–41.

Cottam, R., Ranson, W. and Vounckx, R. (2004b). *Unifying Themes in Complex Systems*, Vol. II, Bar-Yam, Y. and Minai, A. (eds.), Diffuse rationality in complex systems, Westview Press, Boulder, CO, 355–362.

Cottam, R., Ranson, W. and Vounckx, R. (2005). Life and simple systems, *Sys. Res. Behav. Sci.* 22, 413–430.

Cottam, R., Ranson, W. and Vounckx, R. (2006). Living in hyperscale: Internalization as a search for unification, in *Proceedings of the 50th Annual Conference of the International Society for the Systems Sciences*, Wilby J., Allen J. K. and Loureiro-Koechlin C. (eds.), paper #2006-362, pp. 1–22.

Cottam, R., Ranson, W. and Vounckx, R. (2008a). Sapient structures for intelligent control, in *Toward Artificial Sapience: Principles and Methods for Wise Systems*, Mayorga R. V. and Perlovsky L. I. (eds.), Springer, New York, pp. 175–200.

Cottam, R., Ranson, W. and Vounckx, R. (2008b). The mind as an evolving anticipative capability, *J. Mind Theory*, pp. 39–97.

Cottam, R., Ranson, W. and Vounckx, R. (2012). Towards cross-modeling between life and solid state physics, in *Integral Biomathics — Tracing the Road to Reality*, Simeonov P. L., Smith L. S. and Ehresmann A. C. (eds.), pp. 85–95.

Cottam, R., Ranson, W. and Vounckx, R. (2013). A framework for computing like Nature, in *Computing Nature*, Dodig-Crnković G. and Giovagnoli R. (eds.), Springer, Berlin, pp. 23–60.

Cottam, R., Ranson, W. and Vounckx, R. (2017a). Scaled ecosystemic Rosennean complexity. In publication in *Ecol. Complex.*, doi.org/10.1016/j.ecocom.2017.02.001.

Cottam, R., Ranson, W. and Vounckx, R. (2017b). Precognition and the metascalar nature of information, *Eur. Phys. J. ST* 226, 229–242.

Cottam, R., Ranson, W. and Vounckx, R. (2017c). Information: Subjective or objective? *Proceedings* 1, 63.

Damasio, A. (2010). *Self Comes to Mind: Constructing the Conscious Brain*, Pantheon/Random House, New York.

Dennett, D. C. (1991). *Consciousness Explained*, Little, Brown, Boston.

Floridi, L. (2010). *Information — A Very Short Introduction*, Oxford University Press, Oxford.

Floridi, L. (2011). *The Philosophy of Information*, Oxford University Press, Oxford.

Geisen, S., Rosengarten, J., Koller, R., Mulder, C., Urich, T. and Bonkowski, M. (2015). Pack hunting by a common soil amoeba on nematodes, *Environ. Microbiol.* 17, 4538–4546.

Gottesman, D. (2009). An introduction to quantum error correction and fault-tolerant quantum computation, *Proc. Sym. AP.*, pp. 1–46.

Grandin, T. and Johnson, C. (2005). *Animals in Translation*, Harcourt, Orlando.

Kihlstrom, J. F. (1987). The cognitive unconscious. *Science* 237, 1445–1452.

Köhler, S. and Moscovitch, M. (1997). Unconscious visual processing in neuropsychological syndromes: A survey of the literature and evaluation of models of consciousness, in *Studies in Cognition. Cognitive Neuroscience*, Rugg M. D. (ed.), MIT Press, Cambridge, pp. 305–373.

Leahey, T. H. (1997). *A History of Psychology: Main Currents in Psychological Thought.* Prentice Hall, Upper Saddle River.

Lefton, L. A., Brannon, L., Boyes, M. C. and Ogden, N. A. (2005). *Psychology*, Pearson, Toronto.

Lombardi, O., Holik, F. and Vanni, L. (2014). http://philsci-archive.pitt.edu/10911/1/What_is_Shannon_Information.pdf.

Matsuno, K. (2000). The internalist stance: A linguistic practice enclosing dynamics, *Ann. NY. Acad. Sci.* 901, 322–349.

Pattee, H. H. (1973). *Hierarchy Theory — the Challenge of Complex Systems*, George Braziller, New York.

Peirce, C. S. (1931–1958). In *Collected Papers of Charles Sanders Peirce*, Hartshorne C, Weiss P. (eds.), vols 1–6; Burks A. (ed.), Vols. 7–8, Belknap Press, Cambridge.

Rosen, R. (1991). *Life Itself*, Columbia University Press, New York.

Salthe, S. N. (1985). *Evolving Hierarchical Systems*, Columbia University Press, New York.

Salthe, S. N. (1993). *Development and Evolution: Complexity and Change in Biology*, MIT Press, Cambridge, MA.

Salthe, S. N. (2012). Hierarchical structures, *Axiomathes* 22, 355–383.

Schroeder, M. J. (2009). Quantum coherence without quantum mechanics in modeling the unity of consciousness, in *Quantum Interaction: Proceedings of QI 2009*, Bruza P., Sofge D., Lawless W., van Rijsbergen C. J. and Klusch M. (eds.), Springer Verlag, Berlin, pp. 97–112.

Schroeder, M. J. (2012). The role of information integration in demystification of holistic methodology, in Simeonov P. L., Smith L. S., Ehresmann A. C. (eds.), *Integral Biomathics: Tracing the Road to Reality*, Springer, Heidelberg, pp. 283–296.

Shannon, C. E. (1948). A mathematical theory of communication, *AT&T. Tech. J.* 27, 379–423 & 623–656.

Solso, R. L. (1999). *Mind and Brain Sciences in the 21st Century*, MIT Press, Cambridge.

Sperry, R. W., Gazzaniga, M. S. and Bogen, J. E. (1969). Interhemispheric relationships: The neocortical commisures: Syndromes of hemisphere disconnection, in *Handbook of Clinical Neurology*, Vinken P. J., Bruyn G. W. (eds.), Vol 4, North-Holland, Amsterdam, pp. 273–290.

Striedter, G. F., Avise, J. C. and Ayala, F. J., (eds.) (2013). *In the Light of Evolution, Volume VI: Brain and Behavior*, National Academies Press, Washington DC.

Taylor, J. G. (2011). A review of models of consciousness, in *Perception-Action Cycle*, Cutsuridis V., Hussain A. and Taylor J. (eds.), Springer Series in Cognitive and Neural Systems, Springer, New York.

Tononi, G. (2004). An information integration theory of consciousness, *BMC Neurosci.* 5, 42.

Wang, Y. (2012). The cognitive mechanisms and formal models of consciousness, *Intern. J. Cogn. Inform. Nat. Intell.* 6(2), 23–40.

Widhiastuti, H. (2012). The effectiveness of communications in hierarchical organizational structure, *Int. J. Soc. Sci. Human.* 2, 185–190.

Chapter 16

Convergence of Natural Intelligence and Artificial Intelligence

Rao Mikkilineni

Golden Gate University, San Francisco, CA, USA
rmikkilineni@ggu.edu

Natural intelligence has evolved over billions of years to give us two computing models: the "gene" which encodes the process of "life" in an executable form and a neural network that encodes various processes to interact with the environment in real time. Recent advances in computing performance have increased exponentially, allowing us to implement similar computing models, giving us the framework for today's software systems. First, symbolic computing outperforms humans in tasks that can be easily described by a list of formal, mathematical rules or a sequence of event-driven actions such as modeling, simulation, business workflows, interaction with devices, etc. Second, the neural network model allows computers to understand the world in terms of a hierarchy of concepts to perform tasks that are easy to do "intuitively", but are hard to describe formally or a sequence of event-driven actions such as recognizing spoken words or faces.

We examine the role of cognition in symbolic computing and neural networks. We explore the potential for a symbiotic relationship between natural and artificial intelligence to extend the collective consciousness and develop a unified culture that binds the participants to define common goals and execute processes that transcend human and machine capabilities.

1. Introduction

According to Kurzweil (2012), "intelligence may be defined as the ability to solve problems with limited resources, in which key resource is time."

Resources and their optimal utilization have been the preoccupation of all living things from cellular organisms to human beings with a mission to execute well-defined goals even while facing non-deterministic interactions with the environment. Evolution over billions of years has given natural intelligence a way to encode information, store it and use it to execute the "life processes" that define the best practices for sustenance and survival. The result is the emergence of two computing models in the form of the "gene" and the "neural network". These computing models provide the means to capture the knowledge about itself and its environment, processes to define the intent, way to find required resources, ways to execute it with available resources, monitor its progress, identify deviations and take corrective actions.

A cellular organism is the simplest form of life that maintains an internal environment that supports its essential biochemical reactions, despite changes in the external environment. The cell adapts to its environment by recognition and transduction of a broad range of environmental signals, which in turn activate response mechanisms by regulating the expression of proteins that take part in the corresponding processes. The regulatory gene network forms a cellular control circuitry defining the overall behavior of the various cells. The complex network of communication paths and the signaling mechanisms collaborate to create a dynamic, active and temporal representation of both the cell (the observer) and its environment (the observed) with myriad patterns, associations and constraints among their components.

A parallel evolution is taking place in how we encode process information in executable form using general purpose computers and how we use them to analyze vast amounts of data to discern hidden insights and act on them. There are two distinct paths in which various human tasks configure, monitor and control their interactions among themselves and their environment in the form of devices which interface with the physical universe. In this digital universe, first, symbolic computing is exploited to execute those tasks that can be easily described by a list of formal, mathematical rules or a sequence of event-driven actions such as modeling, simulation, business workflows, interaction with devices, etc. Second, the neural network or connectionist computing model is used to analyze vast amounts of data generated in executing these tasks and gaining an understanding of the hidden correlations and insights associated with tasks that are easy to do "intuitively", that feel automatic, but are hard to describe formally or a sequence of event-driven actions such as recognizing spoken words or faces.

In this chapter, first, we analyze the evolution of the symbolic computing model defined by the Turing machine and its stored program implementation by John von Neumann, identify its shortcomings and describe new extensions that mimic the abstractions supported by genes to create self-managing systems. Second, we analyze current advances in artificial intelligence, machine learning (ML) and deep learning, which have evolved by drawing parallels from the neural network in the neocortex. Finally, we speculate on the convergence of both natural intelligence and artificial intelligence where they coexist with a shared conscience and culture and define the necessary requirements for such a convergence.

In Section 2, we describe the evolution of natural intelligence and extract the key abstractions that contribute to self-managing systems and a parallel evolution in symbolic computing. In Section 3, we describe the business drivers and in Section 4, the technology drivers which are forcing the need for creating self-managing software systems. In Section 5, we describe the evolution of Symbolic Computing, which makes it possible to create autonomic software systems that configure, monitor and manage themselves while delivering services that also configure, monitor and manage their interactions with external systems, devices and humans. In Section 6, we describe the evolution of AI, ML and DL systems to manage the real-time interactions of software systems with their environment and gain insights to influence the future course of service evolution. In Section 7, we speculate on the convergence of artificial and natural intelligence to deliver next-generation self-managing business processes. In Section 8, we argue that a collective conscience and culture must evolve to successfully derive the synergy between natural and artificial intelligence. Section 9 concludes with some observations on the future communication, collaboration and commerce services on a global scale with a high degree of resiliency and efficiency.

2. Lessons from Biology

As Ray Kurzweil (2012) points out, the story of human intelligence starts with a universe that is capable of encoding information. This was the enabling factor that allowed evolution to take place. The story of evolution, he observes, unfolds with increasing levels of abstraction that exploit information processing. From atoms to molecules to a very unique and complex molecule DNA, the increasing levels of abstraction of encoding information in reusable form, using it to define a process intent and delivering a means to execute it have resulted in a diversity of self-replicating systems.

The next major evolutionary abstractions, the neuron and the neural network, allowed embedded, embodied, enacted and extended cognition (known as 4E cognition) in a variety of ways (Barrett, 2011) to model, configure, monitor and control the environment using a set of information patterns with a self-identity and optimize the execution of the processes (both of managing its own life and of managing its interactions with the environment). Cognition (Mikkilineni, 2012) here, is the ability to process information, apply knowledge and change circumstances. Cognition is associated with intent and its accomplishment through various information management processes that monitor and control a system and its environment. Cognition is associated with a sense of "self" (the observer) and the systems with which it interacts (the environment or the "observed"). Cognition extensively uses time and history in executing and regulating tasks that constitute a cognitive process. In this section we analyze these biological computing models and derive the abstractions that allow infusing cognition and design self-managing systems with the architectural resiliency of cellular organisms.

2.1. *The gene and the DNA: Blueprint for self-managing systems*

According to Richard Dawkins (1976), a DNA molecule is really a set of plans for building a body and it indirectly supervises the manufacture of a different kind of molecule — protein. "The coded message of the DNA, written in the four-letter nucleotide alphabet, is translated in a simple mechanical way into another alphabet. This is the alphabet of amino acids which spells out protein molecules." The indirect supervision task mentioned by Dawkins is related to the function of the gene, which is a piece of DNA material that contains all the information needed to "build" specific biological structures. Genes thus contain the information to make proteins, the body's building blocks. Proteins make up the structure of the organs and tissues; they are also needed for the body's chemical functions and pathways. Each protein performs a specific job in the different types of cells of the body, and the information for making at least one protein is contained in a single gene. The pattern or sequence of the genes is like a blueprint that tells the body how to build its different parts. The genes constitute the workflow components for building the biological system consisting of a group of cells that act with a single purpose, which is to propagate the equilibrium patterns they have found to survive.

Even the simplest unicellular organism provides a good example of self-management and reproduction to sustain life. Studies in evolutionary developmental biology throw a fundamental insight into the inner workings of how groups of cells are organized and orchestrated to create what the biologist Carroll (2005) calls "endless forms more beautiful". DNA determines when, where and how much of a gene's product is made. With modularity, multifunctionality and redundancy built in its architecture, how does orchestration take place? The orchestration is accomplished by "parallel and sequential actions of tool kit genes — dozens of genes acting at the same time and same place, many more genes acting in different places at the same time, and hundreds of toolkit genes acting in sequence." He goes on to say, "another class of toolkit member belongs to so-called signaling pathways. Cells communicate with one another by sending signals in the form of proteins that are exported and travel away from their source. Those proteins then bind to receptors on other cells, where they trigger a cascade of events, including changes in cell shape, migration, the beginning or cessation of cell multiplication, and the activation or repression of genes."

To summarize, the key abstractions the cell architecture supports are as follows:

- the spelling out of computational workflow components as a stable sequence of patterns that accomplishes a specific purpose;
- a parallel management workflow specification with another sequence of patterns that assures the successful execution of the system's purpose (the computing network);
- a signaling mechanism that controls the execution of the workflow for gene expression (the regulatory network).

These abstractions will play a role in extending the current digital computing models and designing self-managing software systems.

2.2. *The neural network: Real-time reaction to non-deterministic interactions*

Neurons are special purpose cells created by genes that process information received through various sensor cells and control the environment using myriad motor cells. The neuron is the fundamental unit of the nervous system. The basic purpose of a neuron is to receive incoming information and, based upon that information, send a signal to other neurons, muscles or glands. Neurons are designed to rapidly send signals across physiologically long distances. A group of neurons thus is programmed to perform specific functions (sensory and motor) that help an organism to process information, detect patterns and take action based on signals received from within or without. The Hebbian theory of learning summarized as "cells that fire together wire together" has been the premise used to understand "neural nets" and neuronal learning. While the Hebbian unit of learning is a neuron, Ray Kurtzweil on the other hand proposes an assembly (Lego-like) of neurons whose function and structure are predetermined by genes as a fundamental unit of learning, executing information processing, pattern detection and communication with other units of learning. He goes on to say that the purpose of these modules is to recognize patterns, to remember them and to predict them based on partial patterns. In essence, groups of neurons act as basic autonomic computing units with sensors and actuators, processing information based on genetically programmed pattern recognition mechanisms.

The two computing models together define the process intent, an execution path for its evolution, avail the resources required and manage the evolution while making sure deviations are corrected along the path. The result is the assurance that both the organism's survival and its mission are accomplished with efficiency and required resilience.

In the next section, we analyze how we are utilizing machine intelligence to automate various processes that humans use to manage their interactions with each other and their environment and identify how the lessons from biology would guide their evolution.

3. Evolution of Symbolic Computing

In this section, we will discuss the evolution of symbolic computing, its limitations in meeting the demands of current information technologies and new approaches to address these limitations.

3.1. *The Turing machine and its stored program implementation*

The Turing machine (Turing, 1939, 2004) is an abstract model that uses an instruction cycle, {read -> compute (change state) -> write}, to "replace a man in the process of computing a real number (using a paper and pencil) by a machine which is only capable of finite number of conditions." In modern terms, a program provides a description of the Turing machine, and the stored program control implementation in some hardware allows its execution. A universal Turing machine is also a Turing machine but with the ability to simulate a sequence of synchronous Turing machines, each executing its own description. This allows a sequence of programs to model and execute a description of the physical world.

John von Neumann's stored program implementation (von Neumann, 1982) made it possible to encode and execute information in executable form using a digital computer. While Turing showed that any algorithm or process that is composed of tasks which can be precisely described by a list of formal, mathematical rules or a sequence of event-driven actions, such as modeling, simulation, business workflows, interaction with devices, etc., can be executed by a machine, von Neumann showed an implementation for it. The stored computer program held instructions and data in the same memory. This enabled programs to manipulate other programs (or even themselves!) and made it possible to decouple the external representation of a program from the actual instructions executed by the machine. This in turn allowed new forms of abstraction to be created and used by implementing them in software. During the last five decades, many layers of computing abstractions have been introduced to map the execution of complex computational workflows to a sequence of 1 s and 0 s that eventually get stored in the memory and are operated upon by the CPU to achieve the desired result. These include process definition languages, programming languages, file systems, databases, operating systems, etc.

The Turing machine originally started as a static closed system (Goldin and Wegner, 2008) analogous to a single cell. It was designed for computing algorithms that correspond to a mathematical worldview. The computation, in order to successfully execute, must have adequate memory, CPU and time determined by the nature of the algorithm. As Mark Burgin (Burgin and Mikkilineni, 2016) points out, computation started as calculation, or operations undertaken on numbers. It has evolved to computation as information transformation, or operations on multimedia, such as text, audio or video data. More recently, it is transforming into Interactive computation, or computation as interaction where various computing processes interact with each other or with humans or physical devices. During the last seven decades since the first computing machine was introduced, major advances in the design of CPU, memory and their networking have contributed to an exponential growth in the use of general purpose computers to automate many of the human tasks, which has resulted in today's service economy enabled by communication, collaboration and commerce on a global scale almost at the speed of light.

3.2. *Software shortfall in meeting service scale, resiliency and efficiency demand*

With the availability of hyper-scale infrastructure, ubiquitous cloud resources with multi-core servers connected with extremely high bandwidth networks and powerful storage devices, the scale and the scope of software systems delivering various computing services have introduced a new level of requirement to assure non-functional requirements such as availability, performance, security, regulatory data compliance and cost. In addition, these services are often composed of multiple distributed components executed in heterogeneous infrastructures providing the CPU and memory required. A distributed computing system, thus, is a collection of independent software components that are designed to cooperate and deliver a common intent. The characteristics of distributed systems, the motivations to use them, their benefits and the various challenges in designing and implementing those using modern computing systems are extensively discussed in the literature (Jha *et al.*, 2013; Thain *et al.*, 2005; Kshemkalyani and Singhal, 2008; Babaoglu and Marzullo, 1993; Mikkilineni, 2011). It is important to recognize that the distributed computing systems are dynamic systems where software working with hardware delivers information processing services that allow modeling, interaction, reasoning, analysis and control of the environment external to them. The intent of the distributed software components is to execute specific computational workflows implementing well-specified functional requirements in the form of steps to be executed. The software components contain the algorithms (à la the Turing machine) that specify information processing tasks of each component contributing to the execution of common intent. On the other hand, the intent of the hardware components is to provide the computing resources such as CPU, memory, network bandwidth, latency, storage IOPs, throughput and capacity so that the software components pursue their intent without interruption.

The meta-knowledge of the overall intent of the distributed computing system, involving the global context, communication between components, constraints and control abstractions of the software components, the association of specific software component execution to a specific hardware device, and the temporal evolution of information processing and exception handling when the computation deviates from the intent (be it because of software behavior or hardware behavior or their interaction with the environment), is outside the software design and is expressed in non-functional requirements.

Mark Burgin observed (Burgin, 2005) that an Information processing system (IPS) "has two structures — static and dynamic. The static structure reflects the mechanisms and devices that realize information processing, while the dynamic structure shows how this processing goes on and how these mechanisms and devices function and interact." The software contains the algorithms (à la the Turing machine) that specify information processing tasks, while the hardware provides the required resources to execute the algorithms. The static structure is defined by the association of software and hardware devices (function and structure) and the dynamic structure (influenced by the evolution and fluctuations in the structure) is defined by the execution of the algorithms. The meta-knowledge of the intent of the algorithm, the association of specific algorithm execution to a specific device, and the temporal evolution of information processing and exception handling when the computation deviates from the intent (be it because of software behavior or the hardware behavior or their interaction with the environment) are outside the purview of software and hardware design and are expressed in non-functional requirements. Mark Burgin calls this Infware which contains the description and specification of the meta-knowledge that can also be implemented using the hardware and software to enforce the intent with appropriate actions.

Today the Infware is implemented in an *ad hoc* manner and often involves labor-intensive tasks. As the scale of computation increases, so does the complexity of managing the execution of functional requirements. The complexity manifests in the following ways:

1. In a distributed computing environment, multiple computations contribute to the result. Maintaining the availability of the right computing resources (CPU, memory) to the right distributed computing element contributing to the transaction requires orchestration and management of myriad computing infrastructures often owned by different providers with different profit motives and incentives. The resulting complexity in managing application availability, performance and security adds to the cost of computing. For example, it is estimated that up to 70% of current IT budget is consumed in assuring service availability, performance and security. The complexity is compounded in geographically distributed

computing environments which are supported by heterogeneous infrastructures with disparate management systems.

2. In a large-scale dynamic distributed computation supported by myriad infrastructure components, the increased component failure probabilities introduce a non-determinism that must be addressed to meet availability and performance constraints.
3. Fluctuations in the computing resource requirements dictated by changing business priorities, workload variations that depend on service consumption profiles and real-time latency constraints dictated by the affinity of service components demand a run-time response to dynamically adjust the computing resources. Current dependence on myriad orchestrators and management systems cannot be scaled in a distributed infrastructure without either a vendor lock-in or a universal standard that often stifles innovation and competition to meet fast-changing business needs.

Thus, the function, structure and fluctuations involved in the evolution of processes (delivering application intent) are driving a need to search for new computing, management and programming models that address the unification of the management of the computer and the computed. In the next section, we describe a recently introduced computing model that provides an architecture for infusing cognition into computing (the Turing machine) and allows software to become self-aware and self-managing with the architectural resiliency of cellular organisms.

3.3. *DIME computing model and DIME network architecture*

The DIME (Burgin and Mikkilineni, 2014; Burgin *et al.*, 2015; Mikkilineni *et al.*, 2012, 2013; Mikkilineni, 2012; Mikkilineni and Morana, 2014; Burgin *et al.*, 2015a, 2015b), previously discussed at the Turing Centenary Conference (Mikkilineni *et al.*, 2012) in Manchester, is aimed at improving the efficiency and resilience of computational processes by organizing their evolution to model process dynamics under the supervision of intelligent agents who have the knowledge about both the computed and the available pool of computer resources. The computing hardware resources and software functions are arranged to optimally execute processes and evolve their dynamics using a network of DIMEs where each end node can be either a DIME unit or a subnetwork of DIME units executing a workflow. The hardware resources are characterized by their parameters such as the

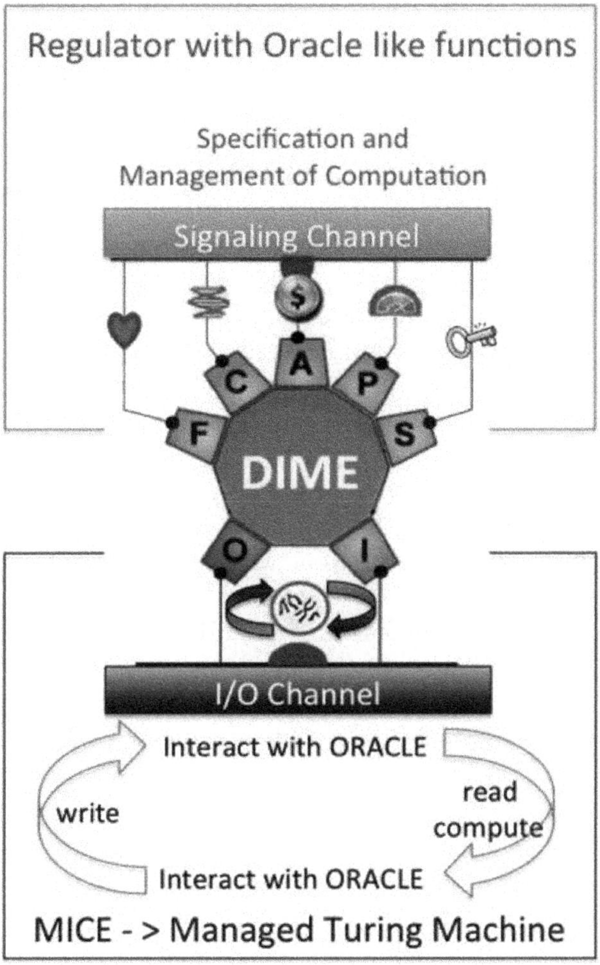

Figure 1. A Distributed Intelligent Managed Element is a managed Turing Oracle machine endowed with a signaling network overlay for policy-based DIME network management.

required CPU, memory, network bandwidth, latency, storage throughput, IOPs and capacity. The efficiency of computation is determined by the required resources, while the expressiveness of the computational process dynamics is established by the structure of the DIME units and connecting hardware units, such as servers or routers, along with their interactions within and with the external world.

Figure 1 shows the DIME implementation.

Here, we describe the managed Turing machine network model using DIMEs with systemic network design, upgrading and maintenance based on three principles: (1) modularity; (2) system representation of each module by grid automata; and (3) utilization of modular operations with networks. Modular approach means division of a complex system into smaller, manageable ones, making implementation much easier to handle.

In its simplest form, a DIME consists of a computing element called Managed Intelligent Computing Element (MICE) which is used to execute the algorithm at hand. Various agents are designed to manage the life processes of the algorithm (such as configuration, monitoring and control) executed in the MICE by knowing the intent of the algorithm, the required resources and various best practices to manage the MICE to sustain its life-process (using the knowledge in the form of an algorithm blueprint). The MICEs are interconnected using a computing (data) channel to communicate and thus execute a networked computing workflow. The regulators are networked using a signaling channel which overlays the computing channel executing a parallel management workflow (Mikkilineni *et al.*, 2012). The regulator can be loaded with the specification and the evolution of the computation that will be executed by the MICE under its control.

Thus, a computation consists of four parts:

1. a blueprint that specifies the resources required, the purpose, run-time behavior and control function (the meta-model of computation);
2. an execution part that specifies the execution of the computation mixed with the communication with a Turing Oracle to accomplish the given purpose;
3. the execution of the Oracle-defined function (with modified instruction cycle as depicted in Figure 1) that uses the knowledge to configure, monitor and control the computer and the computed;
4. a network overlay that allows various DIME management Oracles to collaborate to manage evolution of computation workflow composed of DIME networks.

3.3.1. *Turing machine with an Oracle*

Executing an algorithm, the DIME basic processor P performs the {read -> compute -> write} instruction cycle or its modified version, the {interact with a network agent -> read -> compute -> interact with a network agent -> write} instruction cycle. This allows the different network agents to influence the further evolution of computation, while the computation is still in progress. We consider three types of network agents:

(a) a DIME agent;
(b) a human agent;
(c) an external computing agent.

It is assumed that a DIME agent knows the goal and intent of the algorithm (along with the context, constraints, communications and control of the algorithm) the DIME basic processor is executing and has the visibility of available resources and the needs of the basic processor as it executes its tasks. In addition, the DIME agent also has the knowledge about alternate courses of action available to facilitate the evolution of the computation to achieve its goal and realize its intent. Thus, every algorithm is associated with a blueprint (analogous to a genetic specification in biology), which provides the knowledge required by the DIME agent to manage the process evolution. An external computing agent is any computing node in the network with which the DIME unit interacts.

3.3.2. *Blueprint or policy managed fault, configuration, accounting, performance and security monitoring and control*

The DIME agent, which uses the blueprint to configure, instantiate and manage the DIME basic processor executing the algorithm, uses concurrent DIME basic processors with their own blueprints specifying their evolution to monitor the vital signs of the DIME basic processor and implements various policies to assure non-functional requirements such as availability, performance, security and cost management while the managed DIME basic processor is executing its intent.

3.3.3. *DIME network management control overlay over the managed Turing Oracle machines*

In addition to read/write communication of the DIME basic processor (the data channel), other DIME basic processors communicate with each other using a parallel signaling channel. This allows the external DIME agents to influence the computation of any managed DIME basic processor in progress, based on the context and constraints. The external DIME agents are DIMEs themselves. As a result, changes in one computing element could influence the evolution of another computing element at run time without halting the Turing machine executing the algorithm. The signaling channel and the network of DIME agents can be programmed to execute a process, the intent of which can be specified in a blueprint. Each DIME basic processor can have its own Oracle managing its intent, and groups of managed DIME basic processors can have their own domain managers implementing the domain's intent to execute a process. The management DIME agents specify, configure and manage the subnetwork of DIME units by monitoring and executing policies to optimize the resources while delivering the intent. Figure 2 shows the DIME network architecture implementing a dynamic process.

A DIME network is represented by a grid automaton with such nodes as DIME units, servers, routers, etc. Each DIME unit is modeled by a basic automaton A with an Oracle O. The automaton A models the DIME basic processor P, while the Oracle O models the DIME agent DA.

The Oracle O in a DIME unit knows the intent of the algorithm (along with the context, constraints, communications and control of the algorithm) the basic automaton A is executing under its influence and has the visibility of available resources and the needs of the automaton A as it executes its

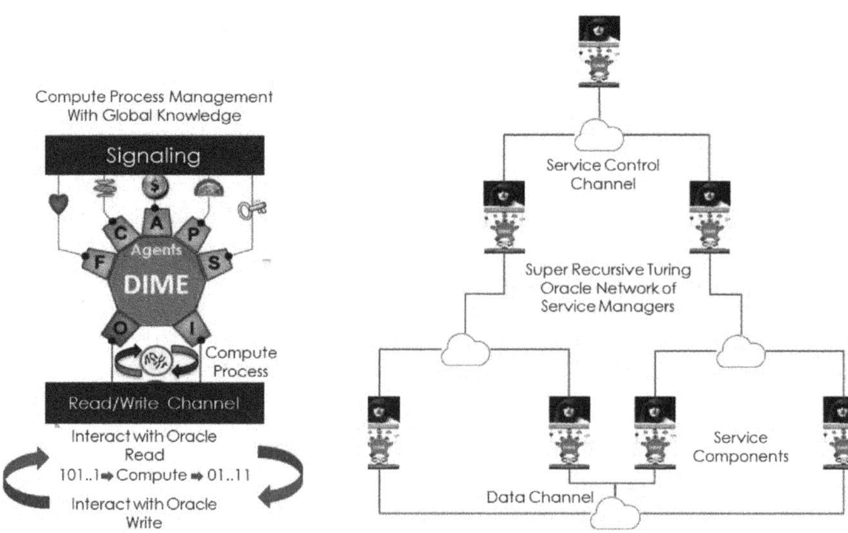

Figure 2. DIME Network Architecture implementing nonfunctional requirements management overlay over the Turing machine network implementing the functional requirements.

function. In addition, the Oracle also has the knowledge about alternate courses of action available to facilitate the evolution of the computation to achieve its intent. Thus, every algorithm is associated with a blueprint (analogous to a genetic specification in biology), which can provide the knowledge required by an Oracle to manage its evolution.

In addition to read/write communication of the basic automaton (the data channel), the Oracles manage different basic automata communicating with each other using a parallel signaling channel. This allows the external Oracles to influence the computation of any managed basic automaton in progress based on the context and constraints, which is much more than a Turing Oracle is expected to do. Indeed, DIME can work not only as a Turing machine with an Oracle but also as a reflexive Turing machine with an Oracle, which is much more efficient.

The Oracle uses the blueprint to configure, instantiate and manage the automaton A executing the algorithm. Utilization of concurrent automata in the network with their own blueprints specifying their evolution to monitor the vital signs of the DIME basic automaton and to implement various policies allows the Oracle to assure non-functional requirements such as availability, performance, security and cost management, while the managed DIME basic automaton is executing its task to achieve its goal and realize its intent.

The external Oracles represent DIME agents, allowing changes in one computing element to influence the evolution of another computing element at run time without stopping the basic automaton executing the algorithm. The signaling channel and the network of the Oracles can be programmed to execute a process whose intent itself can be specified in a blueprint. Each basic automaton can have its own Oracle managing its intent, and groups of managed basic automata can have their own domain managers implementing the domain's intent to execute a process. The management Oracles specify, configure and manage the subnetwork of DIMEs by monitoring and executing policies to optimize the resources while delivering the intent. The DIME network implementing the Oracles is itself managed by monitoring its own vital signs and executing various FCAPS (fault, configuration, accounting, performance and security) policies to assure availability, performance and security.

An Oracle is modeled by an abstract automaton that has higher computational power and/or lower computational complexity than the basic automaton it manages. For instance, the Oracle can be an inductive Turing machine, while the basic automaton is a conventional Turing machine. It is proved that inductive Turing machines have much higher computational power and lower complexity than conventional Turing machines (Burgin, 2005).

3.4. *New class of self-managing systems*

The DIME network architecture mimics the abstractions provided in the gene and the blueprint encodes the service configuration, monitoring and control workflows in executable form based on non-functional requirements. The functional requirements are encoded in executable form in the form of software workflow descriptions. Based on these blueprints, appropriate resource network architecture (RNA) is derived and provisioned to meet both functional and non-functional requirements and their execution simultaneously.

Various implementations of DNA are discussed in detail in Burgin and Mikkilineni (2015), Mikkilineni *et al.*, (2016a, 2016b) and Morana (2017) and various references cited in those papers. The following link is of a video that shows self-managing workloads with auto-failover, auto-scaling, live-migration across multiple cloud infrastructures connected by public or private networks without losing transactions or disturbing user experience:

https://youtu.be/tu7EpD_bbyk

This is a true demonstration of the sentient nature of software infused with cognition where cognition is defined as the ability to process information, apply knowledge and change the circumstance (Mikkilineni, 2012). Cognition is associated with intent and its accomplishment through various processes that monitor and control a system and its environment. Cognition is associated with a sense of "self" (the observer) and the systems with which it interacts (the environment or the "observed"). Cognition extensively uses time and history in executing and regulating tasks that constitute a cognitive process.

4. Evolution of Artificial Intelligence, Machine Learning and Deep Learning

In parallel to the evolution of symbolic computing, the neural network model originally discussed by both Alan Turing (Copeland, 2004) and John von Neumann (Aspray and Burks, 1989) in two different lectures given in 1948, independently, is now at the center of Artificial Intelligence (AI), Machine Learning (ML) and Deep Learning (DL).

Turing while discussing how to organize unorganized machinery says, "If we are trying to produce an intelligent machine, and are following the human model as closely as we can, we should begin with a machine with very little capacity to carry out elaborate operations or to react in a disciplined manner to orders (taking the form of interference). Then by applying appropriate interference, mimicking education, we should hope to modify the machine until it could be relied on to produce definite reactions to certain commands. This would be the beginning of the process. I will not attempt to follow it further now." He is here sowing the seeds for a process to infuse cognition into machines.

Defining the neuron as a black box, von Neumann describes, "McCulloch and Pitts have used these units to build up complicated networks which may be called "formal neural networks." Such a system is built up of any number of these units, with their inputs and outputs suitably interconnected with arbitrary complexity. The "functioning" of such a network may be defined by singling out some of the inputs of the entire system and some of its outputs, and then describing what original stimuli on the former are to cause what ultimate stimuli on the latter." He goes on to say, "McCulloch and Pitts's important result is that any functioning in this sense which can be defined at all logically, strictly, and unambiguously in a finite number of words can also be realized by such a formal neural network. It is well to pause at this point and to consider what the implications are. It has often been claimed that the activities and functions of the human nervous system are so complicated that no ordinary mechanism could possibly perform them. It has also been attempted to name specific functions which by their nature exhibit this limitation. It has been attempted to show that such specific functions, logically, completely described, are *per se* incapable of mechanical, neural realization. The McCulloch–Pitts result puts an end to this. It proves that anything that can be exhaustively and unambiguously described, anything that can be completely and unambiguously put into words, is *ipso facto* realizable by a suitable finite neural network. Since the converse statement is obvious, we can therefore say that there is no difference between the possibility of describing a real or imagined mode of behavior completely and unambiguously in words, and the possibility of realizing it by a finite formal neural network." This is an important insight that brings out the closeness of genetic and brain computing modes and Turing machine-based computing and neural networks. "The two concepts are co-extensive. A difficulty of principle embodying any mode of behavior in such a network can exist only if we are also unable to describe that behavior completely." He asserts that there is an equivalence between logical principles and their embodiment in a neural network, and while in the simpler cases the principles might furnish a simplified expression of the network, it is quite possible that in cases of extreme complexity the reverse is true.

We have come a long way from these observations, and today, we have implementations of the neural networks that analyze vast amounts of data created by the programs that are expressed in words (Turing machine implementations) and address the complexity of analyzing them, extracting correlations and insights, and create actions that are again described in words to take action to produce the interference Turing points out.

Cognition is often used to describe a neural network model that uses a set of algorithms to process large sets of data to discern the underlying patterns and correlations that cannot be easily obtained with conventional computing using symbolic computing. The availability of high-performance multi-core servers, GPUs, connected with high-speed fiber-optics networks, and storage advances with persistent DRAM is accelerating the speed with which these algorithms are executed. However, symbolic computing is still used to execute these algorithms and the so-called "data science workflows" are mostly implemented using conventional computing systems with current software engineering practices. These software engineering practices, which have evolved from memory-starved computers, low-bandwidth networks and complex storage management systems have fallen short of providing the speed, performance, availability, security and data compliance regulations demanded by the computing algorithms driven by ever-changing business priorities. A major factor is global connectivity with large fluctuations in workload demands and available computing resource pools. As the time available to match the resources to the ever-changing workload execution needs becomes shorter, the current state of the art in managing distributed resources falls short.

The limitations of computers that he helped design were very much on John von Neumann's mind, who spent a great deal of time thinking about designing reliable computers using unreliable components (Aspray and Burks, 1989). In the Silliman lectures, and in the Hixon symposium, he touches upon various shortcomings in the computing model discussing how the computers behave differently from cellular organisms. Cellular organisms are autonomic. As von Neumann pointed out, "It is very likely that on the basis of philosophy that every error has to be caught, explained, and corrected, a system of the complexity of the living organism would not last for a millisecond. Such a system is so integrated that it can operate across errors. An error in it does not in general indicate a degenerate tendency. The system is sufficiently flexible and well organized that as soon as an error shows up in any part of it, the system automatically senses whether this error matters or not. If it doesn't matter, the system continues to operate without paying any attention to it. If the error seems to the system to be important, the system blocks that region out, by-passes it and proceeds along other channels. The system then analyses the region separately at leisure and corrects what goes on there, and if correction is impossible, the system just blocks the region off and by-passes it forever. The duration of operability of the automation is determined by the time it takes until so many incurable errors have occurred, so many alterations and permanent by-passes have been made, that finally the operability is really impaired. This is a completely different philosophy from the philosophy which proclaims that the end of the world is at hand as soon as the first error has occurred."

Autonomy in a cellular organism comes from its cognitive mechanisms. Three sources of information and its processing contribute to cognitive behavior:

1. genetic knowledge that is transmitted by the survivor to its successor in the form of executable workflows and control structures that describe stable patterns to optimally create process workflows which deploy the resources available to build the organism and assure its safe keeping in interacting with its environment;
2. the ability to dynamically monitor and control the organism's own behaviour along with its interactions with its environment using the genetic descriptions;
3. using sensory information to develop a history through memorizing the transactions, identifying new associations through analysis and using the insights gained to control both the self and its interaction with the environment.

Evolution of cognition in biology is a widely studied area and throws a great deal of light onto what abstractions contribute to which type of cognitive behavior. Whether they are embodied, embedded, enacted or extended, cognitive architectures play a vital role in providing blueprints for building future intelligent systems supporting a broad range of capabilities similar to those of humans. Cognition refers to the mechanisms by which sentient systems acquire, process, store and act on information from within the system and from the environment with which they interact (Shettlewerth 2010). As Sara Shettlewerth points out, not all agree with such an inclusive definition. Cognitive often is associated with "declarative" rather than "procedural" knowledge. "Declarative knowledge is "knowing that", whereas procedural knowledge is "knowing how", or knowing what to do. The declarative knowledge a chipmunk might gain from moving about its territory could be map-like: "Home burrow is south of that big rock". Or the chipmunk might store information as procedural knowledge such as "Turn left at the rock". The first kind of representation implies more flexible behavior than the second, but in both cases behavior results from processing and storing information about the world. A related distinction is that between first-order and higher order processes, only the latter may be regarded as interestingly cognitive. First-order processes operate directly on perceptual input, as when a stimulus triggers a response or creates a trace in memory. Second-order processes operate on first-order processes, as in evaluating the strength of one's memory for an event.

This insight gives us a path to analyses of what is cognition in software engineering. We can safely say that the Turing machine is a first-order process and cognition is related to second-order processes that operate on the Turing machine. In current software engineering practice, a system's functional behavior is encoded in the form of executable algorithms. A Turing machine uses adequate CPU and memory to execute the algorithms. If adequate CPU or memory is not provided, the execution halts. Implementing second-order processes to assure adequate resources are available in the face of fluctuations in their demand is a cognitive behavior. Analyzing the memory created by the first-order process to alter the behavior that improves the scale, efficiency and resilience of both first-order and second-order processes is also a cognitive behavior.

This gives us a clue to, and a unified computing model that integrates, the two computing models — one which executes processes described in words and the neural networks which deal with descriptions that are too complex to describe in words — and provides a direction for the future information processing methods shown in Figure 3.

The figure shows a unified process for implementing functional requirements and non-functional requirements using managed Turing machines and integrating neural networks to analyze the data created in executing the workflows interacting with the environment consisting of other cognitive

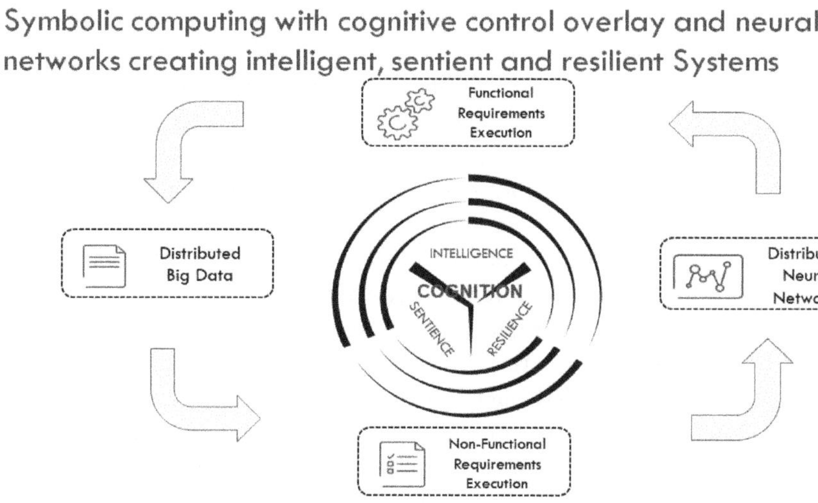

Figure 3. A unified computing model integrating symbolic computing and neural network computing.

processes, humans and devices. The integration of AI/ML/DL processes assists in enhancing the functional and non-functional requirement fulfillment in real time as the computing machines become faster, cheaper and highly scalable.

The integration of neural networks and symbolic computing provides an easy way to deal with complexity as descriptions become more complex to be described in words as von Neumann conjectured.

5. The Role of Conscience and Culture

Everyone agrees that cognitive computing and cognitive systems are game changers (Noor, 2015) "for future engineering systems, as well as for engineering practice and training. They are major drivers for knowledge automation work, and the creation of cognitive products with higher levels of intelligence than current smart products." However, the current focus on cognitive computing invariably gravitates toward "the development of computer systems modeled after the human brain, which has natural language processing capability, learns from experience, interacts with humans in a natural way, and helps in making decisions based on what it learns. All cognitive computing systems are learning systems. They incorporate embedded data analytics, automated management and data-centric architectures in which the storage, memory, switching and processing are moving ever closer to the data. Their way of processing massive amounts of data is neither linear nor deterministic." This approach does not leverage the learning about different forms of cognition such as embedded, embodied, enacted and extended cognition (known as 4E cognition) which is more about coordination of sensory and motor functions to accomplish a purpose. Lyon (2006), for example, defines cognition as the "capacity to infer relations between external circumstance and internal need to facilitate agency whose function is to enable successful action and interaction in a niche. Anderson (2003) offers an even simpler definition of cognition as situated activity. It is not immediately apparent that, under these definitions, cognition depends on the possession of a particular kind of brain or nervous system organization, nor that there are particular kinds of representational or computational processes that qualify as cognitive while others do not.

Natural intelligence improved over a long period of time to include managed computing processes, transcending what genes and neurons do. As history accumulates, it is processed to create new best practices and gain newer predictive insights to configure, monitor and control themselves and their environment in a harmonious way, to assure safety and survival while executing the intent in a most efficient way. Genes transmit the encoded information from the survived to the successor. Neurons organize themselves in the nervous system with or without brain (as in the case of some organisms) to deal with real-time sensory information and motor coordination. In both cases, an overlay control architecture allows the coordination of integrated information from different subsystems. Together, the body (consisting of sensors and motors), the nervous system (including the brain) providing coordination of sensory and motor activities with 4E cognition and an overlay control architecture (mind?) provide integration of information from various quarters.

6. Conclusion

In this chapter, we identified the evolution of intelligence in biological evolution and compared it with current evolution of symbolic computing and neural networks. Both models have been extensively discussed in the literature. Zhong (2017) points out that an intelligent information system, the human information system in particular, should be able to successfully interact with its environment, via acquiring information concerning the problem in the environment and solving the problem based on the information. The system should then possess such functions as information acquiring, transferring and processing, as well as information converts on to knowledge and intelligence for problem solving. All the functions have to be integrated into a process of "information conversion and intelligence creation". What is new in this chapter is the symbiosis of these two separate evolutions of the two computing models to assist each other to take the new computational processes to a higher level of scale, resilience and efficiency. The DIME computing model using Turing Oracle design pattern improves the efficiency, resilience and scale of computations with a control architecture overlay and simplifies the data science workflow. The neural networks help gain new insights to improve both functional and non-functional requirement fulfillment even when the computational processes and their evolution are constantly changing based on their interactions among themselves or with the environment.

Finally, we point out that the insights from two great masters, Alan Turing and John von Neumann, withstand the tides of time and are as relevant today as in 1948. The unified computing model will accelerate our use of these two computing models and also address some of the key issues related to availability, security and ethics in our current information technology implementations.

References

Anderson, M. L. (2003). Embodied cognition: A field guide, *Artif. Intell.* 149(1), 91–130, ISSN: 0004-3702.

Aspray, W. and Burks, A. (1989). *Papers of John von Neumann on Computing and Computer Theory*, MIT Press, Cambridge.

Babaoglu, O. and Marzullo, K. (1993). *Consistent "Global States of Distributed Systems: Fundamental Concepts and Mechanisms" Distributed Systems*, ACM Press, Chapter 4.

Barrett, L. (forthcoming). The evolution of cognition: A 4E perspective, in *The Oxford Handbook of 4E Cognition*, Newen A., De Bruin L. and Gallagher S. (eds.), Oxford University Press, New York, NY.

Barrett, L. (2011). *Beyond the Brain*, Princeton University Press, Princeton.

Burgin, M. (2005). *Super-recursive Algorithms*, Springer, New York.

Burgin, M. and Mikkilineni, R. (2014). Semantic network organization based on distributed intelligent managed elements, in *Proceeding of the 6th International Conference on Advances in Future Internet*, Lisbon, Portugal, pp. 16–20.

Burgin, M. and Mikkilineni, R. (2016). Agent technology, Superrecursive algorithms and DNA as a tool for distributed clouds ang grids, in *Proceedings of the 2016 IEEE 25th International Conference on Enabling Technologies: Infrastructure for Collaborative Enterprises* (*WETICE*), Paris, France, 13–15 June; pp. 89–94.

Burgin, M., Mikkilineni, M. and Morana, G. (2015a). From personal computers to personal computing networks: A new paradigm for computation, in *Proceeding of Future Computing 2015, The Seventh International Conference on Future Computational Technologies and Applications*, pp. 24–30.

Burgin, M. Mikkilineni, R. and Morana, G. (2015b). Intelligent organization of semantic networks, DIME network architecture and grid automata, *J. Embedded Syst.* (to appear).

Carroll, S. B. (2005). *The New Science of Evo Devo — Endless Forms Most Beautiful*, W. W. Norton & Co, New York.

Copeland, B. J. (ed.) (2004). *The Essential Turing*: *The Ideas that Gave Birth to the Computer Age*, Oxford University Press, Oxford.

Dawkins, R. (1976). *The Selfish Gene*, Oxford University Press, Oxford.

Goldin, D. and Wegner, P. (2008). Refuting the strong church-turing thesis: The interactive nature of computing, *Minds and Machines*, 18(1), 17–38.

Jha, S. *et al.* (2013). Distributed computing practice for large-scale science and engineering applications, *Concurr. Comput.*: *Practice Exp.* 25(11), 1559–1585.

Kurzweil, R. (2012). *How to Create a Mind: The Secret of Human Thought Revealed*, Viking Penguin, USA.

Kshemkalyani, A. D. and Singhal, M. (2008). Distributed Computing: Principles, Algorithms, and Systems, ISBN-13: 9780521876346, Cambridge University Press.

Lyon, P. (2006). The biogenic approach to cognition, *Cogn. Process.* 7, 11–29.

Mikkilineni, R. (2011). *Designing a New Class of Distributed Systems*, Springer, New York.

Mikkilineni, R. (2012). Going beyond computation and its limits: Injecting cognition into computing, *Appl. Math.* 3(11A), 1826–1835.

Mikkilineni, R. and Morana, G. (2014). Infusing cognition into distributed computing: A new approach to distributed datacenters with self-managing services, in *Proceeding of the 23rd IEEE International Conference on Enabling Technologies: Infrastructure for Collaborative Enterprises*, Parma, Italy, pp. 131–136.

Mikkilineni R. and Morana G. (2016). Cognitive distributed computing: A new approach to distributed datacenters with self-managing services on commodity hardware, *Int. J. Grid Utility Comput.* (*IJGUC*), 7(2).

Mikkilineni, R., Comparini, A. and Morana, G. (2012). The turing O-machine and the DIME network architecture: Injecting the architectural resiliency into distributed computing, in *Turing-100, The Alan Turing Centenary, EasyChair Proceedings in Computing*, Available online: www.easychair.org/publications/?page=877986046 (accessed on 3 February 2015).

Mikkilineni, R., Morana, G. and Burgin, M. (2015). Oracles in software networks: A new scientific and technological approach to designing self-managing distributed computing processes, in *Proceedings of the 2015 European Conference on Software Architecture Workshops* (ECSAW '15). ACM, New York, NY, USA, Article 11, 8 pages, DOI=10.1145/2797433.2797444, http://doi.acm.org/10.1145/2797433.2797444.

Mikklineni, R., Morana, G. and Keshan, S. (2016). Demonstration of a new computing model to manage a distributed application and its resources using Turing Oracle design, in *2016 IEEE 25th International Conference on Enabling Technologies: Infrastructure for Collaborative Enterprises* (*WETICE*), Paris, pp. 253–254. doi: 10.1109/WETICE.2016.63.

Mikkilineni, R., Morana, G. and Seyler, I. (2013). Implementing distributed, self-Managing computing services infrastructure using a scalable, parallel and network-centric computing model, in *Achieving Federated and Self-Manageable Cloud Infrastructures: Theory and Practice*, Villari *et al.* (eds.), pp. 57–78.

Mikkilineni, R., Morana, G., Zito, D. and Di Sano, M. (2012). Service virtualization using a Non-von Neumann parallel, distributed, and scalable computing model, *J. Comput. Networks Commun.* Available online: www.hindawi.com/journals/jcnc/2012/604018/ (accessed on 3 February 2015).

Morana, G. (2017). Self-managing distributed systems and globally interoperable network of clouds, *Proceedings* 1(3), 188; doi: 10.3390/IS4SI-2017-04024.

Noor, A. (2015). Potential of cognitive computing and cognitive systems, *OpenEng.* 5, 75–88, Available online: http://www.aee.odu.edu/publications/eng-2015-0008.pdf).

Thain, D, Tannenbaum, T. and Livny, M. (2005). Distributed computing in practice: The Condor experience, *Concurr. Comput.: Practice Exp.* 17(2–4), 323–356.

Turing, A. M. (1939). Systems of logic defined by ordinals, *Proc. Lond. Math. Soc., Ser.* 2(45), 161–228.

Turing, A. M. (2004). Copeland, J. B. (ed.), *The Essential Turing: Seminal Writings in Computing, Logic, Philosophy, Artificial Intelligence, and Artificial Life: Plus. The Secrets of Enigma*, The Oxford University Press, New York.

von Neumann, J. (1982). First draft of a report on the EDVAC, in *The Origins of Digital Computers*, Randell, Brian (ed.), 3rd edn., Springer-Verlag, New York.

Zhong, Y. (2017). The law of information conversion and intelligence creation, *Proceedings*, 1, 137, mdpi.com/journal/proceedings.

PART IV

Social, Cultural, Ethical and Legal Aspects of Information

Chapter 17

A Comparative Analysis of Traditional Western and Chinese Music from the Perspective of Complex Information System Theory

Gengxian Cao*,‡ and Donghe Li†,§

*School of Humanities and Social Sciences, Xi'an Jiaotong University, Xi'an 710049, China

†School of Telecommunications, Xi'an Jiaotong University, China

‡494534865@qq.com

§564405722@qq.com

This chapter attempts to discuss the increase of the "disorder" factors in Western modern music at the turn of the 20th century, to analyze the uncertainty factors in its creation. By observing the coherence of concepts between Western modern music and traditional Chinese music, we explore the profound reason for the transformation of creative ideas in Western modern music in the 20th century.

By analyzing the works of contemporary composers and performers which have been adapted to Western classical music, this chapter discusses the change from the original "rigid" principle to the "flexible" characteristic, which appears in the heritage of traditional Chinese music, and views the contemporary value and philosophical significance of Chinese and Western music from the perspective of philosophy.

1. Introduction

Classic scientific theory holds that objects in the world interact and move through linear interaction and the pattern of movement always shows the characteristics of simplicity and rigid determinism. However, as science progresses, especially in the field of fractal and chaos theory, people gradually revealed the nonlinear, complex and non-deterministic property of interaction and movement.

Human understanding progressed from a simply orderly or unorderly way of thinking to a comprehensive way of thinking that integrates both orders and disorders, from unipolar emphasis on determinism or non-determinism to comprehensive integration of determinism and non-determinism. In this duration, people gradually revealed the nonlinear mechanism of things and the uncertainty and complexity in their evolution. In traditional Western music,[1] what composers and theorists believed to be unshakeable principles, such as unipolar pursuit of patterns, orderly appearance of accent and relative low level of change of beat symbol in music, were changed in the creations of some 20th century contemporary Western composers. In traditional Western music rhythm, some orderly phenomenon was developing toward integrating disorders.

Primary creation has always been the focus of traditional Western music in the past hundreds of years. Stave is the carrier of a composer's creation and method of notation that materializes the thoughts of the composer and it comes with a certain rhythm, pitch and speed. Once the stave is generated, it becomes a classic and is not easily changed. However, interestingly, in a 20th century Western contemporary composer's concept, this rigid principle started to shake and even collapsed, adding more uncertainties and allowing greater space for the secondary creation by performers. These types of uncertainties in stave and inheritance have long existed in traditional Chinese music. The characteristics of structure and indication allow more possibilities in outcomes of each inheritance. A music score is followed but is not restricting and each inheritance is different, but with similarities, and shows both certain and uncertain traits.

[1]Western traditional music mainly means European music before the 20th century, which is the music of the 17th, 18th and the first half of the 19th century, compared to the 20th century modern music.

2. From Order to Chaos — Transformation of 20th Century Western Music

Stravinsky is a world-renowned 20th century Western music composer and he is best known for his primitivistic magnum opus, *The Rite of Spring.* This masterpiece vividly portrays the ceremony of Spring Sacrifice by prehistoric Russian people, where living young girls were sacrificed. In the first segment, Return of Spring, the composer attempts to inspire people to think of primitivism. Stravinsky broke the pattern of regularity of traditional Western music by not placing an accent on the first beat but a randomly assigned position in a bar. In traditional Western music, regardless of the beat such as 2/4, 3/4, 4/4, 8/3, 6/8, 9/8, etc., the accent must be placed on the first beat of each bar, making the position of the accent very orderly. Figure 1 shows the position of accent in a 2/4 beat of traditional Western music in red, and we can clearly see the orderly pattern of the beat and accent.

Figure 2 is also a 2/4 beat with the same basic rhythm as Figure 1 but the difference from the previous rhythm is that the accent does not appear orderly, sometimes on hard beat, sometimes on the down hard beat or down soft beat. It appears to be placed in chaos.

The rhythm in Figure 2 comes from the first segment, Return of Spring, in *The Rite of Spring* by Stravinsky. For the sake of ease of observation, we only show the rhythmic pattern of the clip without pitch. It can be seen

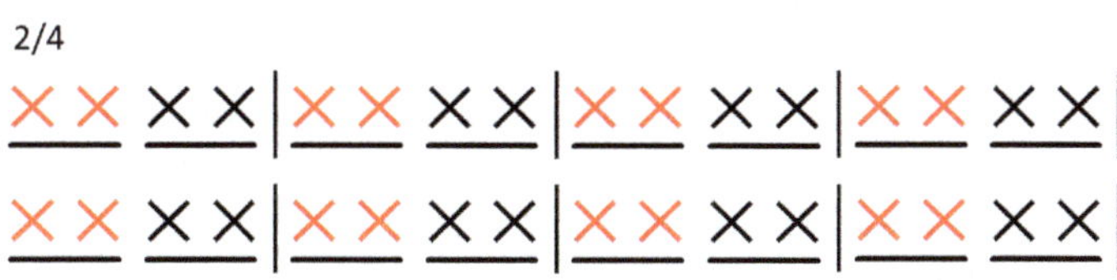

Figure 1. Position of accent in a traditional 2/4 beat.

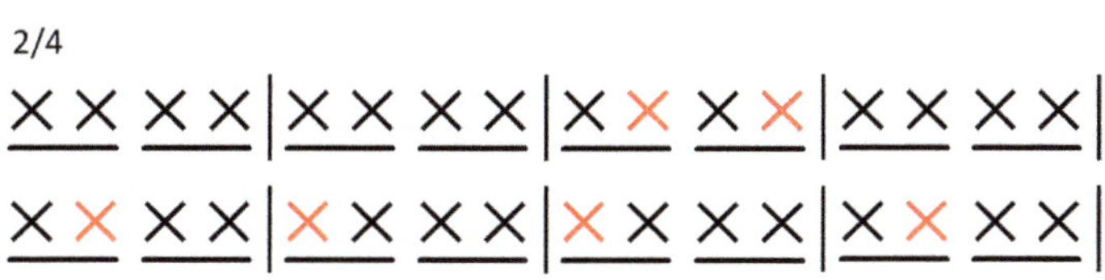

Figure 2. Position of accent in *The Rite of Spring.*

from Figure 2 that the composer did not follow the position of soft and hard beat in traditional Western music, and compared to traditional music, it is difficult to capture the pattern of this rhythm. The accent of traditional Western music has always been very orderly, even for repetitive beats such as 4/4, 6/8 and 9/8. Although there are two or more accents in a bar, the position of these accents still follows a regular pattern, offering listeners a sense of stability, helping them find it easy to recognize the pattern of these accents and to synchronize with the music. It is the extremely ordered pattern of these accents that greatly limits the creativity, imagination and expression of the composers.

The fundamental rhythm in Figures 1 and 2 is the same, both with quaver and the same segmentation of bars, appearing to be in an orderly state. However, there is a great change in the position of accents in Figure 2. It is the change from an orderly state in traditional music to the combination of order and chaos in 20th century music placement of accents. The small change of accent position in Figures 1 and 2 makes the two rhythms audibly distinguished and makes people feel that they are two completely different rhythms.

Tradition is also broken by the composer in the last song of *The Rite of Spring*, "Dance of Sacrifice". In this masterpiece of Stravinsky, very complex variations of rhythm changes have been employed. First the beat symbol is changed frequency. Figure 3 is an extract from "Dance of Sacrifice" and the beat symbol is changed 4 times although there were only 4 bars. The four bars in Figure 3 are respectively 3/16, 5/16, 3/16 and 4/16. Additionally, the first beat of each bar ought to be the position for the hard beat in traditional music, but no sounding notes are placed there by the composer and instead he arranged a pause which takes music time but does not make a sound. Notes with a high pitch are placed on the soft beat of each bar by the composer, the second beat in the first bar, and the second and fourth beat on the second bar. Sound notes are introduced from the second beat in the third bar and although the note with pitch appears in the first beat of the fourth bar, because the liaison links it with the RE in the third bar, there is no hard beat in this bar, making it the same as the previous bars in starting with a soft beat. Such a rhythm is much more complex for performers compared to a

Figure 3. Music clip of the last dance "Devotion of Dance" in "*The Rite of Spring*".

traditional orderly rhythm. The employment of such a complex rhythm is not likely to be a random act by the composer. Then what is the purpose of this?

Stravinsky initially wrote this piece as background music for the ballet *The Rite of Spring.* The last screen of this ballet "Dance of Sacrifice" shows the eventual killing and sacrificial of the chosen young girls on a sacrificial altar. In writing the background music for this ballet, the first task for the composer was to associate the audience to the primitive state of Russia. The second was to express the very complex psychological state of the young girls. The very regular pattern of rhythm of traditional Western music lacks the vivid expression required here.

In prehistorical Russia, young girls who were around thirteen and had just reached puberty in tribes were selected as human sacrifice. Their lives had just started. Perhaps the family and tribe of the chosen girls even found it honorable. However, as the chosen sacrifice, how did the girls think and react? Stravinsky accurately describes the trembling body, bumpy dance steps and moony face of the girls with his music. He bravely broke the rule of singular use of the beat symbol in traditional music and brought in his new approach, showing the girls' internal thinking in the dance as well as his music.

Traditional and regular rhythm patterns are not able to accurately express such complex states of mind. Imagine the girls dancing elegantly a beatific waltz toward their sacrificial altar, or marching in an orderly pace in a march onto the sacrificial altar. Undoubtedly, the regular pattern of beat accent in traditional Western music limits the expression of the complex mind states and stage performances, nor will it associate us with the atmosphere of unruliness and wildness. Therefore, Stravinsky, through this chaotic, unorderly expression, brought listeners back to primitivism audibly and in his best effort showed the psychological and behavioral modality of the girls to be sacrificed. This primitivistic ballet music is creative in its employment of complex rhythms and its characteristic is the collection of a series of minimum beats of varying lengths and the constantly appearing irregular accents at the same time (Shen, 1999, p. 351).

Although Stravinsky broke some music conventions and purposely added chaos in his music to achieve his targeted expression, his music did not completely break away from traditions. His musical construction was still recorded with musical notations with a fixed pitch and his music has traditional proportional rhythmic timings. These are the certainties in music and they form a complex model of both orderly and chaotic existence.

It can be seen from the example of the previous music (analysis of rhythm and music clips in "*The Rite of Spring*") that Western music has the advantages of graceful melody, ordered rhythm and pleasing sound, but its expression is greatly restricted by rhythm and stress. The natural world presents itself in a complex modality where ordered and unordered events both exist. As part of the natural world, the inner world of humans shares this same complex characteristic. For instance, in Buddhism studies, cultivators usually hunker down and reflect on their changing innermost being, observing the constantly changing and intangible inner world. What can be shown by traditional Western music of this complex world and innermost being is a very limited part. This is due to the regular pattern of the beat and the fixed accent position that limit the fields that may be expressed. For this reason, traditional Western music only shows fairy tales, stories of heroes and love stories, for instance the masterpieces of Beethoven in 18th Century and some operas by Wagner from the late 19th Century. In the 19th century, composers started to express natural views with music, such as a series of this type of music by 19th century composer Mendelssohn. However, the modality of traditional Western music cannot express the complexity of the world and inner thinking.

Science provides us with the means to investigate the complex world, and similarly, the sensitive musicians were also starting to break the traditions and orders in traditional Western music that had been in place for hundreds of years, expressing the complexity of the world and the changes in human thinking. They broke traditions, gave up the classics and replaced them with extensions in musical expression.

3. Increase of Uncertain Factors — The Connection between 20th Century Contemporary Musical Concepts and Chinese Traditional Musical Concepts

In the creation of 20th century Western modern music, some composers not only broke the tradition but also started the disorder in the rhythm. Moreover, further increasing the uncertainty factors in the music score became the content considered by modern composers in the 20th century and was put into practice. The musical expression of the "rigid" stave of stature in Western traditional music has changed in the 20th century, where modern composers in the West reduced or even gave up the principle of "rigidity" in their primary creation, and turned to the Chinese "flexible" feature, shifting from the primary creation that focused on composers to focusing on performers or Learner's secondary recreation.

An important feature of music that distinguishes it from other art categories is that musical works need to be recreated twice — playing and singing. "Rigid" music scores have rigorous, absolute pitch, rhythm, speed, strength and other stipulations, and the secondary creation is basically the *reproducible creation*. In the "rigid" musical works, the composer's creation was emphasized. In "flexible" music, the second creation is even more important. The genre style of music art is mainly formed by performers or inheritors. The same tracks can also form different styles of genres. The instrumental and vocal performances of second-degree creations are *expressive recreations*.

The term "rigidity" refers to a modern, invariable rule length resulting from the precise measurement of space in a rigid rod. In music, with the constant frequency and length of the spectrum of the pitch where "rigidity" and "absolute" have the same meaning (Guan, 2014, p. 99), Western traditional music clearly has "rigid" characteristics. Once the composer's works are created, their rhythm, pitch and speed having been stipulated by the composer at the time of creation, if the performers or learners of future generations want to practice the works, they cannot be arbitrarily changed and must respect the intent of the composer. In the teaching of Chinese music majors, when performing traditional Western musical works, students should try to restore their original appearance and express their style. However, most of the existing traditional music works in China do not show the composer's name, and the music score text is only an auxiliary in the inheritance. The scores are written with some backbone sounds. In each performance or transmission, the inheritor is required to perform a secondary creation based on the music score text, and there is a large amount of uncertainty in the secondary creation. As music scores are merely framed and instructive, not all notes and rhythms are written in the score. It is impossible for a performer or a successor to perform literal translations in accordance with the traditional music scores. After improvisations join some notes or change some rhythms, the music becomes more vivid. In traditional Chinese score writings, if different people perform the same musical score, the results will be different; even if the same person performs the same score at different times, different places and different moods, the results will also be different. Therefore, Chinese traditional music has the characteristics of rheology, flexibility and innovation, that is, the characteristics of "flexibility". However, in the 20th century, the modern composers of the West began to attenuate all kinds of stipulations in the primary creation and increased the possibility of the secondary creation, creating "flexible" features of traditional Chinese music.

John Cage was an American composer of the 20th century, of the casual music school. One of his totally non-prescriptive works "4′33 seconds" made him famous. This work was created for piano performance, but the performer did not touch the keys of the piano from the beginning to the end on the stage, but used the opening and closing of piano cober as the division of each segment of music. The composer was influenced by Oriental Zen studies and created this silent work. This work has also become one of the representatives of the casual music school. The elements of music composed of rigid rhythm, pitch and speed in Western music have all fallen apart in this work. He believes that "life is just life itself, everything must go with nature, and there is no deliberate search for order from chaos and accident" (Ma *et al.*, 2011, p. 235).

Chance music is a term used by American composers and is called aleatory music in Europe. It refers to the composer's intention to give up subjective control of certain parameters (such as pitch, rhythm, strength or structure) when he is composing music and adopt some objective methods (such as throwing dice, tossing coins and playing cards). It is determined by playing a gossip, a figure or random data given by a computer program, or by the player performing an improvization when performing. It is in contrast to the tightly controlled sequence music that prevailed in the Western world in the 20th century. It is the composer who brings accidental, uncertain or undesigned elements into the creation or performance of the work. The contingency of creation is that some parts of the work are not written by the composer himself, but are written in a way that is accidental and unexpected in their results. Composers leave occasional techniques in the performance to the performers to decide.

Concert for Piano and Orchestra, p. 57, is another work by John Cage (Figure 4). In this work, he totally gave up the "rigidity" requirement of Western traditional music, and simply marked out in the score the walking curve of music. The composer asks the player to play the music according to the direction of the curve. Based on this, each performer will have

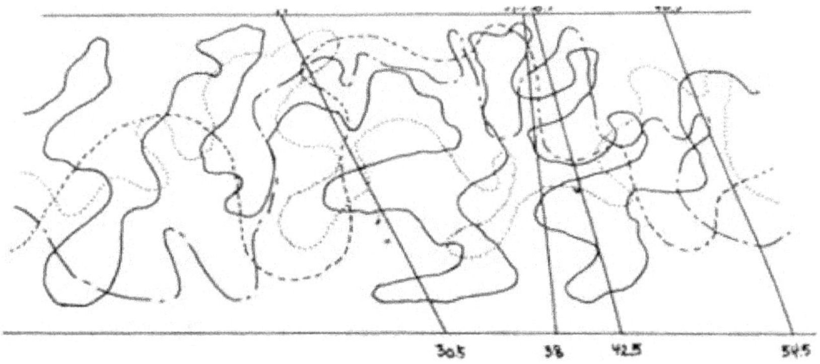

Figure 4. Music score of "*Concert for Piano and Orchestra*, p. 57". © 1960 by Henmar Press Inc. Reprinted by permission of C.F. Peters Corporation.

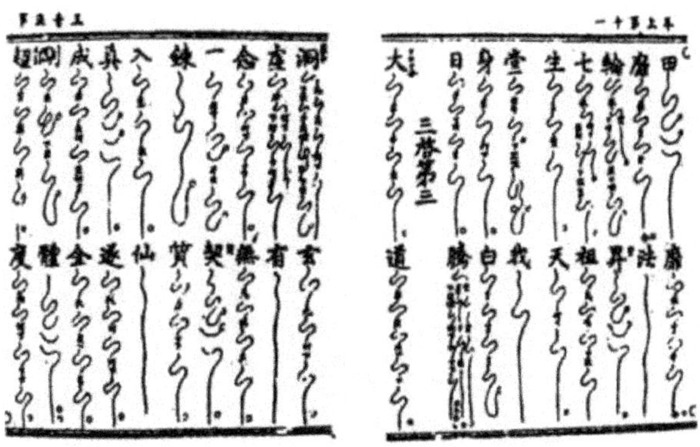

Figure 5. Score of Yu Yin Fa Shi.

different interpretations of the score based on his own feelings and understanding. According to this musical score, performers can play different versions.

Figure 5 is a graph of the Chinese traditional music score "玉音法事 (Yu Yin Fa Shi)" (Wang, 2006). The music is written in ancient China — from top to bottom. The lyrics are written on the top of the score, and the melody tunes are written in the vertical line below the lyrics. From the profile of the vertical line, we can feel that this piece of music is completely twisted and turned. Similarly, everyone who gets this score sings different versions.

The difference between these two musical scores is nearly 500 years old, but they have some similarities in concept. We have not found that the inaccurate quantitative factors in Western modern music notation, suggestive, framework and different degrees of free processing, have long been the essence of Chinese traditional music notation text and inheritance.

Among the existing music types of traditional Chinese music, there are still traditional music scores such as Guqin, Fujian Nanyin, Xi'an Drum Music and so on. Figure 6 is the ancient spectrum of the intangible cultural heritage of Xi'an Drums. The name of the score is 奉金杯 (Feng Jin Bei). In the inheritance of oral teaching, the subjects of the imparter and the imparted person are in an exchange relationship, spreading by word of mouth, and reciprocating. The Song Dynasty folk music used by Xi'an Drum Music only records the skeleton sound of the music, which is the process of the development of the music.

From Figure 6 we can see that it has no rhythm, and the above spectrum words only represent pitches, which have been deciphered by scholars

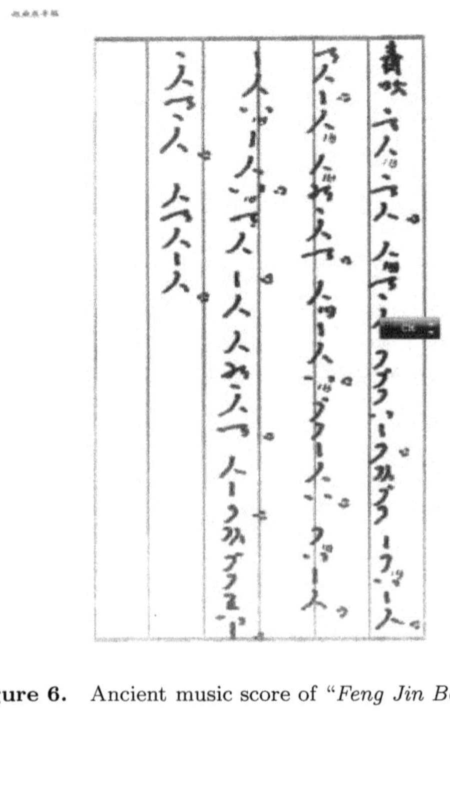

Figure 6. Ancient music score of "*Feng Jin Bei*".

and are fixed pitches. Although there is no accurate rhythm in the music score, a circle is marked on the right side of a spectrum word. The position of the circle is the place where the "board" appears, that is, where the "stress" appears. This is the basic framework of the score. Inheritors can add impromptu rhythms and decorative tones between these sounds, with detailed strengths and weaknesses, based on the position of the "board". In other words, performers have a lot of freedom to express and recreate emotions. This cannot be seen in the spectrum of ancient music, reflected in the inheritance; therefore, different inheritors or different moods can let the song have an ever-changing appearance.

4. Increase of Uncertain Factors in Contemporary Western Traditional Music Performances

The characteristics differ in the scripts of traditional Western music and traditional Chinese music. In traditional Western music, the primary creation by musicians is treaded as the golden standard. Those who perform or practice must closely follow the primary creation of the musicians and not change the music score thus, such a practice is inheritance of certainty and rigidity. Statistically speaking, if each performance of traditional music script is treated as a sample, by putting together all these samples, we will find that they follow the pattern of normal distribution. This is because in each interpretation of music score, the outcomes vary very little if at all. Traditional Chinese music scores, because of their structure and indications, employ elements of both certainties and uncertainties. As a result of the subjective character and implicit experience introduced by performers, the outcome of the same music score varies from one performer to another. If we use musical notations or numbered musical notations that everyone understands to record the performance of the same traditional Chinese music piece by different generations of inheritors and performers and put these recordings together, we should find that they do not follow the pattern of normal distribution. These recordings are in general similar, but also vary in some aspect.

Certain and uncertain acts form a dialectical relationship. Certain acts are expressed in the form of patterns, while uncertain acts appear as a lack or non-existence of patterns. Certain acts provide a stable foundation for the process of development and evolution of music, while uncertain acts are the conditions for variations (Jiang, 2004, p. 6). This explains how a uniform style is kept in traditional Chinese music inheritance even when there is no rhythm or even pitch in its scores while there are variations and creativities in the style.

The method of inheritance of Western traditional music is rigid where the rhythm, pitch and speed and many more follow the original intention of composers rigorously, leaving very little freedom for participation in secondary creation. For this reason, the music score of Western music is still in strict conformity with the original script. For instance, Chopin's music has been published by many presses and performed by many musicians, but the music score kept its original modality and there were only minor differences in playing technique and handling of emotions. This is the rigid method of inheritance in Western music. However, as time progressed, this rigid inheritance method started to fail in satisfying people's demand; thus, some musicians started to perform their own secondary creations of European traditional music according to their own understanding and put in the meaning of their time to the traditional music. There were additions and subtractions on the original score, but when played, it still showed the characteristics of the original piece. This method of inheritance is very similar to what is employed by Xi'an Drum music. Musicians adapt traditional Western music based on their performance abilities, music aesthetics and so on.

"Turkish March" by Mozart is liked by piano learners all over the world for its lively rhythm, smooth melody and moderate technical difficulties, and its music score is followed stringently when people practice and perform. However, some adapted versions of this sonata started to appear in recent years, for instance the polyphony version adapted by the renowned Russian piano performer Arcadi Volodos, Jazz versions by renowned Turkish piano performer Fazil Say and Jazz version by Volodos. We performed a comparison on the third segment of Turkish March between the original version and Jazz Version by Volodos. Following the method described previously, we generated the oscillography of the two versions as shown in Figure 7. It can be seen that the tendency of the two curves is very similar, with Volodos's version being more volatile and embedding more variations. We cannot definitively conclude that these two curves are correlated by just examining the figure, and a mathematical analysis is required to determine their correlation. The computation of their correlation coefficient yields the value of 1, which suggests that they are correlated with 100% certainty. However, despite the high level of correlation, our hearing can still distinguish the differences. By calculating the variance, we find the variance of Volodos's version to be 30.1, while the original is 6.8. This illustrates a higher level of volatility in Volodos's version of Turkish March and it is

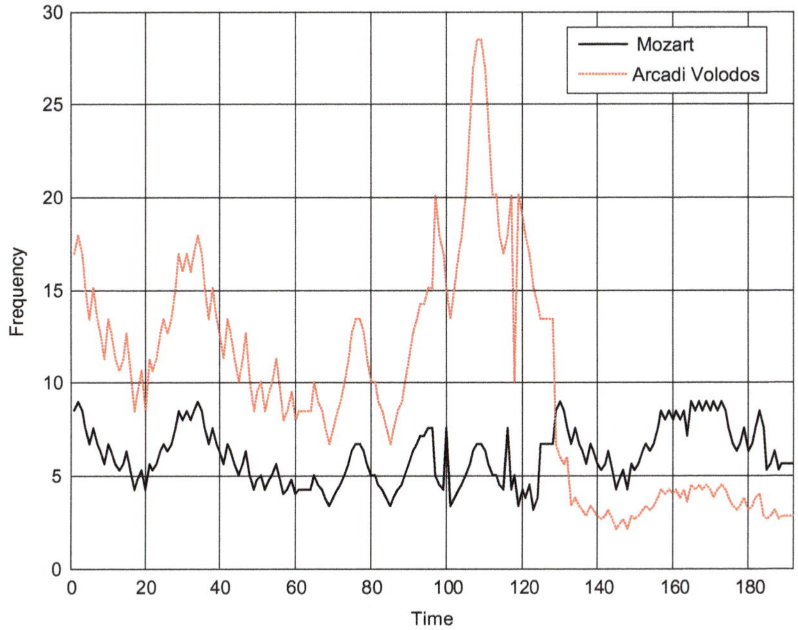

Figure 7. Ancient music score of "*Feng Jin Bei*".

more dynamic in terms of sound. In summary, we can conclude that Volodos's Turkish March and the original Turkish March are the same music, but with different styles and expressions.

Through our analysis above, we find that the inheritance method of oral teaching and inspiring understanding from within practiced in Xi'an Drum Music appeared in traditional Western music. In this method, musicians conduct secondary creation on traditional music with their own understanding. This suggests that compared to the fixed rigid way of inheritance, this method not only ensures traditional music is passed on through generations but also constantly adds contemporary elements into traditional music, keeping it active. It ensures the certainty while reflecting the liberty and creativity in music inheritance. In addition, acoustically adapted music is more refreshing and more capable of attracting attention and inspiring emotions. However, we also find that there are very few adaptable music pieces in traditional Western music. In the vast field of traditional Western music, how many pieces are adaptable and have been adapted? This is because many subjective elements of composers are integrated into the work in primary creation, thus being more difficult to alter. However, this tendency and the examples of traditional Western music adaption just illustrate that the inheritance method of traditional Chinese music is more creative and free, allowing more possibility of variation for the following generations.

5. Conclusion

In conclusion, the new music concept promoted by contemporary Western music connects to the traditional Chinese music. The renowned contemporary philosopher Mr. Zailin Zhang wrote in his article *Postmodernism and Traditional Chinese Culture* (Zhang, 2002/03) that "there exists some covert connection between postmodernism and traditional Chinese culture" and "despite Western postmodernism and traditional Chinese culture originating from completely different social environment and historic period, the two for certain happen to share some commonalities".

The 20th century started with the discovery of a new space-time model and behaviors, and at the same time suggested the beginning of re-recognizing the meaning of space-time and behaviors of Eastern arts (Guan, 2014, p. 107). The establishment of a new musical space-time concept and model must take into consideration the combination of the space-time view and value of both Eastern and Western music, as well as contemporary science and philosophy (Guan, 2014, p. 109).

Compared to traditional Chinese music, traditional Western music shows greater complexity in configuration, orchestration, harmony and so on. For instance, different branches of orchestra cooperate with each other, producing abundant sound and together with harmony show a modality that appears to be complex. However, the realm of its music performance is relatively narrow and simple and lacks creativity and liberty in performers' recreation, thus showing its character of simplicity. Because of this, in the 20th century, when faced with the complexity of nature, humans and the world, Western contemporary musicians added more unordered elements to reflect the complexity of the world together with science.

When compared with its Western counterpart, traditional Chinese music appears to be simple in orchestration and harmony, but does constantly evolve and develop through inheritance. This phenomenon shows a complex modality. The culture behind music is the background and at the same time the resource which supports the aesthetic significance and aesthetic strength. In its inheritance, traditional Chinese music follows the traditional Chinese philosophical theory that man is an integral part of nature. All things in the world are changing unnoticed. They may appear not to have changed but in fact have. This is just like the growth of men and renewal of cells where changes are constant but unnoticed. Ancient Chinese philosophy has different characteristics from the West. Nature is perceived in the West as element association theory whereas Chinese philosophy perceives Nature with Yuanqi metaplasia theory. This view on Nature believes all things are formed with Yuanqi and the formation is not through metaplasia but combination. This is the same as "the Way bears sensation, Sensation bears memory, Sensation and memory bear abstraction and Abstraction bears all the world" claimed by Lao Zi and "metaplasia of all things happen after the Earth and heaven moves" in *Zhouyi*. Qi in Chinese philosophy differs from atom and element in Western sciences. Qi is a continuous body, mutually included and indivisible and an integration of certainty and uncertainty. The script, inheritance and evolution of traditional Chinese music share the same characteristics. Renewed Chinese musician Mr. Xiangpeng Huang commented on the pattern of development of traditional Chinese music over the past thousands of years that "Traditional music following its own principle of oral teaching and inspiring understanding from within, not fixing on a restrictive written script and expelling the possibilities of impromptu development, is gradually changing in an unnoticed manner and this is where its energy lies." In traditional Chinese music, being influenced by cultures and aesthetics, its inheritance itself demonstrates complex modality and reflects diversity, creativity and liberty.

References

Guan, J. (2014). *Comparison of Chinese and Western Music*, Nanjing University Press, Nanjing.

Jiang, W. (2004). Chaos. Thinking of the relationship between fractal and music, *New Voice* 3, 6.

Ma, X. *et al.* (2011). *Music Appreciation*, Xi'an Jiaotong University Press, Xi'an.

Shen, Y. *et al.* (1999). *Compendium of Western Music History*, Shanghai Music Publishing House, Shanghai.

Wang, Y. (2006). *Chinese Traditional Music Scores*, Fujian Education Press, Fuzhou.

Zhang, Z. (2002/03). Postmodernism and traditional culture, *Humanities Magazine*.

Chapter 18

A Reasonable Model of Complexity for the Legal Domain

Cornelis N. J. de Vey Mestdagh
Foundation for Law & ICT, The Netherlands
c.n.j.de.vey.mestdagh@law-and-ict.org

The complexity of the universe can only be defined in terms of the complexity of the perceptual apparatus. The simpler the perceptual apparatus, the simpler the universe. The most complex perceptual apparatus must conclude that it is alone in its universe.

1. Introduction

The concept of complexity has been neglected in the legal domain, both as a qualitative concept that could be used to legally and politically analyze and criticize legal proceedings and as a quantitative concept that could be used to compare, rank, plan and optimize these proceedings. In science, the opposite is true. Especially in the field of Algorithmic Information Theory (AIT), the concept of complexity has been scrutinized. In this chapter, we will first have a quick look at AIT to see what it could mean in this phase of our research in the legal domain.

The conclusion is that there is a difference between (subjective) problem complexity and (objective) solution complexity. In this chapter, we therefore start to develop a model of complexity by describing problem complexity in the legal domain. We use a formal model of legal knowledge to derive and describe the parameters for the description of the problem complexity of cases represented in this formal model.

Further research will focus on refining and extending the formalization of the model of complexity, the comparison of problem and solution complexity for several legal cases using available algorithms and on the validation of the combined model against concrete cases and lawyers' and legal organizations' opinions about their complexity.

2. Complexity in the Legal Domain

As has been said the concept of complexity is hardly developed in the legal domain. Most of the descriptions of concepts related to complexity in legal literature refer to qualitative notions of vagueness (intension of concepts), open texture (extension of concepts), sophistication (level of detail of elements and relations) and multiplicity of norms (competing opinions) — in most cases without explicit reference to the concept of complexity.[1] Complexity arises in all these cases from the existence and competition of alternative perspectives on legal concepts and legal norms.[2] A (subjective) complex concept or norm from a legal point of view is not necessarily a(n) (objective) complex concept or norm from a scientific point of view. If all parties involved agree, i.e., have or choose the same perspective/opinion, there is no legal complexity, i.e., there is no case/the case is solved. The more the parties disagree, the more legally complex the case becomes, even if its quantified information content is not complex.

In science, more exact, quantitative definitions of complexity are common and applied.[3] Complexity is associated with *inter alia* uncertainty, improbability and quantified information content.

---

[1]The epistemology of complexity, specifically the number of variables and their interrelations as a measure of complexity, has been studied extensively in most subdomains of social science (e.g., Hayek, 1967), but not in the legal domain.

[2]For example, Hart (1994), who uses the concept of discretion to characterize hard (complex) cases; and Dworkin (1986), who distinguishes easy from hard cases using the concept of principled interpretation; although fundamentally differing in their opinion about the sources of the decision criteria, they both acknowledge the alternative perspectives that play a role in deciding complex cases (the judge's discretion in the light of the parties' alternative perspectives vs. the judge's principled interpretation in the context of the parties' alternative perspectives).

[3]Legal knowledge, in particular legal decision-making, has been studied applying quantitative methods. However, this research has not been directed at understanding the complexity of legal knowledge. For example, in the field of AI and Law, quantitative methods have been applied to predict new decisions (see the vast literature on legal case-based reasoning) and network analysis of the law has been applied to address the relative importance of legal precedents but not the knowledge theoretical complexity of these precedents. For example, Fowler and Jeon (2008) and Winkels *et al.* (2011).

Despite this discrepancy between the legal domain and the domain of science, in the legal domain complexity is as important as in other knowledge domains. Apart from the obvious human interest of acquiring and propagating knowledge per se, complexity has legal, economic, political and psychological importance: legal, because a coherent concept of complexity helps to analyze and criticize legal proceedings, in order to clarify them, to enable a justified choice of the level of expertise needed to solve legal cases and to reduce unnecessary complexity (an example of reducing complexity by compression is given in the next paragraph); economic, because complexity increases costs and measuring complexity is a precondition for the reduction of these costs (can help in designing effective norms, implementing them effectively, calculating and reducing the costs of legal procedures,[4] planning the settlement of disputes and other legal proceedings, etc.); political, because legal complexity can be an instrument to exert power and can increase inequality; and psychological, because complexity increases uncertainty. A validated model of complexity in the legal domain can help to promote these interests.[5]

3. How to Develop a Model of Complexity in the Legal Domain (Methodology)

In this chapter, I will try to bridge the gap between the intuitive definitions of complexity in the legal domain and the more exact way of defining complexity in science (see Section 4). I will do that on the basis of a formal model of legal knowledge (the Logic of Reasonable Inferences and its extensions) that we introduced before (see Section 5), that was implemented as the algorithm of the computer program Argumentator and that was empirically validated against a multitude of real-life legal cases. The "complexities" of these legal cases proved to be adequately represented in the formal model. In earlier research, I actually tested the formal model against 430 cases, of which 45 were deemed more complex and 385 less complex by lawyers. A first result was that the algorithm (Argumentator) when provided with case facts and legal knowledge was able to solve 42 of the more complex cases and 383 of the 385 less complex cases in exactly the same way as the legal experts did (including the systematic mistakes made by these experts). A second result was that the algorithm when instructed to do so improved the decisions in 30 (66%) of the 45 more complex cases

[4]For example, White (1992).
[5]For example, Schuck (1992); Ruhl (1996) and Kades (1997).

and in 104 (27%) of the 385 less complex cases. This result confirms the relative complexity of the first 45 cases (see Section 7). The selection of these 45 cases thus provides us with the material from which the criteria for the definition of complexity in this chapter could be derived. These criteria are translated to quantitative statements about the formal representation of the cases (see Section 6).

Further research will focus on the fine tuning of this quantitative model by comparing its results with new empirical data (new cases and opinions of lawyers about the (subjective) complexity of cases). Finally, the ability of the fine-tuned model to predict complexity in new cases will be tested. A positive result can be applied to reduce the aforementioned costs of processing complex legal knowledge.

4. Models of Complexity in Science

There are many different definitions of complexity in science. The aim of this research is to develop a quantitative measure of subjective complexity for formal representations of legal knowledge and their algorithmic implementations. In this paragraph, I will therefore refer to definitions of complexity from Algorithmic Information Theory (AIT), which studies the complexity of data structures (representations of knowledge in a computer).

In AIT, the complexity of data structures is equated with its information content. Complexity is postulated to decrease proportionate to the degree of (algorithmic) compressibility of the data structure. To assess the usefulness of AIT, for our practical purpose, i.e., the design of a quantitative model of complexity of legal knowledge, I studied some publications from the domain of AIT. I read that complexity is approached as Algorithmic Probability (e.g., Solomonoff's *a priori* probability), i.e., the higher the probability that a random computer program outputs an object, the less complex this object is considered to be. I read that complexity is approached as Algorithmic Complexity (e.g., Kolmogorov's descriptive complexity), i.e., the shorter the code needed to describe an object (string), the less complex this object is considered to be. This is an interesting approach since it seems to offer a concrete measure of the complexity of certain objects (e.g., of legal judgements) and it associates with the concept of compressibility, which can be transposed as simplification (as opposed to sophistication) to the legal domain. Finally, I read about Dual Complexity Measures in Burgin (2010), which relate AIT to more complex problem structures and distinguish the complexity of the system described (the actual problem and its solution) from the complexity of the description (the algorithm used to describe the problem and its solution). A common and essential aspect of these approaches is the compressibility of the object as a measure of its complexity. In all these cases, the computer program is considered to be an explanation of a (more or less complex) object (or data structure).

My conclusion is that these approaches will be useful when trying to prove certain characteristics of the model of complexity in the legal domain, once developed, but not primarily for the design of the model. I will have to describe the formal model and the algorithm (explanation) first. Just to get a practical insight into the concept of compressibility, I did apply the idea of compressibility to some legal cases (see the following example). However, many of the characteristics of legal cases that make them "complex" according to lawyers are not directly related to compressibility. Moreover, incidentally the most simple "palaver" in the legal domain is meant to be incomprehensible and therefore misses the (semantic and relational) patterns that are needed to be compressible. The conclusion is that this concept only partially covers the problem in the legal domain. I am eager to discuss this with my colleagues in the mathematical domain.

An example of compression using logical equivalence in the legal domain

Objects regulation U.1. Appendix III Decree Indication Chemical Waste reads as follows:

Waste products are ***not*** *considered as chemical waste* [cw] *if they are objects* [o] *that have attained the waste phase* [wp] *of their lifecycle,* ***unless***:

a. This has happened ***before*** *they have reached the user* [ru];
b. This has happened ***after*** *they have reached the user* [ru] *and they are*

1. transformers [1];
...
10. Mercury thermometers [10].

The logical structure of this legal provision is as follows:

not cw is implied by o and wp and not ((not ru) or (ru and (1 or .. or 10))).

Logically equivalent with this formalization of the provision is the following formula:

not cw is implied by o and wp and ru and not (1 or ... or 10)

which is a compression of the original provision.

Interestingly enough, the retranslation of this equivalent formula to natural language is as follows:

Waste products are not considered as chemical waste if they are objects that have attained the waste phase of their lifecycle and they have reached the user and they are not 1. transformers .. 10. Mercury thermometers.

Although this example illustrates that compression can be beneficial because it improves the readability of the regulation, it does not reduce its actual complexity which — in practice — is related to different opinions about the meaning of concepts like "waste products".

5. A Formal Model of Legal Knowledge (Reasonable Inferences)

The first step in developing a model of complexity in the legal domain is to describe the formal characteristics of legal knowledge that are related to the essence of complexity in this domain, i.e., the competition of opinions. In a previous publication (de Vey Mestdagh and Burgin, 2015), we introduced the following model that allows for reasoning about (mutually exclusive) alternative opinions and that allows for tagging the alternatives, e.g., describing their identity and context.

Our knowledge of the world is always perspective bound and therefore fundamentally inconsistent, even if we agree to a common perspective, because this agreement is necessarily local and temporal due to the human epistemic condition. The natural inconsistency of our knowledge of the world is particularly manifest in the legal domain (de Vey Mestdagh and Hoepman, 2011).

In the legal domain, on the object-level (that of case facts and opinions about legal subject behavior), alternative (often contradicting) legal positions compete. All of these positions are a result of reasoning about the facts of the case at hand and a selection of preferred behavioral norms presented as legal rules. At the meta-level, meta-positions are used to make a choice for one of the competing positions (the solution of an internal conflict of norms, a successful subject negotiation or mediation, a legal judgement). Such a decision based on positions that are inherently local and temporal is by definition also local and temporal itself. The criteria for this choice are in most cases based on legal principles. I call these legal principles meta-principles because they are used to evaluate the relations between different positions at the object-level.

To formalize this natural characteristic of (legal) knowledge, we developed the Logic of Reasonable Inferences (LRI, de Vey Mestdagh *et al.*, 1991). The LRI is a logical variety that handles inconsistency by preserving inconsistent positions and their antecedents using as many independent predicate calculi as there are inconsistent positions (Burgin and de Vey Mestdagh, 2011, 2013). The original LRI was implemented and proved to be effective as a model of and a tool for knowledge processing in the legal domain (de Vey Mestdagh, 1998). In order to be able to make inferences about the relations between different positions (e.g., make local and temporal decisions), labels were added to the LRI.

In de Vey Mestdagh and Hoepman (2011), formulas and sets of formulas are named and characterized by labeling them in the form (A_i, H_i, P_i, C_i). These labels are used to define and restrict different possible inference relations (Axioms A_i and Hypotheses H_i, i.e., labeled signed formulas and control labels) and to define and restrict the composition of consistent sets of formulas (Positions P_i and Contexts C_i). Formulas labeled A_i must be part of any position and context and therefore are not (allowed to be) inconsistent. Formulas labeled H_i can only be part of the same position or context if they are mutually consistent. A set of formulas labeled P_i represents a position, i.e., a consistent set of formulas including all Axioms (a perspective on a world, without inferences about that world). A set of formulas labeled C_i represents a context (a maximal set of consistent formulas within the (sub)domain and their justifications, i.e., the world under consideration). All these labels can be used as predicate variables and if individualized, to instantiate predicate variables and consequently as constants (variables as named sets).

Certain meta-characteristics of formulas and pairs of formulas were finally described by labels (e.g., meta-predicates like Valid, Excludes, Prefer) describing some of their legal source characteristics and their legal relations which could be used to rank the different positions externally. The semantics of these three Predicates (Valid, Exclude and Prefer) are described in de Vey Mestdagh and Hoepman (2011). These three predicates describe the elementary relations between legal positions that are prescribed by the most fundamental sets of legal principles (principles regarding the legal validity of positions, principles regarding the relative exclusivity of legal positions even if they do not contradict each other and principles regarding the preference of one legal position over another). It was also demonstrated that the LRI allows for reasoning about (mutually exclusive) alternatives.

In de Vey Mestdagh and Burgin (2015), we have shown that labels can be used formally to describe the ranking process of positions and contexts. With that the thus extended LRI allows for local and temporal decisions for a certain alternative, which means without discarding the non-preferred alternatives like belief revision does and without using the mean of all alternatives like probabilistic logic does. This extended the LRI from a logical variety that could be used to formalize the non-explosive inference of inconsistent contexts (opinions) and naming (the elements of) these contexts to a labeled logical variety, in which tentative decisions can be formally represented by using a labeling that allows for expressing the semantics of the aforementioned meta-predicates and prioritizing (priority labeling). In de Vey Mestdagh and Burgin (2015), we illustrated the use of these labels by examples.

In the next paragraph, the extended LRI is used to identify the quantitative parameters of complexity in the legal domain.

6. A Formal Model of the Complexity of Legal Knowledge (Parameters for a Reasonable Calculation of Complexity)

The processing of legal knowledge takes place in successive phases. Each phase is characterized by its own perspectives and associated parameters of complexity. Roughly, first the different parties in a legal dispute take their positions, then the positions are confronted and a decision is made and finally the decision is presented.

The complexity of the dispute differs from phase to phase: again roughly, from intermediate *initial complexity* (the separate positions), to high *problem complexity* (the confrontation and decision making), to low *solution complexity* (the decision itself). The separate positions are basically consistent and their contents can each be processed within a separate single logical variety. When the dispute starts, complexity increases, because the shared axioms of the dispute have to be calculated and the positions are by definition mutually inconsistent and several calculi within the logical variety have to be used to calculate the joint process of the dispute and to decide between different hypotheses within the dispute. Ultimately, the decision principles included in the different positions have to be used to rank the different consistent solutions. The dispute ends by presenting the highest ranking consistent (local and temporal) decision, representing a concurring opinion or a compromise. The complexity of this result is reduced again, because it can be (re)calculated within a single consistent variety. In the following, these phases are described in more detail and the related parameters of complexity in terms of the formal model introduced above.

6.1. *Phases and parameters of legal knowledge processing*

In a particular case, the complexity of the case can be quantified on the basis of the case elements and relations presented by all parties. The processing takes place in five phases:

(1) At the start of legal knowledge processing, the case can be described as follows: A number of sets n (the number of parties involved) of labeled formulas $\mathrm{H}_{i,l}$ representing the initial positions of each of the parties in a legal discourse, i.e., hypotheses$_i$ of parties$_l$ about the (alleged) facts and applicable norms in a legal case.
(2) The next step is determining the intersection between these sets $\mathrm{H}_{i,l}$, which defines A_i representing the agreed case facts and norms, and determining the union of all complements, which defines H_i. (A_i, H_i) represents the initial case description.
(3) The third step is calculating all possible minimal consistent positions P_i that can be inferred from (A_i, H_i) applying a logic, e.g., the LRI, a logical variety that allows each position to be established by its own calculus. If these calculi differ, this adds to the complexity of the problem. In earlier publications, we assumed all the calculi to be the same (predicate calculus).
(4) The fourth step is calculating all maximal consistent contexts (possible consistent worlds) C_i on the basis of (A_i, H_i, P_i).
(5) The last step is making a ranking of these contexts on the basis of the application of the meta-norms (decision criteria) included in them. A formal description and an example of this process can be found in de Vey Mestdagh and Burgin (2015).

6.2. *Parameters for a reasonable calculation of complexity*

Each step in this process is characterized by its own parameters of complexity. In legal practice, different procedures are used to determine and handle (reduce) complexity in these different phases:

(1) In the first phase, a direct, static measure of complexity is commonly applied: the number of parties and the number of hypotheses. This is

a rough estimate of the number of different positions (interpretations, perspectives, interests).

(2) In the second phase, a direct, relative measure of complexity is commonly applied: the number of A_i and its relative size to H_i. The larger the relative size of A_i, the less complex a case is considered to be, because there is supposed to be more consensus.

(3–4) In the third and fourth phases, all positions P_i and contexts C_i are derived:

Given the resulting set of labeled formulas (A_i, H_i, P_i, C_i) representing the legal knowledge presented in a certain case, the problem complexity of this set can be defined as follows:

(a) The subset A_i (agreed case facts and norms) is by definition included in each P_i and C_i so its inclusion as such is not a measure of complexity as it reflects absolute consent.

(b) The elements of the subset H_i are by definition not all included in each P_i and C_i, so the relative size of the inclusion of its elements is a measure of complexity as it reflects relative consent. If there is more conformity, there is less complexity. It is even possible that certain elements of the subset H_i are not included in any P_i and C_i. The number of these "orphaned" elements can also contribute to the complexity of a case, because they represent antecedents without consequents (a justification for an absent conclusion) or consequents without antecedents (a decision is proposed without justification). Orphaned elements can be the result of incompletely presented positions or — worse — be smoke screens.

(c) The relative size of the fraction of subset A_i in (A_i, H_i) — relative to the fraction of A_i in other cases — is a measure of complexity as it reflects the size of shared (consented) knowledge in a legal dispute. This holds, even if the size of A_i is manipulated by one or more of the parties involved (as a winning strategy or for billing reasons), because the other parties have to take the A_i into consideration.

(d) The relative size of the fraction of subset H_i in (A_i, H_i) — relative to the H_i in other cases — is a measure of complexity as it reflects the size of the disputed knowledge in a legal dispute. This holds even if the size of H_i is manipulated by one or more of the parties involved (as a winning strategy or for billing reasons), because the other parties have to take the H_i into consideration.

(e) The relative size of the subset P_i (relative to the P_i in other cases) is a measure of complexity as it reflects the number of different minimal positions that can be taken logically in this specific case. The size of P_i can only be manipulated indirectly (through the respective sizes of A_i and H_i).

(f) The relative size of the subset C_i (relative to the C_i in other cases) is a measure of complexity as it reflects the number of different consistent contexts (possible decisions) that can be distinguished in this specific case.

(5) In the fifth phase, ranking of the contexts takes place. The number of rankings depends on the inclusion of ranking meta-norms in the respective contexts. For example, if C_i includes a ranking meta-norm that concludes to Prefer(C_i, C_j) and C_j includes a ranking meta-norm that concludes to Prefer(C_j, C_i), there are (at least) two rankings. Meta-norms that are agreed upon are part of A_i, meta-norms that are not agreed upon are part of H_i. The process of applying the meta-norms is fully recursive, since the objects of the meta-norms are other (meta-)norms, which are themselves also part of (A_i, H_i). This means that the determination of the complexity of the *application* of the meta-norms is included in the previous phases. In this phase, only the resulting number of rankings is established and can be considered to be an independent measure of complexity.

7. Validation of the Model of Complexity

The model of parameters for a reasonable calculation of complexity of legal knowledge as described in the previous paragraph is based on prior theoretical and empirical research into — among other things — the complexity of legal knowledge (de Vey Mestdagh, 1997, 1998). In this research, a theory of legal knowledge has been developed and translated into a formal model of legal knowledge. This model — the logic of reasonable inferences (LRI) — is introduced in Section 5. The LRI has been implemented as a general computer model of legal knowledge, the legal knowledge system (algorithm) Argumentator. A total of 430 environmental permit law cases and knowledge of environmental permit law, administrative and court rulings, principles, policies and practices have been formally represented in

Argumentator resulting in an expert system for environmental permit law (ESM). The relative complexity of the 430 cases has been established on the basis of legal expert judgements. The opinion of the experts was that 45 cases (concerning secondary aluminum smelters) were of a relatively complex nature and 385 cases (concerning petrol stations) of a relatively less complex (more general) nature. This has been verified by applying the expert system ESM to these cases. The ESM was enabled (provided with more complete general data and knowledge) to improve on the human judgements in the 430 cases. The test results have shown that in the complex cases, 66% of the human judgements were improved by the expert system (of which 20% were full revisions), while in the less complex, general cases only 27% of the human judgements were improved by the expert system (of which only 2% were full revisions).

The testing was conducted in two ways. First, an investigation was made of which rules the expert system applies and which conclusions it draws if it starts with the same data as the human decision makers did in the cases from legal practice (*descriptive mode*). This investigation made it possible to draw conclusions about *the expert system as a model of the actual administration of the law* by comparing the norms applied and the conclusions drawn in the decision-making process of the human experts vs. that of the ESM. Second, an investigation was made of which rules the expert system applies and which conclusions it draws if it not only considers the case data but is also enabled to apply all the known norms and independently collect additional data (*supplementary mode*). This investigation made it possible to draw conclusions about *the expert system as a (normative) model of the formal and informal rules in force within Environmental Permit Law.* The expert system and the tests were evaluated by 18 independent assessors. In de Vey Mestdagh (1997), the test research and the independent assessments are described.

In *descriptive mode*, the ESM successfully (re)constructs 425 of the 430 examined decisions. So, only 5 mismatches were encountered, 3 of them in complex cases and 2 of them in the less complex cases. In all 3 mismatching complex cases, an invalid rule was applied in practice (7% of the 45 complex cases) while a valid rule was applied by the ESM. In just 1 of the 2 mismatching less complex cases, an invalid rule was applied in practice (0.3% of the 385 cases). In the other case, an — unpublished — valid local rule was applied. The test in descriptive mode validates the theoretical basis and its formalization and implementation. The legal expert system is a valid model of the use of legal knowledge in legal practice because it applies the same knowledge in a large majority (425) of the cases. It also shows a higher occurrence of the application of invalid rules (mistakes made) in legal practice in more complex cases.

In *supplementary mode*, in the complex cases 30 (66%) of the 45 decisions were amended by the ESM: 9 (20%) of the 45 examined decisions were revised (contradicted); furthermore 11 (24%) of the 45 examined decisions turned out to be subject to competition of legitimate (contradicting) alternatives, which were not taken into consideration in these cases, and 10 (22%) of the 45 examined decisions qualified for legitimate extensions.

In the less complex cases — as a consequence of the erroneous or non-application of valid rules — according to the ESM, 104 (27%) of the 385 decisions should be amended: 7 (2%) of the 385 decisions should be revised, 34 (9%) of the decisions are certainly subject to the competition of legitimate alternative (contradicting) decisions which were not taken into consideration, 31 (8%) of the decisions are subject to the competition of legitimate alternative (contradicting) decisions depending on the content of unknown additional data and 32 (8%) of the decisions should be extended.

In both complex and less complex cases, the legitimacy of the decisions proposed by the ESM has been judged to be in accordance with Dutch law by the 18 independent assessors.

In conclusion, in supplementary mode, the ESM is enabled to apply more complete knowledge of Environmental Permit Law (formal and informal object-level and meta-level rules). As a consequence, it improves the quality of a large number of the complex and less complex legal cases from practice. The difference between complex and less complex cases is reflected in the much larger improvement in complex cases (66%) than in less complex cases (27%). The improvements consist of (1) revision of decisions by applying valid rules instead of the erroneous or non-application of valid rules or the systematic application of invalid rules (also detected in descriptive mode), (2) weighing up legitimate alternatives (rules) by applying decision rules (meta-rules) instead of only applying the routine alternatives and (3) the application of valid and legitimate additional rules. In all cases, the number of normative opinions (rules and meta-rules) considered in making the improved decision is larger than that considered in making the original decision.

The distinction between the complex and less complex cases can be quantified by the differing numbers of normative opinions taken into consideration. The test of the ESM in supplementary mode thus confirms the assessment of complexity by the independent assessors. The complex cases are characterized by higher counts of the parameters distinguished in the previous paragraph. The number of parties and the number of hypotheses (rules and meta-rules) and with that the number of elements of H_i are higher in the complex cases. In an average complex case, the number of parties is 4 (3 licensing parties with diverging policies, principles and practices, 1 licensee). In an average less complex case, the number of parties is 2 (1 licensing authority, 1 licensee). The number of rules and meta-rules is relative to the number of regulations and licensing parties. The number of regulations is much larger in the complex cases (see list of regulations and description of policies, principles and practices in the original publication). The test results show that the number of (contradicting) alternatives and with that the number of elements of P_i and C_i are higher in the complex cases. Finally, the number of meta-rules and with that the number of rankings are higher in complex cases.

Further validation research is needed to refine the model of parameters for a reasonable calculation of complexity of legal knowledge as described in the previous paragraph. The relative weight of the counts of the parameters described will be varied against the available dataset of legal cases. The results will also be correlated with other variables that are available to gain further insight into possible parameters of complexity. Examples of these variables are number of submitted documents, length of procedure, number of appeals, spending power of the parties involved, level of expertise of the lawyers involved, etc.

8. Conclusion: A Reasonable Model of Complexity and Further Research

In this chapter, we have explored the concept of complexity in the legal domain. A first conclusion is that the concept has not been studied explicitly in the legal domain. Only indirectly as a qualitative concept (vagueness, open texture, etc.) and hardly ever as a quantitative concept. However, a quantitative model of complexity in the legal domain has — apart from its scientific meaning *per se* — legal, economic and political implications. It will allow us to improve the quality and efficiency of legal proceedings. Algorithmic Information Theory offers several approaches to the quantification of complexity that inspired the approach chosen in this chapter. It induced the thought that in the legal domain a distinction between subjective problem complexity and objective resolution complexity is necessary and that a quantifiable model of complexity based on the formal representation of subjective legal knowledge should be the first step in developing a model of complexity in the legal domain.

In this chapter, we have given a description of a formal representation of legal knowledge (the extended Logic of Reasonable Inferences) and we have described the quantitative parameters of complexity for this model. The result of this we would like to call *a Reasonable Model of Complexity*, because it is based on the Logic of Reasonable Inferences (LRI) and because it inherits its relative, perspective-bound character. Complexity is specifically relative to the number of perspectives combined in the knowledge under consideration.

Further research will focus on extending the model of complexity to resolution complexity, using — among others — available algorithms (e.g., Argumentator, the computer program we developed to implement the LRI). It will also use the available dataset of 430 environmental law cases that have been described and analyzed before and that have already been represented in Argumentator.

References

Burgin, M. (2010). Algorithmic complexity of computational problems, *J. Comput. Inform. Technol.* 2(1), 149–187.

Burgin, M. and de Vey Mestdagh, C. N. J. (2011). The representation of inconsistent knowledge in advanced knowledge based systems, in *Knowledge-Based and Intelligent Information and Engineering Systems*, Koenig, A., Dengel, A., Hinkelmann, K., Kise, K., Howlett, R. J. and Jain, L.C. (eds.), Vol. 2, Springer Verlag, pp. 524–537.

Burgin, M. and de Vey Mestdagh, C. N. J. (2013). Consistent structuring of inconsistent knowledge, *J. Intell. Inform. Syst.* 1–24.

de Vey Mestdagh, C. N. J., Verwaard, W. and Hoepman, J. H. (1991). The logic of reasonable inferences, in *Legal Knowledge Based Systems, Model-based Legal Reasoning*, Breuker, J. A., de Mulder, R. V. and Hage, J. C. (eds.), Vermande, Lelystad, pp. 60–76.

de Vey Mestdagh, C. N. J. (1997). *Juridische Kennissystemen, Rekentuig of Rekenmeester? Het onderbrengen van juridische kennis in een expertsysteem voor het milieuvergunningenrecht* (PhD thesis), p. 400, series Informatica en Recht, nr. 18, Kluwer, Deventer (in Dutch).

de Vey Mestdagh, C. N. J. (1998). Legal expert systems. Experts or expedients? in *The Law in the Information Society, Proc. of the Fifth International Conference of the Italian National Research Council*, Ciampi, C. and Marinai, E. (eds.), Florence, Italy, Istituto per la documentazione giuridica del Centro Nazionale delle Richerche, Firenze, p. 8.

de Vey Mestdagh, C. N. J. and Hoepman, J. H. (2011). Inconsistent knowledge as a natural phenomenon: The ranking of reasonable inferences as a computational approach to naturally inconsistent (legal) theories, in *Information and Computation, Essays on Scientific and Philosophical Understanding of Foundations of Information and Computation*, Dodig-Crnkovć, G. and Burgin, M. (eds.), World Scientific, New Jersey, pp. 439–476.

de Vey Mestdagh, C. N. J. and Burgin, M. (2015). Reasoning and decision making in an inconsistent world: Labeled logical varieties as a tool for inconsistency robustness, in *Intelligent Decision Technologies*, Neves-Silva, R., Jain, L. C. and Howlett, R. J. (eds.), Vol. 39, Springer; Smart Innovation, Systems and Technologies, pp. 411–438.

Dworkin, R. (1986). *Law's Empire*, Belknap Press, Cambridge, MA.

Fowler, J. H. and Jeon, S. (2008). The authority of supreme court precedent, *Soc. Networks* 30(1), 16–30.

Hart, H. L. A. (1994). *The Concept of Law*, Oxford University Press, New York.

Hayek, F. A. (1967). The theory of complex phenomena: A precocious play on the epistemology of complexity, *Studies in Philosophy, Politics and Economics*, Routledge & Kegan Paul, London, UK, pp. 22–42.

Kades, E. (1997). The laws of complexity & the complexity of laws: The implications of computational complexity theory for the law, *Faculty Publications*, Paper 646. Available online: http://scholarship.law.wm.edu/facpubs/646.

Ruhl, J. B. (1996). Complexity theory as a paradigm for the dynamical law-and-society system: A wake-up call for legal reductionism and the modern administrative state, *Duke Law J.* 45(5), 849–928.

Schuck, P. H. (1992). Legal complexity: Some causes, consequences, and cures, *Duke Law J.* 42(1), 1–52.

White, M. J. (1992). Legal complexity and lawyers' benefit from litigation, *Inter. Rev. Law Econom.* 12, 381–395.

Winkels, R., Ruyter, J. and Kroese, H. (2011). Determining authority of dutch case law, in *The Twenty-Fourth Inter. Conf. Legal Knowledge and Information Systems. JURIX 2011*, Vol. 235, pp. 103–112.

Chapter 19

Reform and Innovation: The Social Sciences Research in Big Data Times

Ouyang Kang

Institute of State Governance, Institute of Philosophy,
Huazhong University of Science and Technology,
Wuhan, China
kouyang@mail.hust.edu.cn

The goal of this chapter is to analyse the impact of the information overflow and big data on the contemporary society. The author attempts to explain the nature of big data and the opportunities and challenges this provides to us, promoting the reform and innovation of humanities and social science research.

1. Introduction

We are already living in the information age and Big Data times. But, many of us have not been ready for the greatest change of our times. We should deeply understand the nature of big data and the opportunities and challenges it provides us, and consciously promote the reform and innovation of humanities and social science research.

2. Full Understanding of Big Data and its Multidimensional Significance

What is big data? Wikipedia defines it as follows: "big data is large, high-speed, and/or changing the information assets, it needs to handle the new way to promote stronger decision-making ability, insight and optimization process." In China, the File to Promote the Development of Big Data Action Program (2015, State Council) pointed out: "Big data is based on the capacity of a large, multi type, high access speed, high application value as the main features of the data set." Victor Mayer–Schoenberger and Kenneth Cukier proposed the "4V" of big data: (large) Volume, Velocity (great) generation, Variety (fast modal variety) and Value (great value and low density) in their book, *Big Data.*

In our opinion, the emergence of big data is a positive product of human development and development of the law on the road to explore the world.

In the direct sense, big data is compared to "small data" or traditional data, which refers to the first data humans may obtain from nature, in contrast to which this forms in large numbers, with various types of high-speed access to social and human activities and values of data to be set up, forming a huge database, to change the way the world is and the things in it with comprehensive mass data in a detailed manner, helping users to realize the outside world to suit their needs in a more comprehensive, detailed, in-depth and accurate way.

In a wide sense, big data means data acquisition, storage, processing, transmission and application of the related information technology facilities, system and operation mode in order to build a data-oriented world. People may get to know and grasp the reality of the world through data processing, without direct contact with the world, and change the reality of the world; to reconstruct the real world through perception and operation data of the world, they form the interaction between virtual world and real world, show the natural, social and human digital process by which modern information technology and the network world reached a new stage and level.

In a broader sense, big data implies a high degree of information age people consciously or unconsciously relying on modern information network system and data of the mode of production, mode of living, communication, thinking and governance, whereby big data not only changes people's lives but also is a comprehensive guide to a revolutionary change in line with the times to enhance the human way of life.

In the broadest sense, big data refers to a kind of reliance on science and technology for the construction of the society by which mankind itself is transformed in terms of how it views nature, society, human values and methodology, marking the human understanding of the world and transforming the world with self-understanding, self-transformation into a more comprehensive and clear level, and a more consciously effective stage.

3. Exploring the Special Opportunities and Challenges of Big Data

In general, the diversity of social phenomena and the complexity of the corresponding humanities and social sciences present a huge group of subjects, which are very different disciplines, and the relationship and correlation of these different disciplines with big data are also very different. For example, from the overall perspective, the traditional social sciences, such as economics, management science, sociology, political science, law and pedagogy, have a strong dependence on data, showing the significance of data more fully, while literature, history, art and philosophy pay more attention to people's spiritual life, value philosophy, the spiritual world, and dependence on data is low, showing the significance of data is relatively weak, and the impact of the era of big data is relatively small here. Big data technology is still in development and cannot simply replace the traditional data technology of small systems; there are many problems in big data research, and we should not simply cover the humanities and social science research with the traditional big data research.

Yet, even as big data in the ascendant has given us a lot of problems and challenges in the humanities and social sciences research, it also provides a lot of opportunities for innovation.

First of all, from the "assumption" or from "data start to begin our study?", the bold hypothesis, "careful", has always been the basic mode when we are engaged in the research of humanities and social sciences. The era of big data, massive data and even "data" has become our very multidisciplinary research field and their characterization a premise. A lot of irrefutable facts require us to provide necessary descriptions and interpretations.

Secondly, it provides a choice between "individual sample" or "massive data" as the object of study. Because of data collection difficulties, past research in humanities and social sciences had to depend on the sampling survey, collected information was always individual samples, and therefore one had to increase the sampling ratio, but the largest proportion of the sample cannot be the exhaustive object, therefore all the conclusions came with inference properties. In the era of big data, it is possible to design a collection of things through the whole sample, provide full data, yielding a more solid and reliable object foundation and prerequisite for the humanities and social science research.

Again, the possibility of big data begs the question: Is "emotional contact" or "behavioral statistics" more reliable? The humanities and social sciences research by many researchers use direct contact as a basis for gathering reliable data, such as direct dialogue, face-to-face interviews, telephone interviews, etc., which are aided by perceptual intuition. But, people's innermost thoughts, full of rational and irrational reasoning, accompanied by whimsy, dispersion, and change of individual characteristics, often directly influenced by the interaction between feeling and reality, especially relating to some sensitive topics, will produce some special shield and so may not always provide data that are true and reliable. The behavior of large data statistics, data for fine-grained segmentation, the full range of long-time cumulative collection, especially gathered from a large number of people by way of "mouse" or "network voting", the value orientation showing the beauty of likes and dislikes, the evaluation criteria showing the individual soul, outlining the psychological state of the overall society, generally to depict the spirit of the times, may help achieve the objective of a more reliable and rational basis for the study of the humanities.

Again, the availability of big data provides us with a choice between using "entity law" or "Statistics Law" for the purpose of the study. In the past, the humanities and social science research has been mainly based on local observations of the social phenomena and on physical exploration. Big data provides social variables statistics, can measure objects and calculations, by tracking the society, with "people" as the smallest unit, and the overall social inclusion accumulation shows high precision tracking, real social society and social organizations at a relatively low cost of implementation. On the whole, with all the data as the foundation, as is the inherent law of nature, the better the social development, evolution trend and future direction, for example. The research on the modernization of national governance system, with the help of big data, can try to establish the evaluation system and international comparison to draw a scientific conclusion.

4. Humanities and Social Sciences Research into the Era of Big Data

Big data era has arrived. The U.S. government started the construction of the information highway in 1993. President Obama announced in March 2012 an investment of $2 trillion for the "big data research and development program", focusing on maintaining data in large strategic aspects of the world, giving the Chinese priority. The State Council issued an important file in August 2015 named Platform For Action For Big Data Development in order to promote the development of big data operations into social practice. Big data will be included in the national strategy. The problem is, for humanities and social sciences to consciously enter the era of big data, it is important to promote the reform and innovation of humanities and social science research.

First of all, conscious learning and effective use of complex thinking need to be encouraged. Big data world is an extremely complex world, requiring complex thinking. In modern times, we may tend to produce a simple, linear, causal perspective and reduction research theory. But, there are many uncertainties in the real-world mutation of non-equilibrium, nonlinear, fuzzy and other complex phenomena that are closely related, resulting from the dissipative structure theory and the complexity of modern science. The human world is a more complex emotional world and value world, there are very interactive and complex emotions intertwined, full of uncertainty and risk, raising the need to learn from the ideas and methods of complex science.

Secondly, there is the need to consciously study and use the big data technology. The application of big data technology can be divided into three types based on the level of proficiency in the use of complex information in modern information technology and data processing equipment in the humanities and social sciences to interpret scientific and rational implications of the significance of modern humanities and social sciences information. Big data technology can be used to lead humanities and social sciences research and the social reality of healthy development, leading to the development of the digital social civilization. Through the big data technology applied to the humanities and social sciences research, aided by improvements in the natural science and engineering technology, they can be compared to that of the "hard sciences", which will greatly enhance the humanities and social sciences of China, providing the contemporary world with explanation and influence.

Again, people are encouraged to consciously update and develop themselves. First of all, to master the technology and go beyond technology, we should try our best to learn and use science and technology, walking in the forefront of the times, while at the same time strive to maintain the humanities and social sciences thought, the height, the theoretical character and value orientation, and the reflection and criticism. Secondly, the use of big data should be based on professional requirements. Beyond the age of big data, one cannot do without any professional approach, but without rigidly adhering to any professional requirements, we consciously stand in contemporary disciplines, both differentiated and highly integrated, in the background of the research. Thirdly, our research needs to be based on the industry, but must also have the courage to go beyond the industry. The era of big data is not the result of any one industry and also will not limit itself to any one industry. We must consciously stand as one with the contemporary industry structure and study the degree of differentiation in a highly integrated background. Finally, the adaptation of big data is based on the self and the courage to go beyond the self. The biggest feature in the era of big data, object world, in the multidimensional interactive data and data technology world, is that people consciously tune themselves to integrate the complex system of nature and society — those with talent in complex data can truly become the masters of the world, become the masters of humanities and social sciences research, and eventually become a master of self.

Chapter 20

The *Differentia Specifica* of Interhuman Communications: Luhmann and the Sociological Reflection of Information Theory

Loet Leydesdorff

University of Amsterdam, Amsterdam, The Netherlands
loet@leydesdorff.net

Unlike the coding of biological evolution in terms of DNA, cultural evolution is driven by further development of codes in communication. These codes are cultural constructs which structure how we give meaning to information. They are not hardwired. Complementary to the information flow, they can be modeled in terms of generating *redundancy*: the observables are provided with meaning from the perspective of hindsight and with reference to horizons of meaning. Orders of expectations can be developed, which operate as feedback on the arrow of the entropy flow by providing envisaged options. Cultural evolution thus incurs on the biological evolution. The codes operate as selection mechanisms on the phenotypical variation of the observables.

1. Introduction

When "something" is communicated, the exchange can first, and in the abstract, be described as information: the distribution of what was communicated has changed. The expected information value I of the message that the prior distribution $(\Sigma_i\, p_i)$ was changed into the posterior distribution

($\Sigma_i q_i$) can be formulated as follows (Shannon, 1948; Kullback and Leibler, 1951):

$$I = \sum_i q_i \log_2(q_i/p_i). \tag{1}$$

When the two-base logarithm is used, I is measured in bits of information. Theil (1972, p. 59f.) has proven that I is necessarily equal or larger than zero. I is also called the Kullback–Leibler divergence (Leydesdorff, 1991; Theil, 1972).

Note that I is asymmetrical in p and q. When the values of all prior probabilities (p_i) are equal to the posterior ones (q_i), no information is generated since $\log(1) = 0$. This indicates that no exchange has taken place in the dimension under study. Specification of this dimension provides the information with meaning. However, I itself is dimensionless and not yet meaningful. I is derived from Shannon's H ($= -\Sigma_i p_i \log_2(p_i)$) which has a thermodynamic interpretation. Using the H in Gibb's formulation of entropy ($S = k_B * H$), Shannon chose to define information as probabilistic entropy. Since k_B is a constant (with dimensionality Joule/Kelvin), the development of information follows the second law of thermodynamics and can therefore only be positive (Krippendorf, 2009). H itself is a statistic of the distribution; it remains meaningless without the specification of a system of reference.

Shannon distanced himself from the question of meaning processing by declaring that the "semantic aspects of communication are irrelevant to the engineering problem." Weaver, however, suggested that Shannon's analysis "has so penetratingly cleared the air that one is now, perhaps for the first time, ready for a theory of meaning" (Shannon and Weaver, 1949, p. 27), but he added that "(i)nformation must not be confused with meaning" (Shannon and Weaver, 1949, p. 8).

In summary, Shannon refrained from addressing questions concerning "what" is communicated; this is a theory of communication (of information) from a sender to a receiver via a medium. Thus, the focus is on "how" communications can be expected to generate how many bits of information. The specification of "what" is substantively communicated when the system under study operates, however, raises the question of the meaning of the communication. For example, at the biological level, molecules are exchanged, whereas exchanges of atoms are typical for chemistry. This specification of the substance makes reference to a *differentia specifica*. Shannon's theory itself, however, is only a mathematical theory of information; as a general theory, it is void since it is dimensionless. Specification of one or more systems of reference enables us to develop the respective special theories of information.

2. Aim of the Chapter

The aim of this chapter is the specification of what is communicated in interpersonal communications. Human communications are reflexive and, in this respect, different from the communication of energy and momenta among particles or the exchanges of molecules in biological processes. I shall argue that codes constructed in and by the communication can feedback as controls of the communication or operate as feedforward loops enhancing the communication. The codes of communication do not "exist," but can be expected to operate as the eigenvectors of a communication matrix. Since communication is multidimensional, a multidimensional array can be assumed in which differently codified processes of communication can operate in parallel and disturb one another, resulting in a "fractal manifold" of communications (Ivanova and Leydesdorff, 2014; Leydesdorff *et al.*, 2017).

3. *Differentia Specifica*: "What is Communicated?"

The communication is specific in terms of "what is exchanged" when the system operates. For example, momenta and energy are communicated in classical physics so that they remain conserved when elastic particles (e.g., "billiard balls") collide. Note that in the ideal case, the thermodynamic entropy is zero, but the probabilistic entropy in the exchange can be at its maximum because of the frictionless exchanges. For example, if two particles have a distribution of momenta of 3:7 before a head-on collision, this distribution will change in the ideal case into 7:3. Consequently, $I = 0.7 \log_2(0.7/0.3) + 0.3 \log_2(0.3/0.7) = 0.86 - 0.37 = 0.49$ bits of information are generated.

Other systems are dissipative, for example, living systems exchange molecules with their environments. In the case of *autopoiesis,* the system achieves closure in terms of specific operations (Varela *et al.*, 1974). The closure determines what can be communicated and what not. For example, the exchange of molecules is typical for a biological operation, while the exchange of atoms prevails in chemistry. Thus, the various system levels can be distinguished information-theoretically in terms of "what" specifically is communicated while as exchange processes, they can be modeled as generating bits of information.

Using the mathematical theory of communication — which is devoid of meaning and therefore neutral — as a heuristic, the models developed in one substantive domain can be explored and tested as *hypotheses* in another. However, there is no *ex ante* correspondence between one and another exchange mechanism; one expects the systems to be also *specifically* different. However, the specification of the communication is consequential since defining the system of reference. The complexity of "what" can be communicated determines the relevant selection environments; selection can only operate on dimensions which are relevant to the selected substances.

For example, Varela *et al.* (1974) developed *autopoiesis* as a model of the organization of living systems. Maturana further developed this model into a *biology* of language and knowledge (Maturana, 1978). The model can be related to von Foerster's more formal model of self-organization (von Foerster, 1960). In a study entitled "The Autopoiesis of Social Systems," (Luhmann, 1986a), Luhmann proposed to generalize the *autopoiesis* model using a parallel line of reasoning as above, that is, developing a general theory by abstracting from the biological content. He formulated as follows:

> [...] if we abstract from life and define autopoiesis as a general form of system-building using self-referential closure, we would have to admit that there are non-living autopoietic systems, different modes of autopoietic reproduction, and general principles of autopoietic organization which materialize as life, but also in other modes of circularity and self-reproduction. In other words, if we find non-living autopoietic systems in our world, then and only then will we need a truly general theory of autopoiesis which carefully avoids references which hold true only for living systems (Luhmann, 1986a, p. 172f.).

In my opinion, such a general theory would be a mathematical theory devoid of meaning (Baecker, 1999). As noted above, a mathematical theory can sometimes be used to mediate among substantive theories. Luhmann wished to formulate the relations between two of the latter by distinguishing between "consciousness" at the individual level and interhuman "communications" as two different systems of reference. He formulated this as follows:

> This kind of approach is usable only if we are prepared to accept [...] that social systems, and even psychic systems, are not living systems. The concept of autopoietic closure itself

> requires this theoretical decision and leads to a sharp distinction between meaning and life as different kinds of autopoietic organization; and meaning-using systems again have to be distinguished according to whether they use consciousness or communication as modes of meaning-based reproduction.

The program of studies could then be specified as follows:

> On the one hand, then, a psychological and a sociological theory have to be developed which meet these requirements; on the other hand, the concept of autopoiesis has to be abstracted from biological connotations. Both tasks are clearly interdependent. The general theory of autopoietic systems forms the foundation of the theories of psychic and social systems; the general theory itself, however, is meaningful only if this implementation succeeds, because otherwise we would be unable to determine which kind of attributes are truly general (Luhmann, 1986a, p. 173).

In other words, Luhmann specifies a hierarchical model with a "general theory of autopoietic systems" as "the foundation" of special theories. I argued above that such a theory cannot carry substantive theories. Each relation between different operations has to be carefully articulated and investigated since the systems operate in parallel with potential couplings and uncouplings in co-evolutions (McLuhan, 1964).

However, this is not the place to elaborate on the autopoiesis model in general. For our purpose, it suffices to note that Maturana (1978, p. 49) claims that one can specify when "the behavior of the organism becomes indistinguishable from the behavior of an observer; the second-order consensual domain that it establishes with other organisms becomes indistinguishable from a semantic domain." In other words, the organisms behave as if they were able to use a language for their communication. However, the biologist cannot specify "what" is communicated since the translation of a linguistic model into a biological one is metaphorical. Maturana's (2000) subject of study in the biology of language and knowledge consequently is "languaging" (that is, behavior) and not "language" (Maturana, 2000).

Even if dolphins and monkeys were able to use a language for their communication, we would not have access to this language because autopoietic systems are operationally closed. We can interpret "monkey speech," for example, as signaling danger. However, the biologist herself using biological discourse is a *super-observer* to be distinguished analytically from the languaging animals under study. Maturana (1978, p. 56ff.) formulated this as follows:

> Human beings can talk about things because they generate the things they talk about by talking about them. That is, human beings can talk about things because they generate them by making distinctions that specify them in a consensual domain, and because, operationally, talking takes place in the same phenomenic domain in which things are defined as relations of relative neuronal activities in a closed neuronal network.

From the biological perspective, not language itself, but "languaging behavior" is considered as the system of reference. Language, however, is generated and reproduced by languaging, and therefore, a second-order domain is attributable not to individual agents but to their interactions, that is, to their languaging as a first-order attribute of these agents.

What is constructed by the languaging constructors? What are the dynamics of language and culture? The cybernetic hypothesis is that the next-order system is constructed bottom-up, but once constructed, tends to take over control top-down (Ashby, 1958; Craver and Bechtel, 2007). In this case, the language is expected to structure the use of language by languaging agents. Unlike languaging among communicators, however, interhuman communication cannot be observed — as Luhmann (1995a, p. 164) emphasized — since the next-order layer is no longer (meta- or epi-)biological (Luhmann, 1995a). The cultural one is a domain *sui generis* (Durkheim, 1912; Hemmung, 1926; Parsons, 1968; Freud, 1926).

Cultural phenomena emerge on top of the biological ones and then tend to take over control. Language enables us to carry culture as interpersonal expectations, which feedback from envisaged futures on the historical information flow *against the arrow of time*, that is, by adding redundancy. This redundancy is not communicated because then it would be information.

4. The *Differentia Specifica* of Meaning Processing

Meaning is provided with reference to "horizons of meaning" (Husserl, 1929; 1962). Meaning cannot be communicated, but it can be shared on the basis of communication. The sharing presumes communication of information and these parallel interactions generate a nonlinear dynamics of information and redundancy. (The information plus the redundancy add up to the maximum entropy at each moment.) Redundancies are measurable as not-yet-realized options. Increasing redundancy decreases the relative information and increases the maximum entropy — $\log(N)$ — of the systems under reconstruction (Brooks and Wiley, 1986, p. 46).

Luhmann operationalized Husserl's "horizons of meaning" as "symbolically generalized codes of communication" (Luhmann, 1986b). The codes developed in the communication serve the generation of redundancy by spanning a phase space of possible combinations that is reflected in the communications. Following Parsons (1968), Luhmann used the expression "symbolically generalized media of communication" (Luhmann, 1974). However, Distin argued convincingly that this phrase can be considered a misnomer (Distin, 2012). The codes — Distin uses "meta-representations" — can be functionally differentiated and then carry more complexity (Distin, 2010). Not the (sub)systems, but the codes are differentiated and then potentially generalized (Luhmann, 1982).

As against "languaging" animals, humans have the capacity to translate among codes of communication — Latour (1988a, pp. 169ff.) invented the word "infra-reflexively" — thus to entertain more than a single meaning (Latour, 1988a). Luhmann (2002a, p. 182) noted that the codes are operationally coupled in the communication in addition to being structurally coupled as specializations of the communication (Luhmann, 2002a). Variation by languaging is only operationally or loosely coupled to language (Simon, 1973a).

In contrast to Parsons, who referred to *linguistic* codes, Luhmann wanted to explain codes of communication as binary in analogy to the model of genetic code, that is, DNA (Künzler, 1987, p. 323). At several places (e.g., Luhmann, 1985, p. 2; 2002b, p. 116), Luhmann formulated as an example of binary codes that "A woman is pregnant or she is not pregnant; she cannot be somewhat pregnant" (Luhmann, 1985; 2002b). However, this example is biological. This type of binary unambivalence — either/or — is not provided in the social domain: what can be right in one context can be wrong in another. "Pregnancy" can have very different meanings in social contexts. For example, one can be pregnant with ideas. Analogously, the code of true/false operates in the sciences as a heuristic (Simon, 1973b). Some scientific statements, for example, can be true more than others.[1]

[1]Luhmann distinguishes between binary codes and the programs using the codes.

5. From Observations to Expectations

The sciences can be considered as systems of rationalized expectations. Rationalized expectations are coded in scholarly discourse, that is, the context of validation, and expectations can be tested against observations (Hempel and Oppenheim, 1948; Popper, 1959). With reference to George Spencer Brown's *Laws of Form* (Spencer, 1969), Luhmann (1993), however, followed a biologically inspired focus on observations (Luhmann, 1993; Von Foerster, 1982). His sociological theory unintentionally thus became untestable with sociological methods and statistics.

From a biological perspective, observations have a status more fundamental than expectations (Fuchs, 2004; Gumbrecht, 2006; Leydesdorff, 2006). Such a focus on observables leads in sociology to historicism and descriptive instead of inferential statistics. Dodig-Crnković argues that theoretical biology nowadays has abandoned reflexively the focus on observing in favor of computational modeling of the construction (Dodig-Crnković, 2014).

In the social sciences, the focus is on expectations which can be tested against observations. Spencer Brown (1969) has been misread as a legitimation to focus irreflexively on observations as if they were a given. Spencer Brown's theory is mathematical and strictly empty when he argues that the first moment is "drawing a distinction." From this perspective, an observation can be considered as such a mark (p. 76). Marking this distinction by identification, one generates not a substantive observation, but only a yet empty observational category. An observational report is needed for filling the box with a value.

Whereas animals such as predators are able to observe their prey, they are not able to report their observations. Observational reports enable us to take the next step and thus feed the results of observations reflexively into a cultural evolution. Observational categories enable us to specify expected values before proceeding to the measurement. For example, one can cross-table distinctions. When the distinctions along the axes are theoretically relevant, the specification of categories in the boxes invites measurement.

6. Expectations as Foundational

One advantage of this turn to expectations is the possibility to test expectations against observations. The statistical significance of observations can then also be specified, for example, by using chi-square statistics. Observations are no longer an epistemological foundation (Popper, 1959; Kauffman, 1980). But, from this perspective, observations are at variance and hence include error terms. The specification of uncertainty lifts the observation out of the domain under study into the domain of communication among "super-observers," that is, scholars observing on the basis of expectations.

In my opinion, Luhmann (1995b) could have stayed closer to Husserl (1962) who argued that the priority given by positivists to observing an assumingly observable "reality" has led to *The Crisis of the European Sciences.* Husserl called for a "concrete philosophy of science" (Husserl, 1929, pp. 159f.) which should begin with the recognition that the sciences develop in terms of intersubjective expectations. The latter can discursively be codified into testable hypotheses. The sciences are not *res extensa* to be observed, but *res cogitans* — the substance to which we have access reflexively — with a dynamic very different from that of observables or languaging behavior.

7. From *Harmonie Préétablie* to Post-established Engineering

Husserl's (1929) reference to Descartes in the title *Cartesian Meditations* is pertinent. Huygens formulated thus already in the 17th century in response to Descartes:

> Against Cartesius' dogma, that the nature or notion of a body should consist in extension alone, I have a notion of space that differs from the notion of a body: space is what may be occupied by a body (Huygens, 1888–1950; Elzinga, 1972).

Note the "may be" in this formulation. In opposition to Descartes, Huygens and his contemporaries had achieved an understanding of *empty* space (Newton) and *infinite* time (Leibniz) which allowed them to test abstract theories experimentally. Huygens would call this method "*ratione atque experientia*": by means of both reason and experiments (Elzinga, 1972). However, the arrow of the inference was reversed: space was no longer considered exclusively as a consequence of the extension of matter, as had been the case in the earlier mechanistic philosophy. The specification of an expected value is given priority in relation to possible observations.

Discursive knowledge is constructed not from observations, but from arguments. Observations enter the discourse after translation into observational reports, since observations themselves are not discursive before being articulated (De Zeeuw, 1993; Pask, 1975). This discursive articulation is selected with reference to the codes available in the discourse; the codes function as both selection and coordination mechanisms. In differently codified systems such as law, the economy, love, etc., distinctions are validated with reference to varying expectations; but the codes develop among human beings alongside each other ("heterarchical") in terms of expectations.

The rule of law, for example, is not a given, but remains a construct of expectations as much as the knowledge base of an economy. All cultural phenomena process meanings on top of the facts. The facts are observables and thus at variance. Phenotypical variation can be tested against theoretically constructed "genotypes". The theorizing codes the distinctions at the supra-individual level.

Expectations can proliferate much faster than realizations (Weinstein and Platt, 1969) and hence the redundancy — the not-yet realized options — can increase much faster than the entropy (Ehresmann, 2014). In my opinion, we need a calculus of redundancy supplementing Shannon's calculus of information (Deacon, 2012; Leydesdorff *et al.*, 2018; Bar-Hillel, 1955). The dynamics of cultural evolution are very different from the dynamics of biological evolution because the codes are not given as DNA, but constructed and potentially at variance (Langton, 1989, p. 22f.).

While there is not a single code, one can hypothesize an alphabet of them (Simon, 1973a, p. 19ff.). These constructed "genotypes" can be functionally different and can "infrareflexively" be translated into one another (Latour, 1988a, p. 169ff.). Latour (1988b, p. 164) added: "There is no preestablished harmony, Leibniz notwithstanding, harmony is *post*established locally through tinkering (Latour, 1988b)." The tinkering with differently codified communications may lead to the engineering of new technologies when codes can be recombined (Arthur, 2009; Sahal, 1985).

References

Arthur W. B. (2009). *The Nature of Technology*, Free Press, New York, NY.

Ashby, W. R. (1958). Requisite variety and its implications for the control of complex systems, *Cybernetica* 1(2), 1–17.

Baecker, D. (ed.), (1999). *Problems of Form* (German title: *Die Kalkül der Form.* Frankfurt a.M.: Suhrkamp ed, Stanford University Press, Stanford, CA.

Bar-Hillel, Y. (1955). An examination of information theory, *Philos. Sci.* 22(2), 86–105.
Brooks, D. R. and Wiley, E. O. (1986). *Evolution as Entropy*, University of Chicago Press, Chicago.
Craver, C. F. and Bechtel, W. (2007). Top-down causation without top-down causes, *Biol. Philos.* 22(4), 547–563.
De Zeeuw, G. (1993). Improvement and research: A Paskian evolution, *Syst. Res.* 10(3), 193–203.
Deacon, T. W. (2012). *Incomplete Nature: How the Mind Emerged From Matter*, Norton & Company, New York, London.
Distin, K. (2010). *Cultural Evolution*, Cambridge University Press, New York, NY.
Distin, K. (2012). Symbolically generalized communication media: A category mistake? *Construct. Found.* 8(1), 93–95.
Dodig-Crnković, G. (2014). Info-computational constructivism and cognition, *Construct. Found.* 9(2), 223–231.
Durkheim, E. (1912). *Les formes élémentaires de la vie religieuse*, Félix Alcan, Paris.
Ehresmann, A. C. (2014). A mathematical model for info-computationalism, *Construct. Found.* 9(2), 235–237.
Elzinga, A. (1972). *On a Research Program in Early Modern Physics*, Akademiförlaget, Gothenburg.
Freud, S. (1926). *Die Frage der Laienanalyse*, Internationaler Psychoanalytischer Verlag, Leipzig.
Fuchs, P. (2004). *Der Sinn der Beobachtung*, Weilerswist, Velbrück.
Gumbrecht, H. U. (2006). *Luhmanns intellektuelles Vermächtnis, Merkur.* 60(8), 696–706.
Hemmung, F. S. (1926). *Symptom und Angst*, Internationaler Psychoanalytischer Verlag, Leipzig.
Hempel, C. G. and Oppenheim, P. (1948). Studies in the logic of explanation, *Philos. Sci.* 15, 135–175.
Husserl, E. (1929). *Cartesianische Meditationen und Pariser Vorträge* [Cartesian Meditations and the Paris Lectures, translated by Dorion Cairns], Strasser, E. b. S. (ed.), The Hague, Martinus Nijhoff, 1960.
Husserl, E. (1962). *Die Krisis der Europäischen Wissenschaften und die Transzendentale Phänomenologie*, Biemer, W. (ed.), Den Haag, Martinus Nijhoff [1935/36].
Huygens, C. (1888–1950). *Oeuvres Complètes*, The Hague, Nijhoff.
Ivanova, I. A. and Leydesdorff, L. (2014). Rotational symmetry and the transformation of innovation systems in a triple helix of university-industry-government relations, *Technol. Forecast. Soc. Change.* 86, 143–556.
Kauffman, L. H. and Varela, F. J. (1980). Form dynamics, *J. Soc. Biol. Struct.* 3(2), 171–206.
Krippendorff, K. (2009). Information of interactions in complex systems, *Int. J. Gen. Syst.* 38(6), 669–680.

Kullback, S. and Leibler, R. A. (1951). On information and sufficiency, *The Annals Mathe. Statist.* 22(1), 79–86.

Künzler, J. (1987). *Grundlagenprobleme der Theorie symbolisch generalisierter Kommunikationsmedien bei Niklas Luhmann, Zeitschrift für Soziologie* 16(5), 317–333.

Langton, C. G. (1989). Artificial life, in *Artificial Life*, Langton, C. G. (ed.), Addison-Wesley, Redwood, etc., pp. 1–47.

Latour, B. (1988a). The politics of explanation: An alternative, in *Knowledge and Reflexivity: New Frontiers in the Sociology of Knowledge*, Woolgar, S. and Ashmore, M. (eds), Sage, London, pp. 155–177.

Latour, B. (1988b). *The Pasteurization of France*, Harvard University Press, Cambridge, MA, and London.

Leydesdorff, L. (1991). The static and dynamic analysis of network data using information theory, *Soc. Networks* 13(4), 301–345.

Leydesdorff, L. (2006). The biological metaphor of a (Second-order) observer and the sociological discourse, *Kybernetes* 35(3/4), 531–546.

Leydesdorff, L., Johnson, M. and Ivanova, I. (2018). Toward a calculus of redundancy: Signification, codification, and anticipation in cultural evolution, *J. Assoc. Inform. Sci. Technol.* 69(10), 1181–1192.

Leydesdorff, L., Petersen, A. M. and Ivanova, I. (2017). Self-organization of meaning and the reflexive communication of information, *Soc. Sci. Inform.* 56(1), 4–27.

Luhmann, N. (1974). *Einführende Bemerkungen zu einer Theorie symbolisch generalisierter Kommunikationsmedien, Zeitschrift für Soziologie* 3(3), 236–255.

Luhmann, N. (1982). Systems theory, evolution theory, and communication theory, in *The Differentiation of Society*, Holmes, S. and Larmore, C. (eds.), Columbia University Press, New York, pp. 255–270.

Luhmann, N. (1985). *Einige Probleme mit "reflexivem Recht", Zeitschrift für Rechtssoziologie* 6(1), 1–18.

Luhmann, N. (1986a). The autopoiesis of social systems, in *Sociocybernetic Paradoxes*, Geyer, F., Zouwen, Jvd., (eds.), Sage, London, pp. 172–192.

Luhmann, N. (1986b). *Intersubjektivität oder Kommunikation: Unterschiedliche Ausgangspünkte soziologischer Theoriebildung, Archivio di Filosofia* 54(1–3), 41–60.

Luhmann, N. (1993). *Die Paradoxie der Form*, in Kalkül der Form Suhrkamp, Baecker, D. (ed.), Frankfurt a. M., pp. 197–212.

Luhmann, N. (1995a). *Social Systems*, Stanford University Press, Stanford, CA.

Luhmann, N. (1995b). *Die neuzeitlichen Wissenschaften und die Phänomenologie*, Picus, Vienna.

Luhmann, N. (2002a). How can the mind participate in communication? in *Theories of Distinction: Redescribing the Descriptions of Modernity*, Rasch, W. (ed.), Stanford University Press, Stanford, CA, pp. 169–184.

Luhmann, N. (2002b). *Einführung in die Systemtheorie, hrsg. v. Dirk Baecker*, Carl-Auer, Heidelberg.

Maturana, H. R. (1978). Biology of language: The epistemology of reality, in *Psychology and Biology of Language and Thought Essays in Honor of*

Eric Lenneberg, Miller, G. A. and Lenneberg, E. (eds.), Academic Press, New York, pp. 27–63.

Maturana, H. R. (2000). The nature of the laws of nature, *Syst. Res. Behav. Sci.* 17(5), 459–468.

McLuhan, M. (1964). *Understanding Media: The Extension of Man*, McGraw-Hill, New York.

Parsons, T. (1968). Interaction: I. Social interaction, in *The International Encyclopedia of the Social Sciences*, Sills, D. L. (eds.), McGraw-Hill, New York, pp. 429–441.

Pask, G. (1975). *Conversation, Cognition and Learning*, Elsevier, Amsterdam.

Popper, K. R. (1959). *The Logic of Scientific Discovery*, Hutchinson, London [1935].

Sahal D. (1985). Technological guideposts and innovation avenues, *Res. Policy* 14, 61–82.

Shannon, C. E. (1948). A mathematical theory of communication, *Bell Syst. Techn. J.* 27(July and October), 379–423 and 623–656.

Shannon, C. E. and Weaver W. (1949). *The Mathematical Theory of Communication*, University of Illinois Press, Urbana.

Simon, H. A. (1973a). The organization of complex systems, in *Hierarchy Theory: The Challenge of Complex Systems*, Pattee, H. H. (ed.), George Braziller Inc., New York, pp. 1–27.

Simon, H. A. (1973b). Does scientific discovery have a logic? *Philos. Sci.* 40(4), 471–480.

Spencer, B. G. (1969). *Laws of Form*, George Allen and Unwin, London.

Theil, H. (1972). *Statistical Decomposition Analysis*, North-Holland, Amsterdam.

Varela, F. G., Maturana, H. R. and Uribe, R. (1974). Autopoiesis: The organization of living systems, its characterization and a model, *Biosystems* 5(4), 187–196.

von Foerster, H. (1960). On self-organizing systems and their environments, in *Self-Organizing Systems*, Yovits, M. C. and Cameron, S. (eds.), Pergamon Press, London, pp. 31–50.

von Foerster, H. (1982). *Observing Systems* (with an introduction by Francisco Varela, ed.) Intersystems Publications, Seaside, CA, 331 + xvil pp.

Weinstein, F. and Platt, G. M. (1969). *The Wish to be Free: Society, Psyche, and Value Change*, University of California Press, Berkeley.

Chapter 21

Li Erqu's Theory on Epistemology and Gong Fu Ontology

Jun Liu

Department of Philosophy,
Xi'an Jiaotong University, China
liujun1216@xjtu.edu.cn

Li Erqu, one of the three greatest Confucians from the beginning of the Qing Dynasty, asserts in his theory that everything in the world keeps changing and evolving through the self-generation process. He focuses his attention on the first primitives and their movements in the world, which shows that the thought process and generative theory in ancient Chinese philosophy are not only rich but also profound. He inherits and develops the Confucian Theory on epistemology and Gong Fu Ontology accomplished by Zhang Zai, who saw people as being one with the universe and put forward the idea that one's heart makes appropriate and fast responses to the changing environment outside. Li Erqu, therefore, believes that the real Confucian has to build his morality inside and struggle to become a ruler in real life at the same time. This was based on the old version of the book *Great Learning*, whose core Li Erqu believes is "Mingtishiyong", the first step of learning is "Gewu", and the guiding principle is "Huiguozixin". The mode of interpretation mentioned above reflects that Li Erqu combines Neo-Confucianism with Xin Xue, which was proposed by Liu Jiuyuan of the South Song dynasty. To an extent, Li Erqu's epistemology of *Great Learning* happened to coincide with some theory of information construction in contemporary information philosophy on this point.

1. Introduction

According to the history of Confucianism, *Great Learning* is a work that is more fundamental among the systems of the Four Books in China (compared with the other three, the Analects, the Mencius and the Doctrine of the Mean), therefore scholars pay more attention to it.

The academic differences on this point between Zhu Xi and Wang Yangming mainly focus on the understanding of the version of *Great Learning* and the philosophical problems derived from it, and gradually became the philosophical topic that the later generations of scholars couldn't go around. Li Erqu (1627–1705) was born in Shanxi province and was known as the backbone of Guanxue (a school of Confucianism created by Zhang Zai in the North Song dynasty). Li Erqu had established his theoretical system of the study of human nature based on ontology (Gong Fu Ontology) which inherits and develops Zhang Zai's Confucian theory. Besides, based on the old version of *Great Learning*, Li Erqu absorbed the rational core of Wang Yangming's thought. To an extent, his annotation of *Great Learning* is to clarify the ideas from an overall perspective, so as to enrich the theory through further studies.

Following are the three real innovation points:

1. the core of *Great Learning* is "Mingtishiyong";
2. the first step of learning is "Gewu";
3. the guiding ideology is "Huiguozixin".

This innovation is indeed objective and valuable, coinciding with some theory of information construction in contemporary information philosophy on this point.

2. Theory of Process and Generative Theory

The most remarkable feature of Chinese philosophy is that it does not have logic, epistemology and methodology. Through the understanding and communication of the language, a dialogical civilization can be realized. The present research on this point in China mostly uses only methods of history and paleography, and focuses on the explanation of words in ancient books and from the historical narrative paradigm, which prevents experts from carrying out an in-depth research around the work of *Great Learning* as a whole and ignores its precious value in the whole history of Confucian thought.

Therefore, it is a good way to carry out research on *Great Learning* based on the background of information philosophy and recreate the value of Confucian classics through a special form of language and text. The world is expressed through language, so that by changing the traditional methods of hermeneutics and the concept of classic, we can seek another new way of interpreting China's classics.

3. Combination of Neo-Confucianism and Xin Xue Proposed by Liu Jiuyuan in the South Song Dynasty and Developed by Wang Yangming in the Ming Dynasty

The understanding of the theory of "Gewu" (theory of knowledge acquisition, a way for Chinese ancient scholars to investigate and research things) by Zhu Xi and Wang Yangming is totally different. Zhu Xi explains that "Gewu" means "Zhi" (toward), while Yangming approaches it from the angle of Gongfu Ontology. The former prefers knowledge acquisition from outside, while the latter concentrates on improving the morality within. Here, one can note a real innovation point in Li Erqu's theory.

Li Erqu combines Zhu Xi's theory of knowledge acquisition ("knowledge from the study of the physical world") and Wang Yangming's thoughts on the rediscovery of intuitive knowledge.

The value of Li Erqu's combination of these two thoughts is as follows. Firstly, it widens the scope of Wang Yangming's theory. Secondly, it limits Zhu Xi's theory of knowledge acquisition, so that in the theory of Gongfu Ontology, knowledge acquisition gives way to moral cultivation, which makes Zhuxi's theory directly beneficial to the Confucian Ideology. In addition, it shows the sustainable development of Chinese civilization.

4. Moral Epistemology

Western experts prefer to discuss about the problem of morality, of where human morals come from, which is still an important issue in China. Actually, the problem is central in moral epistemology. From the sociological angle in western countries, the best explanation of the depth of moral disagreements and social diversity that they reflect is one and the same thing, while the evolutionary theory offers a familiar and widely accepted answer that human morals are in essence, despite their modern variations, Darwinian adaptations. In China, scholars are also engaged in the same issue, but in different ways. For example, Li Erqu focuses on how to improve one's own morality. In his view, the realization of life lies in the integration of both external factors and internal ones. When the goodness which man is born with is shadowed, some faults occur that include improper words and actions, arising from selfish motives inside. Therefore, one significant step for people to take in terms of moral epistemology is to correct one's own faults so as to go back to one's good nature which existed at the very beginning.

5. Conclusion

Like the western experts, the Chinese philosophers still pay attention to the relationship between moral civilization and practice. So in Li Erqu's theory, he considers the Confucianism theory of Xinxing as the best in terms of practical utility, which is the origin based on which scholars express their ambition and willingness as annotations from the classics. Firstly, Li Erqu's thoughts on *Great Learning* reflect that his views have reflexive practice as their finial aim. At the same time, in general, we should also emphasize that the interpretation of *Great Learning* by Li Erqu shows he is not limited by bias between different schools of thought. Although such relative fair research is still not thoroughly carried out in philosophy, the direction and path of thinking that he has displayed in his work seem to be a stream surging intensified in the early Qing Dynasty when thoughts of Zhuxi and Wang Yangming were increasingly in dispute. Additionally, his respect of and construction based on the study of Confucian classics gradually changed the academic atmosphere in the Ming Dynasty, which is considered to be the reason for the revival of the study of Confucian classics later in the history of Chinese Confucianism. Li Erqu's thoughts coincide with some relevant ideas of information construction theory in the course of human cognition in the contemporary information philosophy.

Acknowledgments

This project is funded by the financial grant from Shaanxi Provincial Postdoctoral Foundation (2017BSHEDZZ134) and Shaanxi Provincial Social Science Fund (2017C006), and specially supported by the 2018 National Social Sciences Foundation of China (18FZX011).

References

Li, J. and Liu, J. (2013). Zhuxi's investing things to obtain knowledge: One possible technological rationality, *Stud. Dialect. Nat.* 3, 79–83.

Steup, M. and Neta, R. (2020). Epistemology, in *Stanford University Encyclopedia of Philosophy*, Zalta, E. N. (ed.) Metaphysics Research Lab, Stanford, CA.

Wu, K. (2010). *Information, System and Complexity Problems of the Ancient Philosophy*, Vol. 5, Commercial Press, pp. 201–206.

Wu, K. (2013). Segmentation of existence field and the "whole new evolutional" significance of the philosophy of information, *J. Humanit.* 5, 1–6.

Chapter 22

Law as Information and the Impact of Information Technologies

Ugo Pagallo
University of Turin, Italy
ugo.pagallo@unito.it

1. Introduction

The law has been considered in many ways throughout the centuries. Aristotle, Rousseau, and Hegel, although for different and even opposite reasons, conceived the law as a second nature; Austin, as a command; Kelsen, as a set of norms; Dworkin, as interpretation; etc. This chapter adopts the level of abstraction of the law as information. By assuming information as that which is made of a data, namely, that which is made of a lack of uniformity in the real world if, and only if, such data are additionally "well formed" and "meaningful" (Floridi, 2009, p. 21),[1] the level of abstraction of the law as information sets the interface that makes an investigation of some crucial aspects of the legal phenomenon possible. From a methodological viewpoint, the idea is to compose a set of features representing the observables and variables of the analysis, the result of which provides a model for the field (Pagallo, 2013, pp. 28–29; Durante, 2017).

[1]Together with different types of data, such as analog, binary, and digital data, there are of course multiple kinds of information. As stressed in Section 2, attention is drawn here to the kind of semantic information, which has to do with the content of the information that can be further distinguished in an instructional and factual manner.

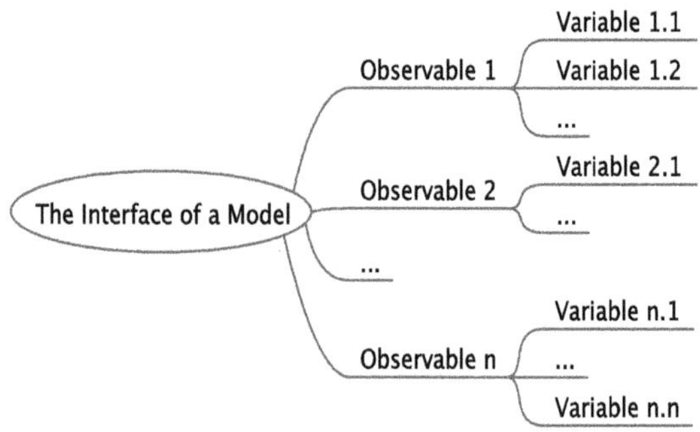

Figure 1. Levels of abstraction.

The approach of this chapter is illustrated with a figure on the interface of the model, its observables, and variables, namely, the illustration of Figure 1.

This stance on the law as information is not new (Pagallo and Durante, 2016a). A classical text like Thomas Hobbes's *Leviathan* makes the point clear. In Chapter 26 on "Of Civil Laws", the aim is to define what "men are therefore bound to observe, because they are members, not of this or that Commonwealth in particular, but of a Commonwealth" (Hobbes, 1999). Here, Hobbes presents the law as "a command, and a command consisteth in declaration or manifestation of the will of him that commandeth" (*op. cit.*). Because of this signaling function, the law as information appears as a set of rules or instructions for the determination of every legal subject of the system. This information must be expressed "by voice, writing, or some other sufficient argument of the same". Since this set of commands is enforced through the threat of physical sanctions, "nor is it enough — according to Hobbes — the law be written and published, but also that there be manifest signs that it proceedeth from the will of the sovereign.... There is therefore requisite, not only a declaration of the law, but also sufficient signs of the author and authority" (Hobbes, 1999).

Contrary to the tenets of Hobbes's legal philosophy as information, however, we should widen our perspective in terms of law, information, and reality. In light of a new figure — see Figure 2 — three variables of the analysis follow as a result. While Hobbes's stance on law as information can be summed up as legal information "for" reality, i.e., the law as a command, we can also refer to the legal phenomenon in terms of legal information "about" reality, and legal information "as" reality. In the first case, legal information concerns knowledge and concepts that frame the representation and function of a given system, and inform us about the different states of reality. This view is closely related to people's ordinary experience every time they visit a lawyer's office or a clerk to the justices. When they ask for semantic content information, which is alethically qualifiable, what people hope is that such a lawyer, clerk, or attorney, will speak the truth. On the other hand, in the case of legal information "as" reality, the focus is on "the possibility that data might be meaningful independently of an intelligent producer/informer" (Floridi, 2009, p. 32). As patterns of physical signals, which are neither true nor false, this is a popular view in the field of evolutionary psychology, according to which basic human concepts would be wired in our brains, thus grounding cultural evolution. The law is correspondingly understood as an emergent property of social interaction, for "man did not adopt new rules of conduct because he was intelligent. He became intelligent by submitting to new rules of conduct" (Hayek, 1981, p. 163).

Against this threefold connection between law, information, and reality — as shown by another figure, i.e., Figure 3 — three sets of observables of the analysis with their variants are under scrutiny in this chapter. They

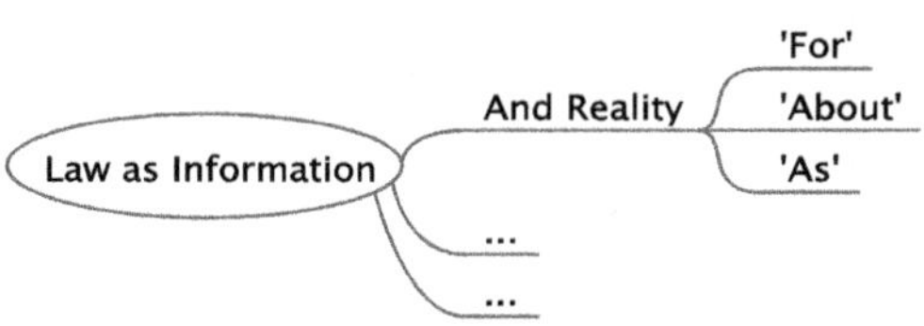

Figure 2. Law, information, and reality.

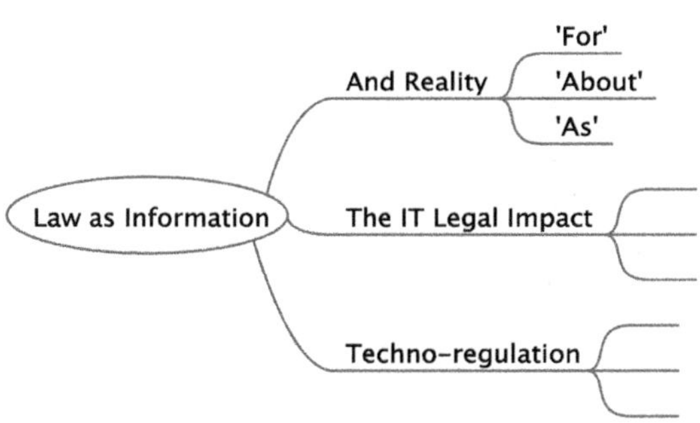

Figure 3. Three observables with their variables.

concern the different schools of jurisprudence, the legal impact of information technologies on this traditional framework, and how legal systems have reacted to this impact through the means of techno-regulation.

More particularly, Section 2 examines the variables of Figures 2 and 3, i.e., legal information "for" reality, "about" reality, and "as" reality. This stance sums up the different schools of jurisprudence, such as legal positivism (i.e., legal information "for" reality), legal sociology (i.e., legal information "about" reality), and the tradition of natural law (i.e., legal information "as" reality). On this basis, Section 3 draws attention to the ways in which the information revolution and the use of information technologies, such as computers, the Internet, or the Internet of Things (IoT), may affect both requirements and functions of the law, namely, what the law is supposed to be (requirements), and what it is called to do (functions). Whereas, over the past centuries, human societies have been related to information and communication technologies (ICTs), but have mainly been dependent on technologies that concern energy and basic resources, today's societies are increasingly dependent on ICTs and, furthermore, on information and data as a vital resource (Pagallo, 2015a). Cases that scholars address as a part of their everyday work in the fields of information technology (IT)-Law, i.e., data protection, computer crimes, digital copyright, e-commerce, and so forth, illustrate the several ways in which this flow of information in current ICT-driven societies jeopardizes traditional assumptions of legal and political thought. The new realities of an information society have not only increased the difficulty of law enforcement on, for example, the Internet but this new scenario also concerns the illegitimate condition where states claim to unilaterally regulate extraterritorial conduct by imposing norms on individuals who have no say in the decisions affecting them.

Law and technology, however, may not only clash but also reinforce each other. While Section 3 examines some crucial aspects of the impact of ICTs on the law — e.g., the shortcomings of the traditional idea of the law as a set of rules, or commands, enforced through the threat of physical sanctions on the Internet — Section 4 dwells on the observable of techno-regulation. Here, the focus is on legal information "for" reality embedded into the design of ICTs and of further convergent technologies, such as robotics and artificial intelligence (AI) in IoT. The interplay between law and technology should not be thought of as a one-way movement of social evolution from technology to law. Rather, law and technology should be deemed as regulatory systems that may not only compete but also strengthen each other. Article 8 of WIPO's 1996 *Copyright Treaty* and Article 14 of the twin *Performances and Phonograms Treaty* are good examples of current techno-regulation. These provisions represent the legal umbrella for the adoption of such automatic techniques as digital right management (DRM) in the private sector. Shifting from the traditional stance on punitive or positive sanctions that have to follow terms and conditions of legal accountability to effects that follow natural causes or technical instructions, e.g., DRM, techno-regulation, and the use of self-enforcing technologies, is an instance of what this chapter dubs as today's re-ontologization of the law.

The conclusions of this chapter aim to make clear how the re-ontologization of the law and the new challenges of techno-regulation, as a novel form of legal automation "for" reality interact with the previous variables of the analysis, i.e., legal information "as" reality and "about" reality. The contention of the chapter will be that such a legal impact of information technologies does not only regard the upgraded normative side of the law by design and through further mechanisms of legal automation but this impact also has to do with the emergence of spontaneous orders and new sources of transnational law in cyberspace. In the first case, new forms of legal information "as" reality have indeed emerged on the Internet, e.g., online communities rapidly developing their social norms. In the second case, attention should be drawn to the ways in which legal systems have evolved as a matter of information "about" reality, e.g., private actors and the market as a regulatory source of current systems. In order to grasp the issue of and then take sides on today's re-ontologization of the law, let us first understand how the variables of the analysis on legal information and reality interact.

2. Legal Information and Reality

The first variable of Figures 2 and 3, i.e., legal information "for" reality, is the most common and important one in the legal domain: the law is conceived as a set of rules or instructions for the determination of other informational objects, this is not only the common ground for every kind of legal positivism, normativism, or imperativism whatsoever but this stance also casts light on some aspects of constitutionalism, institutionalism, and the natural law tradition as well. What all such views have in common is the regulatory aim of the law to govern social and individual behavior. As a sort of instructional information, the law is meant to (contribute to) bring about something. This normative perspective can be properly illustrated with the work of one of the most prominent legal scholars from the 1900s, Hans Kelsen. Although Kelsen would only admit a kind of instructional information, i.e., semantic information — as opposed to environmental information, from which to deduce the norms of the law — his definition of the law captures the essence of legal information "for" reality: "if A, then B" (Kelsen, 1949). The formula shows what should be, rather than what is. The reference is to either punitive or positive sanctions (B) that have to follow terms and conditions of legal accountability (A), rather than effects (B) that follow natural causes (A).

The second variable of Figures 2 and 3, i.e., legal information "about" reality, is elucidated by work on legal sociology, where attention is drawn to the semantic content of the information that refers to the states of the world, and which is alethically qualifiable. Since focus is on what the law is, rather than what it should be, this kind of analysis shows what some different approaches to the law have in common. Reflect on, say, Max Weber's sociology of law in his masterpiece *Economy and Society* and some variants of the American (as opposed to the Scandinavian) legal realism. In the first case, Weber offers a general analysis of the types of political authority, of the role of law in securing political legitimacy, of the character of modern law, and the development of the modern state, up to the sociological significance of legal history. The aim of this research is to grasp the different ways in which the law actually functions *vis-à-vis* the "ideal types" of the model (Weber, 2005). In the case of American realism, the law appears as a sort of prophecy of what the courts will do in fact (Holmes, 1963). We even have nowadays smart AI systems that are particularly good in predicting the outcomes of a court, namely, legal information "about" reality.

Still, we should not miss a critical connection between legal information "about" reality and legal information "for" reality. Some undisputed champions of legal positivism admit this sort of overlap. In Chapter 21 of *Leviathan*, for instance, Hobbes concedes, "the obligation of subjects to the sovereign is understood to last as long, and no longer, than the power lasteth by which he is able to protect them" (Hobbes, 1999). As a result, scholars have to pay attention to the level of legal information "about" reality, namely, whether the law is actually effective, in order to ground the obligation of the subjects and whether the command of a sovereign as legal information "for" reality makes sense. Likewise, in his *General Theory of Law*, Kelsen sees a similar connection between these levels of legal information, because "a norm is considered to be legally valid ... on the condition that it belongs to a system of norms, to an order which, on the whole, is efficacious" (Kelsen, 1949, p. 42).

In addition to the cumbersome relation between the principle of legality and the legitimacy and efficacy of legal systems, another open issue concerns the different ways in which these levels of analysis on legal information "for" reality, and "about" reality, relate to legal information "as" reality. For example, in order to explain why the subjects shall obey the sovereign's commands, advocates of legal information "for" reality, such as Hobbes and Kelsen, need some kind of anthropology. As to the philosophy of Hobbes, scholars have stressed time and again how the attempt to make the political obligation of the social covenant legitimate ends up in a sort of naturalistic fallacy. All in all, Hobbes grounds the moral principles of his doctrine on a number of empirical propositions about human nature (Hampton, 1990). In the case of Kelsen, his theory of the basic norm as a legal assumption, which is required to make the validity of the norms intelligible, prevents such naturalistic fallacy on the basis of a moral formalism: "You shall obey to the constitution" (Bobbio, 1993). Of course, further accounts of the connection between legal information "for" reality and legal information "as" reality do exist, e.g., the manifold variants of the natural law tradition. In this context, suffice it to mention three aspects of this connection. They regard the instructional information of the law, its environmental dimension, and how this latter environmental information may compete — and even clash — with legal information "for" reality.

First, legal information "as" reality may refer to instructional information, namely, information that is meant to bring about something, and regards the content of the information, i.e., semantic information, rather than environmental information. From this viewpoint, the assumption is that the law can be found in the nature of the individuals, of the world, of the things, i.e., the German "Natur der Sache", or in Nature as such, as in Kant's 1795 essay on the "perpetual peace" (Kant, 1996). An illustration of this stance on legal information "as" reality is given by the Roman maxim, according to which "nobody is (legally) bound to that which is (physically) impossible". Yet, key differences exist between modern theorizations of natural law and the state of nature, as in Locke, or in Kant, and the ancient school of Plato and Aristotle, up to the Christian tradition of the Middle Ages (e.g., Thomas Aquinas). Despite all these different instances of natural law, a common feature persists, namely, the idea that some basic principles and norms of the law could be deduced from an examination of being *qua* being.

Second, attention should be drawn to the ways in which today's network theory has deepened the stance on the environmental information of the law as an emergent property of social interaction, e.g., the theses of Friedrich Hayek on spontaneous orders and the *kosmos* side of the law. In the phrasing of the first volume of *Law, Legislation and Liberty*, "one of our main contentions will be that very complex orders, comprising more particular facts than any brain could ascertain or manipulate, can be brought about only through forces inducing the formation of spontaneous orders" (Hayek, 1982, p. 38). Recent empirical evidence in the field of network analysis confirms this idea on the emergence of spontaneous orders, in such fields as jurisprudence and how academic studies quote each other. In order to grasp this environmental information, the analysis starts with the construction of the network, which requires the identification of the nodes, e.g., judicial decisions and scholarly citations, together with the explicit annotation of the references among them, namely, the edges of the network. By considering the flow of information among the source nodes, scholars have found out, time and again, a power law distribution in such legal networks (Agnoloni and Pagallo, 2016). This is to say that a small number of cases, such as the decisions of the courts, or scholarly papers, receive a large number of citations, while most cases and papers have few citations or none at all. Although some coefficients of the network, such as its diameter and average path length, may vary, the power law distribution of information is at work in such different networks as the case law of the US Supreme Court (Fowler and Jeon, 2008), of the EU Court of Justice (Malmgren, 2011), or of the Italian Constitutional Court (Agnoloni and Pagallo, 2015). From this latter perspective on the environmental information of the law, there is no difference between the series of concentric rings visible in the wood of a cut tree trunk, which allows us to estimate the age of that tree. As a kind of environmental information, even legal information can be meaningful independently of an intelligent producer/informer.

Third, the connection between legal information "as" reality and legal information "for" reality may regard the competition between regulative systems. Going back to Hayek's work, his distinction between *kosmos* and *taxis*, i.e., evolution vs. constructivism, has to do with the potential conflict between spontaneous orders and human political planning. In the phrasing of *Rules and Order* (1973), i.e., the first volume of *Law, Legislation and Liberty* (1982), "these rules of conduct have thus not developed as the recognized conditions for the achievement of a known purpose, but have evolved because the groups who practiced them were more successful and displaced others.... The problem of conducting himself successfully in a world only partially known to man was thus solved by adhering to rules which had served him well but which he did not and could not know to be true in the Cartesian sense" (*op. cit.*, 8). The conflict between legal information "as" reality (Hayek's *kosmos*) and "for" reality (*taxis*) has been the subject of a lively debate since ancient times, e.g., between advocates of the natural law tradition and of the positive laws of the community. Still, leaving aside this kind of debate, it seems fair to affirm that such competition between regulative systems is here to stay. In addition to Hayek's views, however, we should widen our perspective and consider the role that technology plays in this context.

In general terms, we may say that every regulatory system claims to govern social behavior by its own means, and can even render the claim of another regulatory system superfluous. Therefore, in addition to systems that spontaneously emerge from the complexity of the environment, such as Hayek's social norms, or the forces of the market, we may wonder about the regulatory means of technology and to what extent technology — as a regulatory system — can affect the *taxis* side of the law. After all, there are several cases in which the legal intent to regulate the process of technological innovation has failed. Think of the legal umbrella for the adoption of such automatic techniques as DRM in the private sector, i.e., Article 8 of WIPO's 1996 *Copyright Treaty* and Article 14 of the twin *Performances and Phonograms Treaty*, which were mentioned in Section 1.[2] Such provisions enable copyright holders to monitor and regulate the use of their protected artefacts and still, two decades after such international agreements, legal information "about" reality suggests that these legal rules have fallen short of coping with the dynamics of technological innovation, e.g., the introduction of *Cascading Style Sheets* (CSS)-technology to protect DVD artefacts, soon after followed by its DeCSS antidote. In this cat-and-mouse game, the interplay between law and technology should not thus be grasped as a one-way movement of social evolution from law to technology. Rather, *pace* advocates of both legal positivism and legal information "for" reality, a key component of the legal challenges in an information society concerns the other way around. The attention should be drawn here to how the regulatory tools of technology aim to govern social and individual behavior by their own means. This entails a competition with further regulative systems, such as the law, up to the point that technology may even affect requirements and functions of the legal system.

[2]The EU e-money directive 46 from 2000 is another good example. Soon after its implementation, further forms of online payments, such as PayPal, forced the Brussels legislators to intervene, finally amending themselves with the new directive 110 from 2009.

Section 3 examines how information technologies may impact observables and variables under scrutiny so far, i.e., legal information "for" reality, "about" reality, and "as" reality. On this basis, Section 4 will then scrutinize how legal systems have addressed this very impact.

3. The Impact of Information Technologies

Today's information revolution is affecting our understanding about the world and about ourselves: we are interconnected informational organisms that share with biological entities and engineered artefacts "a global environment ultimately made of information", i.e., what Luciano Floridi calls "the infosphere" (Floridi, 2014). A crucial feature of this new environment is illustrated by how the information revolution is transforming principles and concepts of the law, much like the approach of experts to legal information. This section examines three different ways in which we can appreciate this legal impact of information technologies. Figure 4 represents these ways as the new variables of the analysis.

First, focus should be on how the flow of information on the Internet tends to blur conventional boundaries of national legal systems. Some stress that the traditional representation of the legal and political order, as grounded on the principle of national sovereignty, is questioned because there are no clear national boundaries in cyberspace: "border-crossing events and transactions, previously at the margins of the legal system and of sufficient rarity to be cabined off into a small corner of the legal universe ... have migrated, in cyberspace, to the core of that system" (Post, 2002, p. 1380). Others claim that principles and provisions of the legal tradition would be capable of tackling the set of issues emerging with

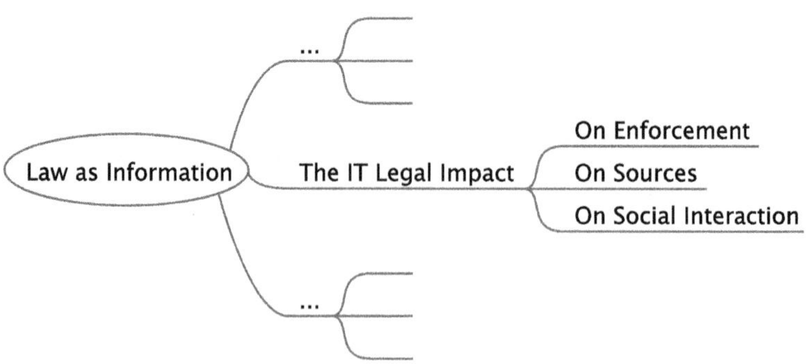

Figure 4. The threefold impact of information technologies on the law.

the information revolution, because the transnational legal impact of the Internet, for example, should be conceived as "identical to transnational activity mediated by other means, such as mail or telephone or smoke signal" (Goldsmith, 1998, p. 1240). Even if we embrace this latter claim, however, two further issues remain open. On the one hand, the canonical idea of the law as a set of rules enforced through the menace of physical sanctions, e.g., Kelsen's approach to legal information "for" reality, is currently challenged by such cases as spamming, phishing, digital frauds, or cyber attacks: despite harshening criminal laws like, say, the *CAN-SPAM Act* passed by the US Congress in 2003, no threat of sanctions seems to limit spamming. On the other hand, cases that scholars address as a part of their everyday work in the fields of information technology (IT)-Law, such as data protection, computer crimes, or digital copyright, cast light on how the flow of information on the Internet has not only magnified the troubles with the enforcement of the law but has also often lead to the illegitimate condition where states claim to unilaterally regulate extraterritorial conduct by imposing norms on individuals who have no say in the decisions affecting them. Remarkably, a general disagreement on how we should grasp the legal institutions of an information society, and how such institutions should be built, and even conceived of, persists in today's debate (Murray, 2007; Post, 2009; Solum, 2009; Reed, 2012; Pagallo, 2015a,b).

The second variable of Figure 4 has to do with legal information "about" reality, and the fact that today's system is tripartite, rather than bipartite. In addition to the traditional sources of national and international law, in which the only relevant actors used to be the sovereign states, the system includes the sources of transnational law and the agency of non-state, or private (as opposed to public) actors (Pagallo, 2015b). The new scenario triggered by the information revolution, and the interplay between the forces of the market and of social norms, can be properly illustrated with the set of rules regulating such a legal system as the global eBay trading community (Goldsmith and Wu, 2006, pp. 130–145; Schultz, 2007). This legal system works through a web of contractual obligations, so that, theoretically speaking, "these contracts are justifiable under the relevant state laws, and so it might be thought that the ultimate source of authority for the community's rules is in fact state law. This is not how these systems work in practice, however" (Reed, 2012, p. 90). The primary source of the system is indeed represented by eBay's rules, because the sums involved are often too small to justify litigation, especially when sales take place cross-border, while "the content of the rules has evolved over time to align with the expectations of eBay users" (*ibid*). In addition, nation states have seen their powers shrink even in more crucial fields of legal interaction, such as the Internet's naming and numbering system. The decisions of the Internet corporation for the assignment of names and numbers (ICANN), which is formally a company registered under the law of California, affect and shape the whole infrastructure of the system via a top-down and distributed approach, i.e., by ensuring one single system for the Internet's naming and numbering, while leaving responsibility to regional registries. At the core of the Internet infrastructure, nation states have thus no veto power and rather they should be considered as one actor of the Internet governance. Since this rearrangement of the legal sources entails a redistribution of power, it is no surprise that sovereign states have attempted to divest ICANN of its authority (Pagallo, 2015b).

The third variable of Figure 4 concerns the manifold ways in which the information revolution and the use of ICTs have affected social interaction. Some insist on the challenges to the rule of law and the democratic institutions, since the threat is that "controls over access to content will not be controls that are ratified by courts; the controls over access to content will be controls that are coded by programmers" (Lessig, 2004, p. 152). Yet, as a matter of fact, it should be admitted that the use of ICTs has also strengthened social interaction. Although some have devoted themselves to nurture such myths, as a new direct online democracy, a libertarian cyberspace, or even a digital form of communism (e.g., Bauwens, 2005), information technologies have favored new forms of cooperation and self-governance (Benkler, 2011). Most online communities have, or rapidly develop, their own social norms (Murray, 2007; Schultz, 2007; Reed, 2012), or subject themselves, through their choice, to the rule of the authority that sets the rules, insofar as members of the online communities generally deem such rules as suitable (Post, 2009; Reed, 2012; Pagallo, 2015b). Thus, in addition to the traditional tools of legal constructivism, such as norms and rules of national and international legislators, social norms are a key part of the system. Furthermore, from this latter perspective, we can appreciate a new sort of ICT-driven competition between regulatory systems, namely, between Hayek's *kosmos* side of the law, or legal information "as" reality, and the regulation of lawmakers as legal information "for" reality.

The relevant distribution of power triggered by the information revolution and the worldwide use of ICT technologies — down to the massive spread of AI apps, in the Internet of everything — does not mean that current trends in the system should be taken for granted. There are several examples of how the regulatory tools of technology may clash with social norms and social expectations, e.g., privacy and data protection issues (Pagallo, 2017a). In addition, current advancements of technology have obliged legislators, policy makers, and governance actors to forge more sophisticated ways to think about legal enforcement through the mechanisms of design, codes, and architectures. Such a shaping is not necessarily digital: the installation of speed bumps in roads as a means to reduce the velocity of cars is a good example of such a legal design. However, scale does matter, in that many impasses of today's legal and political systems are increasingly tackled by embedding normative constraints into ICTs — as well as robots, AI systems, and further artificial agents — through the design of interfaces, self-enforcing technologies, default settings, and more. The next section explores how this form of legal information "for" reality affects the requirements and functions of the law.

4. The Re-ontologization of the Law

The final observable of our analysis concerns "techno-regulation", as shown in Figure 5. The variables correspond to the three different meanings of the formula related to technology as a regulatory system, to technology as the subject matter of legal regulation, and to technology as a new means of legal information "for" reality.

First, we may refer to the regulatory tools of technology and how they interplay with the rules of further regulatory systems, such as the law (e.g., Hayek's notion of *taxis*), the forces of the market, and of social norms

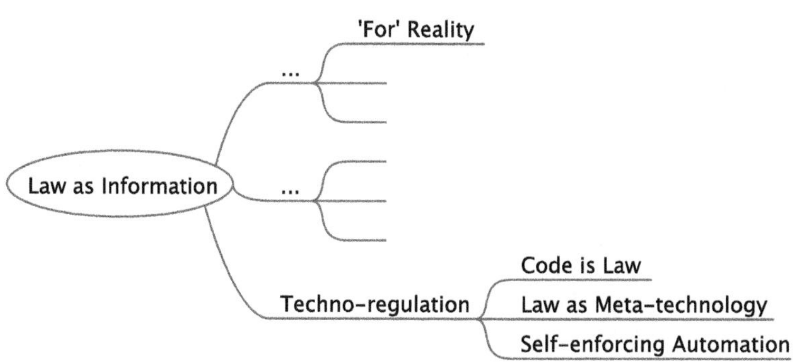

Figure 5. Three roads to techno-regulation.

(e.g., Hayek's *kosmos*). This is the level of abstraction that Lawrence Lessig and others made popular since the late 1990s (Reidenberg, 1998; Lessig, 1999). As mentioned in Section 2, several examples clarify the ways in which the regulatory tools of technology can compete with — and even prevail over — the rules of lawmakers, the forces of the market, or of social norms. As an illustration of such competition, think again about the cat-and-mouse game in the field of copyright protection with the use of DRMs and their technological countermeasures.

The second variable has to do with the idea of the law as a "meta-technology". Going back to Kelsen's formula, "if A, then B", this means that the law aims to govern social and individual behavior "by means of a specific technique" (Kelsen, 1949, p. 26). Once such a technique regulates other techniques and, moreover, the process of technological innovation, we can thus conceive the law as a form of meta-technology (Pagallo, 2013, pp. 147–152). Such regulatory aim of the law operates through both the primary and secondary rules of the system. In the first case, lawmakers may pursue different legislative goals, such as (a) the achievement of particular effects; (b) functional equivalence between online and offline activities; (c) non-discrimination between technologies with equivalent effects; and (d) future-proofing of the law that should neither hinder the advance of technology nor require over-frequent revision to tackle such a progress (Koops, 2006). Others propose to differentiate between (a) technological indifference, i.e., legal regulations which apply in identical ways, whatever the technology, as occurs with the right to authorize communication of a work to the public in the field of copyright law; (b) implementation neutrality, according to which regulations are by definition specific to that technology and yet they do not favor one or more of its possible implementations, e.g., the signature of e-documents; and (c) potential neutrality of the law (Reed, 2012). Yet, in addition to the primary rules of the law, the aim to govern the process of technological innovation also involves the secondary rules of the law. More specifically, the reference is here to the rules of change that create, modify, or suppress the primary rules of the system. Some legal techniques, such as the rules of competitive federalism, legal experimentation, or rules of procedural regularity, offer mechanisms of legal flexibility that allow us to properly address the challenges of the information revolution at a meta-regulatory level (Pagallo, 2017b).

The third variable of Figure 5 brings us back to the idea of embedding legal constraints into ICTs through the design of interfaces, self-enforcing technologies, default settings, and so forth. Some countries, like China, have built up systems of filters and re-routers, detours, and dead-ends to keep Internet users on the state-approved online path. Other states, such as France or South Korea, have endorsed the so-called "three strikes" doctrine, as a part of the graduated system which ends up with the disconnection of the user's Internet after three warnings of alleged copyright infringements. The aim to embed legal safeguards into places and spaces, products, and interfaces of social interaction presents however some risks of its own. In *The Design with Intent Method* (2010), it is noteworthy that Lockton, Harrison, and Stanton describe up to 101 ways in which products can influence the behavior of their users. In legal terms, the aim can be either to encourage the change of social behavior, or to decrease the impact of harm-generating conducts, or preventing that such harm-generating conducts may even occur. As to the first aim of legal design, e.g., nudging (Thaler and Sunstein, 2009), this approach has been quite popular over the past decade, especially in the US and UK, where administrations often adopt it as part of their policies in the public health sector and related fields, from nutrition habits to organ donation. As to the aim of design to decrease the impact of harm-generating conducts, contemplate the field of data protection in Europe, e.g., the principles of data protection by design and by default. In this latter case, the default configuration of ICT interfaces can ensure that values of design are appropriate for novice users and, still, the system improves efficiency (Kesan and Shah, 2006). Likewise, think of security measures, such as reCAPTCHA, that aim to prevent automated programs from abusing online services (von Ahn *et al.*, 2008). The third aim of legal design is finally devoted to the prevention of harm-generating conducts, as occurs with filtering systems on the Internet that provide a complete packet inspection and Information eXtraction ("IX"), together with DRMs in the field of intellectual property protection, or some versions of privacy by design. Although the automation of legal information "for" reality through the use of self-enforcing technologies is increasingly popular among policy makers and legislators, several reasons suggest why "techno-regulation" is highly problematic in this context. Such reasons concern pillars of the rule of law and the traditional view of the legal system as a set of rules or instructions for the determination of the subjects, i.e., the Kelsenian formula "if A, then B".

The automation of the law through the use of self-enforcing technologies, or legal enforcement by design, can in fact be understood in accordance with effects (B) that automatically follow technical instructions (A), rather than sanctions (B) that should follow terms and conditions of legal accountability (A), i.e., that which is, rather than that which should be. Because legal interaction has increasingly turned into a matter of access to, and control and protection over, information in digital environments, a fourfold threat should be addressed properly (Pagallo and Durante, 2016b). First, legal automation impinges on human dignity, leaving no room for human reason and a choice to disobey (Caulfield and Brownsword, 2006; Van den Berg and Leenes, 2013). Second, legal automation wipes out the exceptions that may be taken into account in each case: what is imperiled concerns "the public understanding of law with its application eliminating a useful interface between the law's terms and its application" (Zittrain, 2007, p. 143). Third, individual behavior can unilaterally be determined on the basis of automatic techniques, rather than by individual choices on levels of access and control over information: the risk concerns hence new forms of paternalism, because the more the personal choices are wiped out by the legal automation, the bigger the danger of modeling social conduct through design (Pagallo, 2012). Four, there is the technical difficulty of applying to a machine concepts traditionally employed by lawyers, through the formalization of norms, rights, or duties: legal safeguards do present highly context-dependent notions that raise a number of relevant problems when reducing the informational complexity of a legal system where concepts and relations are subject to evolution. Some insist on the nature of legal regulation and the flexible ways in which we often have to interpret it, so that "the idea of encoding legal norms at the start of information processing systems is at odds with the dynamic and fluid nature of many legal norms, which need a breathing space that is typically not something that can be embedded in software" (Koops and Leenes, 2014, p. 167). Others stress the difficulties of achieving the aim to prevent harm-generating conducts from occurring through the use of self-enforcing technologies. Doubts cast by "a rich body of scholarship concerning the theory and practice of 'traditional' rule-based regulation bear witness of the impossibility of designing regulatory standards in the form of legal rules that will hit their target with perfect accuracy" (Yeung, 2007, p. 106).

In light of such threats and challenges of legal automation, however, we should not overlook the reasons why the use of self-enforcing technologies is increasingly popular among lawmakers (and private companies as well). Who would not aim to prevent harm-generating behaviors from occurring in such cases as cyber-terrorism, child pornography, and the like? Moreover, the use of some technological measures such as, e.g., full packet inspections and IX, could overcome traditional troubles with the enforcement of the law in digital environments, together with possible conflicts among jurisdictions, divergent interpretations of statutes and legal acts, finally repositioning the role of nation states at the center of the sources that govern today's legal systems. Notwithstanding the pros of legal automation and their popularity among national and international lawmakers, we should not forget what was previously mentioned about the interplay between law and technology. The public enforcement of the law through the regulatory tools of technology does not occur in a sort of normative vacuum, but competes and even clashes with further forms of social regulation. The upgrading of the normative side of the law by design, i.e., from punitive or positive sanctions that have to follow terms and conditions of legal accountability to the effects that follow natural causes and technical instructions, goes hand in hand with the ways in which the law has evolved in the information era. This evolution concerns both legal information "about" reality and legal information "as" reality.

5. Conclusions

The chapter has adopted the level of abstraction on the law as information, which conveniently summarizes the different schools of jurisprudence with their variants. The aim was to ascertain whether and to what extent the information revolution has affected this traditional framework, and how today's legal systems have reacted to this impact. By examining the law as legal information "for" reality, "about" reality, and "as" reality, the outcome of the analysis presented a series of double-edged swords, particularly the following:

- As to the variable of legal information "for" reality, the chapter scrutinized the troubles of the law with the enforcement of its rules on the Internet and how legal systems have reacted by embedding normative constraints into technology. Compare the first variable of Figure 4 on the difficulties of law enforcement on the Internet with the last variable of Figure 5 on self-enforcing automation and its impact on some tenets of the rule of law.
- As to the variable of legal information "about" reality, focus was on the rearrangement of today's legal sources and the subsequent redistribution of powers that nation states intend to counteract. In light of the different ways in which the law may aim to discipline social and individual

behavior, as illustrated in Section 4, the contrast is illustrated here by the second variables of Figures 4 and 5. Focus is on the realignment of current legal sources and the legal aim to govern the process of technological innovation, i.e., the law as meta-technology.

- As to the variable of legal information "as" reality, the chapter insisted on how the use of ICTs can strengthen social interaction, favoring the emergence of spontaneous orders and new forms of cooperation and self-governance. Yet, as shown by several examples, such as mass surveillance, privacy, or data protection, the impact of technology also works the other way around. This tension can be elucidated by the last variable of Figure 4 on today's role of social norms *vis-à-vis* the first variable of Figure 5 on technology as a regulatory system.

Against this framework, should the conclusion be that technology is thus simply "neutral", i.e., a bare means to obtain whatsoever end?

The short answer is "no". What the series of double-edged swords intend to show is that today's information revolution should be comprehended as a set of constraints and affordances that transform, or reshape, the environment of people's interaction and their institutions. This profound transformation reverberates throughout the different informational dimensions of the legal phenomenon, and the latter interplays with the regulatory tools of technology, the forces of the market, and of social norms. Whether the focus is on the troubles with the enforcement of the law and its re-ontologization, or on the redistribution of power that follows the re-arrangement of the legal sources, or on the dynamics of social interaction *vis-à-vis* restraints and opportunities triggered by the use of information technologies, the alternatives brought about by this legal impact have to be grasped as a series of open questions that are likely to draw the attention of experts for quite a long time. The information revolution is affecting requirements and functions of the law as information, and its threefold connection with reality. No legal thought which ignores this profound transformation can survive at the present time.

References

Agnoloni, T. and Pagallo, U. (2015). The power laws of the Italian constitutional court, and their relevance for legal scholars, in *Legal Knowledge and Information Systems — JURIX 2015: The Twenty-Eighth Annual Conference*, A. Rotolo (ed.), IOS Press, Amsterdam, pp. 1–10.

Agnoloni, T. and Pagallo, U. (2016). The case law of the Italian constitutional court between network theory and philosophy of information, *Informatica e diritto* XXV(1), 139–152.

Bauwens, M. (2005). *P2P and Human Evolution: Placing Peer to Peer Theory in an Integral Framework*, Available online: http://integralvisioning.org/article.php?story=p2ptheory1 (last accessed April 20, 2018).

Benkler, Y. (2011). *The Penguin and the Leviathan: The Triumph of Cooperation Over Self-interest*, Crown, New York.

Bobbio, N. (1993). *Thomas Hobbes and the Natural Law Tradition*, D. Gobetti (ed.), University of Chicago Press, Chicago.

Caulfield, T. and Brownsword, R. (2006). Human dignity: A guide to policy making in the biotechnology era? *Nature*, 7, 72–76.

Durante, M. (2017). *Ethics, Law and the Politics of Information. A Guide to the Philosophy of Luciano Floridi*, Springer, Dordrecht.

Floridi, L. (2009). *A Very Short Introduction to Information*, Oxford University Press, Oxford.

Floridi, L. (2014). *The Fourth Revolution*, Oxford University Press, Oxford.

Fowler, J. H. and Jeon, S. (2008). The authority of supreme court precedent, *Social Networks*, 30, 16–30.

Goldsmith, J. (1998). Against cyberanarchy, *University of Chicago Law Review*, 65(4), 1199–1250.

Goldsmith, J. and Wu, T. (2006). *Who Controls the Internet: Illusions of a Borderless World*, Oxford University Press, New York.

Hampton, J. (1990). *Hobbes and the Social Contract Tradition*, Cambridge University Press, Cambridge.

Hayek, F. A. (1981). *Law, Legislation and Liberty*, Vol. 3: *The Political Order of a Free People*, University of Chicago Press, Chicago.

Hayek, F. A. (1982). *Law, Legislation and Liberty*, Vol. 1: *Rules and Order*, University of Chicago Press, Chicago.

Hobbes, T. (1999). *Leviathan*, R. Tuck (ed.), Cambridge University Press, Cambridge.

Holmes, O. W. Jr. (1963). *The Common Law*, M. De Wolfe Howe (ed.), Little, Brown, Boston.

Kant, I. (1996). *Practical Philosophy*, M. J. Gregor (ed.), Cambridge University Press, Cambridge.

Kelsen, H. (1949). *General Theory of the Law and the State*, A. Wedberg (ed.), Harvard University Press, Harvard.

Kesan, J. P. and Shah, R. C. (2006). Setting software defaults: Perspectives from law, computer science and behavioural economics, *Notre Dame Law Review*, 82, 583–634.

Koops, B. J. (2006). Should ICT regulation be technology-neutral? in *Starting Points for ICT Regulation: Deconstructing Prevalent Policy One-liners*, B. J. Koops *et al.* (eds.), The Hague: TMC Asser, pp. 77–108.

Koops, B.-J. and Leenes, R. (2014). Privacy regulation cannot be hardcoded: A critical comment on the "privacy by design" provision in data protection law, *International Review of Law, Computers & Technology*, 28(2), 159–171.

Lessig, L. (1999). *Code and Other Laws of Cyberspace*, Basic Books, New York.

Lessig, L. (2004). *Free Culture: The Nature and Future of Creativity*, Penguin Press, New York.

Lockton, D., Harrison, D. J. and Neville, A. S. (2010). The design with intent method: A design tool for influencing user behaviour, *Applied Ergonomics*, 41(3), 382–392.

Malmgren, S. (2011). *Towards a Theory of Jurisprudential Relevance Ranking: Using Link Analysis on EU Case Law*. Master of Laws degree at Stockholm University under the supervision of C. Magnusson Sjöberg.

Murray, A. D. (2007). *The Regulation of Cyberspace: Control in the Online Environment*, Routledge-Cavendish, New York.

Pagallo, U. (2012). Cracking down on autonomy: Three challenges to design in IT law, *Ethics and Information Technology*, 14(4), 319–328.

Pagallo, U. (2013). *The Laws of Robots: Crimes, Contracts, and Torts*, Springer, Dordrecht.

Pagallo, U. (2015a). Good onlife governance: On law, spontaneous orders, and design, in *The Onlife Manifesto: Being Human in a Hyperconnected Era* L. Floridi (ed.), Dordrecht: Springer, pp. 161–177.

Pagallo, U. (2015b). The realignment of the sources of the law and their meaning in an information society, *Philosophy & Technology*, 28(1), 57–73.

Pagallo, U. (2017a). The group, the private, and the individual: A new level of data protection? in *Group Privacy: New Challenges of Data Technologies*, L. Taylor, L. Floridi and B. van der Sloot (eds.), Springer, Dordrecht, pp. 159–173.

Pagallo, U. (2017b). LegalAIze: Tackling the normative challenges of artificial intelligence and robotics through the secondary rules of law, in *New Technology, Big Data and the Law. Perspectives in Law, Business and Innovation*, M. Corrales, M. Fenwick and N. Forgó (eds.), Springer, Singapore, pp. 281–300.

Pagallo, U. and Durante, M. (2016a). The philosophy of law in an information society, in *The Routledge Handbook of Philosophy of Information*, L. Floridi (ed.), Oxon & New York, pp. 396–407.

Pagallo, U. and Durante, M. (2016b). The pros and cons of legal automation and its governance, *European Journal of Risk Regulation* 7(2), 323–334.

Post, D. G. (2002). Against "against cyberanarchy", *Berkeley Technology Law Journal*, 17(4), 1365–1383.

Post, D. (2009). *In Search of Jefferson's Moose: Notes on the State of Cyberspace*, Oxford University Press, New York.

Reed, C. (2012). *Making Laws for Cyberspace*, Oxford University Press, Oxford.

Reidenberg, J. R. (1998). Lex informatica: The formulation of information policy rules through technology, *Texas Law Review*, 76(3), 553–594.

Schultz, T. (2007). Private Legal Systems: What Cyberspace Might Teach Legal Theorists, *Yale Journal of Law and Technology*, 10, 151–193.

Solum, L. B. (2009). Models of internet governance, in *Internet Governance: Infrastructure and Institutions*, L. A. Bygrave and J. Bing (eds.), Oxford University Press, New York, pp. 48–91.

Thaler, R. H. and Sunstein, C. R. (2009). *Nudge: Improving Decisions about Health, Wealth, and Happiness.* Rev. and expanded ed. Penguin Books, New York.

van den Berg, B. and Leenes, R. E. (2013). Abort, Retry, Fail: Scoping Techno-regulation and other Techno-effects in *Human Law and Computer Law: Comparative Perspectives*, M. Hildebrandt and J. Gaakeer (eds.), Springer, Dordrecht, pp. 67–88.

von Ahn, L., Maurer, B., McMillen, C., Abraham, D. and Manuel, B. (2008). reCAPTCHA: Human-based character recognition via web security measures, *Science*, 321(5895), 1465–1468.

Weber, M. (2005). *Economy and Society*, Ch. Camic *et al.* (eds.), Stanford University Press, Stanford.

Yeung, K. (2007). Towards an understanding of regulation by design, in *Regulating Technologies: Legal Futures, Regulatory Frames and Technological Fixes*, R. Brownsword and K. Yeung (eds.), Hart, London, pp. 79–108.

Zittrain, J. (2007). Perfect enforcement on tomorrow's internet, in *Regulating Technologies: Legal Futures, Regulatory Frames and Technological Fixes*, R. Brownsword and K. Yeung (eds.), Hart, London, pp. 125–156.

Chapter 23

Information and Openness

Tomáš Sigmund

University of Economics in Prague, Department of System Analysis,
nám. W. Churchilla 1938/4, 130 67 Praha 3, Czech Republic
sigmund@vse.cz

This chapter attempts to answer the questions to what extent information is unchangeable and to what extent it can be changed. It analyses the three aspects of information — its syntactic, semantic and pragmatic dimension. It concludes that on all three levels information is changeable and should serve more as a trigger than as a hard fact. Information is not an objective and independent concept. It is influenced by the society we live in, by the culture and its historical development. States of affairs we would like to change, changeable and unchangeable components of the world are all expressions of power over the world.

1. Introduction

Information is an answer to a question of some kind. That reflects the fact that to be aware of information, one has to be prepared for its acceptance. The acceptance of information is not a trivial act; it is the result of a conflict in which an individual meets the world. It is the process of coming to terms with the brute world.

First comes the question of why something is so and so, the wish that something be such and such follows, then comes recognition of the different state of affairs in reality which have to be changed and finally reconciliation with some unchangeable facts.

The distinction between the wish, the fact that has to be changed and the unchangeable fact is not stable. It often depends on one's power, perspective and presuppositions that are however necessary as they reflect involvement and interest in the world. The development of modern science, which breaks our common understanding and shows that the everyday world as we normally understand it can look very different, is just one example. However, science thinks it has come to a final true description of the world and doesn't concede how relative its stance can be. If the results aren't stable, at least the scientific procedures and the ways of achieving the evidence or procedures used in experiments are stable in science. However, we may doubt the scientific stability. As an argument in the support of the constructivist claim in science, we can state Kuhn's conception of scientific revolutions (1962), Duhem Quine's thesis (1998) or Wittgenstein's rule-following theory (2002), all of which stress the importance of background assumptions. Awareness of their influence depends on the level of abstraction and so on the kind of information we use.

In human history, various theories have been developed that claim to have found a universally valid differentiation of these three principles of the wish, changeable fact and unchangeable fact, or at the very least they show their limits. For example, man's wishes are limited because of his nature; what can be changed is limited by the rules governing the world and some stable parts of the world can't be changed at all. That division can have powerful consequences. Thinking is always related to power. Even in pure mathematics, those who have the power to differentiate between right and wrong are more powerful than those who have to conform. Teachers are more powerful than students. They may supress those diverting from them.

Understanding and expressing something as unchangeable means the person doesn't have power to change it and doesn't even try. Defining something as something implies that other objects don't belong in this category and must be treated differently. We can also express the problem of conforming in terms of the powerful prevailing limits and fighting with the world as differentiation between identity and non-identity. Identity means accepting the state of affairs, non-identity means fighting against it.

Information is a unity of three aspects: the object the information is about, the code the information is presented in and the subject which receives the information. In relation to Ch. S. Peirce's semiotics, we may designate the three approaches to information arising from these three aspects as syntax, semantics and pragmatics. Syntax studies the internal relations of signs carrying information, semantics deals with the relation of signs to its objects and pragmatics with the implications and consequences the coded content of information has on its recipient. Similar differentiation can be found in N. Weaver's article "The Mathematics of Communication" (1949). Weaver also differentiates three aspects of information. He calls the syntactic level technical as for him it deals with the quantification of information. Semantic problems consider meaning and truth (relation of information to the world) and pragmatic-level problems are called "influential" in his text — they concern the impact and effectiveness of information on human behaviour. In all these three aspects, the interplay of wish, changeable and unchangeable facts can be identified.

2. Syntactics

Syntactics is the set of rules, principles and processes that govern the structure of information. In order to do that we need a set of rules governing the syntactic relations of information. The rules can be qualitative as is the case, e.g., in linguistics, or quantitative as is the case, e.g., in programming languages.

One approach attempting to quantify the relations of information is based on combinatorics (Kolmogorov, 1968). Combinatorics defines the entropy which quantifies the probability of the outcome of a random process of variable x as $H(x) = \log_2 N$. If x gets a value a, we remove the entropy and get an identical quantity of information. If a variable x can have a definitive value a of the set consisting of N elements ($x = a$), we have combinatorial information. The total combinatorial information contained in the list of all binary strings is $k * N = k * 2^k$. A k-digit binary string ($N = 2^k$) contains $k$bits of information. The combinatorial approach doesn't reflect the various probabilities of events.

The second approach to the syntactics of information was introduced by Shannon in his theory of information transmission (1948). Shannon studied quantification, storage and communication of information. He didn't consider the content of information, just its syntactic properties. Shannon focused on the coding of information and considered the probability of occurrence of letters in a message. For him, information shows how many binary digits are necessary to code information. If the probability of the ith value a of X is p_i, the Shannon entropy H in bits is given by $H = -\sum_i^s p_i \log_2 p_i$. If we have an alphabet with the frequency of occurrence of individual letters in a message of length n satisfying the inequality $\chi = \sum_1^s p_i \log_2 p_i \leq h$, we can make an asymptotic estimate

$H = \log_2 N \sim nh$. h is the entropy of the ensemble X. T is the set of typical messages. Probability is related to the fact that with large n, x will fall in a small subset of outcomes called the typical set, which consists of typical messages.

The probabilistic measure of information doesn't in turn reflect the complexity of the system, but is a measure of uncertainty of a given state or of an event. Both complex and simple repetitive structures have identical amounts of information. That is why Kolmogorov suggested his algorithmic information or the relative complexity defined as the minimal length of the program p for obtaining y from x (Duch, 1993).

Are these theories dealing with objective features of information or are they results of cultural processes? We will use Wittgenstein's rule-following argument to highlight some points. Syntactic structures represent the rules governing the relations within information sets. Rules should be applicable to all relevant instances. Knowing the rule includes knowing how to use it. Wittgenstein begins by an example: "we get [a] pupil to continue a series (say "+2") beyond 1000 — and he writes 1000, 1004, 1008, 1012" (Wittgenstein, 2001). What do we do when the pupil after being corrected claims he went on in the same way? Wittgenstein therefore asks how we learn rules, how we follow them, what the standards are, where they come from and in the decision-making whether the rule has been followed correctly or not. He investigates the options, whether the rules have a mental representation, whether intuition is applied and how rules are taught and enforced. Unfortunately, Wittgenstein's questions are not answered positively as that would suggest there is an external authority (or rule) which would give us a certain answer. Wittgenstein wants to avoid such ways of thinking and claims there is nothing but actual application of the rule. In Philosophical Investigations (2001, p. 201) Wittgenstein claims, "This was our paradox: no course of action could be determined by a rule, because every course of action can be made out to accord with the rule. The answer was: if everything can be made out to accord with the rule, then it can also be made out to conflict with it. And so there would be neither accord nor conflict here." The famous interpretation of the argument comes from Kripke (1982) according to whom there are no facts that could determine what can be considered as following the rule, and no reasons that could justify somebody in saying that he is actually following a rule. Only the consent of others, their acceptance may privilege one way of following the rule over another (Biletzki and Matar, 2016).

Wittgenstein criticized systematic knowledge because the use of language and understanding meaning stem from our common and shared practices, the forms of life, not from agreement in opinions. Any formulation of the rule faces the risk that it would be misunderstood and using other rules to determine how to use the original rule leads to a vicious circle. This is a so-called sceptical solution according to which we have just a conventional solution for rule following. Michael Dummett (1959) considers Wittgenstein to be a real conventionalist who thought that there is no need to follow the rule one way rather than the other. Staten (1984) interprets Wittgenstein in the Derridean way according to which there is a difference between the rule and its application that makes any theory explaining what it is to follow a rule an impossibility.

Interpreters agreeing on the sceptical interpretation of Wittgenstein's rule-following differ as to whether the resources for the solution can be provided by an individual who can decide himself about the practices involved in rule following (McGinn, Blackburn) or by a community (Kripke, Bloor). Kripke himself refuses the communitarian solution as it in his opinion doesn't meet the criteria set by the sceptical problem. Bloor on the other hand considers the solution appropriate.

Another interpretation of Wittgenstein's rule-following argument was offered by McDowell (1993). McDowell thinks Wittgenstein rejects the paradox for its connection of understanding and interpretation. The paradox presupposes that in order to understand something, we must have an interpretation of it. If we agree, we finish either in scepticism or in relativity. In this reading, Wittgenstein wants to get rid of any theory of meaning. This reading is known as quietism as it denies that Wittgenstein has anything to say on big philosophical theories. Putnam and Diamond are further supporters of this interpretation.

Both these perspectives (scepticism and quietism) see in rule following something problematical and use man or society as a way out of the situation. In case of scepticism, the way out is just an emergency solution — not solving the problem, but suggesting a way how to live with it. The quietist solution is interesting as it suggests that man can do things without being able to explicate them properly. That idea is supported by Dreyfus and Taylor (2015) who promote the idea that there are things we know, but we don't know that we know them. In their perspective, man has a privileged access to the world, he understands it, but his understanding is not reducible to a set of rules. Reduction of understanding to a set of rules is the result of a mechanistic worldview which considers perception to be the result of the influence of the outer world on a sensor in the human body. This perspective results in understanding man, and the human brain, as a digital computer. This perspective is often useful and successful, but can't be considered universal. Perceptions are not atomistic, but get sense because they belong to a context, within the world that gives them some sense. Many actions man performs are performed intuitively; activities and perceptions are implicitly involved in many relations which give them their meaning. The net of relations can be partly explicated, but never in its totality. The net is transformed with every new perception (Dreyfus and Taylor, 2015). Many activities we perform have such a character; paradigmatic examples are sportsmen who act almost intuitively or experts. They know what to do, but are not able to explicate the reasons and relations in full detail. Our everyday activities have this implicit character, too. They are based on many presuppositions, many relations and are context dependent. We are usually not aware of these hidden features. In the area of semantics, which deals with the relation of information to the facts in the real world, the importance of the human aspect has been more recognized.

Syntactic information in the form of explicit rules is therefore not as stable as it may seem. The rules can be either results of individual decisions, social processes or are generalized from one or more examples without respecting the specificity of every case where many inexplicable relations and experiences play a role.

3. Semantics

For Shannon, the less probable the situation, the more information is garnered when we learn that it will happen. This leads to the conclusion that contradiction loads maximum, i.e., infinity of, information content. Tautologies on the other hand would contain no information and so deductions from them would also have no information content. However, our intuition tells us that the value of information depends not only on its probability but also on its accuracy, on its truth. One way to evaluate the truthfulness of information is the number of instances it can be applied to. So probability is not the only way to characterize information content. We must consider semantics, too (D'Agostino and Floridi, 2009).

Semantics deals with the interpretation of information. When thinking about the way to signify information and how such information may be interpreted, the role of the cultural context has been widely recognized. In many cases, the context decides the meaning (e.g., meaning of pronouns). Further, the meaning of signs does not consist in their attachment to things, but in the way they function in every situation and context. So meaning is not full in itself, the element of context co-determines it. Gärdenfors (2000) gives the example of the word "red", the dictionary definition of which would be "the colour of fresh blood, rubies, human lips, the tongue, maple leaves in the autumn, post-office pillar boxes in Great Britain". However, the world "red" is also used in phrases like red wine, red hair, red skin the colour of which would normally be called by a different colour name. Red wine is called red in contrast to white wine which refers to the contextual space of contrasting elements. We can also think of body language, rituals and other social codes bearing meaning or contributing to it.

Semantic content of information can be either a fact or an instruction. Instructual information does not describe a situation, a fact or state, it brings it about. There are many types of instructions which have been studied and classified in the speech act theory. Austin (1962) claims that only a few sentences of our language have truth value. Every language has a lot of performative sentences which are not used to describe something, but to perform an action; they have a performative function. Speech acts can be analysed on three levels: (1) Locutionary act — the actual performance of an utterance including phonetic, verbal, semantic and related aspects of any meaningful utterance. (2) Illocutionary act — the real intended meaning or force of an utterance as a social action. (An example may be promising or commanding. We do something by saying something as when we order somebody to do something or when two people are pronounced husband and wife.) (3) Perlocutionary act — the actual effect of the utterance.

The factual information is declarative and may be true or untrue. With the concept of truth, the situation becomes more complex as we not only deal with the interpretation of information but also its relation to the world. Frege differentiated between sense and reference. A sign (e.g., a name) expresses the sense (meaning) of a thing and designates its reference (the object). For Frege, the reference is the truth value of the proposition. From the logical perspective, the situation is more complicated if we consider one can analyse categorical propositions from extensional or intensional perspective. This distinction refers to Frege's differentiation between sense (Sinn) — meaning in the intuitive sense — and meaning (Bedeutung) — object referred to. R. Carnap then suggested to replace Frege's concepts of meaning and sense with extension and intension. The extension of an individual expression is the object for which it stands, the extension of a predicate is a function assigning truth value to objects for which the predicate is true and the extension of a sentence is its truth value. Expressions are in this interpretation mapped on their extensions, names on individuals, predicates on sets of individuals and sentences on their truth value. Frege's sense was presented as the way in which the term presents its referent, but Carnap modified it and said that intension is what a term shares with the terms that are logically equivalent to it, i.e., their equivalence can be proven just by the laws of logic.

So extension is the referent of a term and the intension is the meaning or semantical aspect of the term which determines its extension. Carnap introduced in 1947 the concept of state description, which shows some similarities with possible worlds. Thus, he prepared the way for subsequent understanding of intension as the set of possible worlds in which the sentence or term is true (Peregrin, 2006). It means that at least in some cases, the denotate (object) of the statement is not the criterion of the truth of the statement. In these cases, the meaning of the terms is decisive, too. In the sentence "I am not sure the author of *Waverly* is Walter Scott", the extension of both Walter Scott and author of *Waverly* is the same person. However, if we say "I am not sure Walter Scott is Walter Scott", the sentence can't be true as nobody can fail to know that $a = a$. So at least in some cases, truth value is not the result of the combination of lower truth values and so is not compositional. In some cases, we must consider the meaning of the sentence — its intension. As we have seen, intension can be explicated as the set of possible worlds in which the sentence is true and so the meaning is not absolute, but is context or world or culture relative.

Regarding language (and these ideas can be extended to any symbolic system), we mustn't forget the ideas of postanalytical philosophy, which refuses the correspondence theory of language according to which language corresponds to the building blocks of the world and true sentences correspond to the factual structures of the world. The building blocks of the world are individual things with their properties and relations. Some postanalytical thinkers like the late Wittgenstein, N. Goodman, D. Davidson, R. Rorty, etc., realized the world has structure only through language. The world is thus a language construct. It doesn't mean that the world is made of language, but with the help of language. Language does not correspond to the world, but is a way of coming to terms with the world. The consequence is not that anything we say is true, but that the structure of the world depends on the language we choose. Different languages put their users into different worlds. Interpretation of the world is done in language which allows something, but restricts something else (Peregrin, 1988).

The semantic content of information is so dependent on the speaker and his culture as it depends on the signifying system he uses and on his intentions. In the case of performative sentences, the subject may even influence if the information is true or not by realizing it.

4. Pragmatics

With the perlocutionary act, we got close to the pragmatic aspects of information. A perlocutionary act considers the consequence of the speech act like persuading, explaining, inspiring, etc., the listener. It may be considered the effect of an illocutionary act expressed in the locutionary act. It focuses on the effect on the addressee. I may say the sentence “I am rich” with the perlocutionary effect on the listener consisting in his feeling impressed, seeking help, etc.

For the pragmatic aspect of information, the value of information is very important. Pragmatics as such studies the way context of the utterance, pre-existing knowledge, intent of the speaker, etc., influence implications of the information.

In general, pragmatic information is the content received by the addressee. Pragmatic information measures the information received, not contained, in the message. Pragmatic information may be different for different recipients. Weizsäcker claims (2014) that the highest information acceptance is achieved with good balance between novelty and confirmation. The increase of these two components increases the potential for pragmatic information. These two components are complementary. Without novelty, there is nothing new and there is no event. Without confirmation, which contains concepts of redundancy and physical reliability, there is chaos. Too much confirmation means knowing everything about nothing, too much novelty means knowing nothing about everything. If there is too much novelty in the information, we strive for repetition, i.e., confirmation to keep the balance. If there is too much repetition of the known content, some change at the expense of confirmation is welcome to break the boring repetition.

Weizsäcker (2014) adds that there are aspects of reality where no disturbance or novelty is welcome. An example may be the introduction of a new letter to the alphabet, which is usually improper or the way our physical body fights against the intrusion of new molecules. “It is usually on the highest hierarchical levels, such as meaningful speech, animal behaviour or population genetics, that novelty and confirmation ought to be equilibrated to ensure maximum information production" (Weizsäcker, 2014, p. 40). The code and the channel belong usually to confirmation; messages carry novelty and produce novelty. The receiver is exposed to the message and normally does not understand the full message, as his transmission capacity is limited. Receivers usually highlight a part of the message, disregard another or add something to the original message and so produce further novelty. In a special case, the receiver may kill the information by not forwarding it and not even storing it. Information changes its meaning in the course of its spreading and growing. The receiver in whom the message combines previous knowledge with the satisfaction of understanding will carry the message further. The innovative aspects are necessary for solving new or modified tasks; however, one has to count with errors, bad luck and problems.

The effect of information on the subject can be changed by the subject. The effects of information on him depend on his knowledge, values, etc. He also needn't consider all the novelty of the information. The confirmative part can be evolved and cultivated in a number of ways and the more advanced the receiver, the more the potential of the information he will use.

In reality, the availability and accessibility of information didn't lead so much to the development and empowerment of man. Postman (1985) referring to the ideas of M. McLuhan claims that television is a medium that excludes a high level of rational argument and satisfies the entertainment needs only. Television news is for him just a form of entertainment programming using music, catchy pictures, etc. It is not only the subject but also the medium that changes the effect of the information.

Postman's student Lance Strate examines Postman's thesis in the contemporary media including the Internet, web 2.0 and social media (2014). He proves that people are still unable to distinguish between amusement and important matters. Strate doesn't consider the social media to enlarge the public sphere because they stress the novelty and immediacy of the news at the expense of their informativeness.

Hjarvard (2013) shows in his book how medialization has spread out through various social spheres (politics, religion, play, habitus), that media have become an important institution in society and culture and have influenced human interactions.

5. Conclusion

The problem with information consists in the fact that we often consider it too rigid, that we consider some rules governing it, ways of interpretation and effects as given and unchangeable while they are the results of social processes, individual decisions that were forgotten or power structures. Information should be rather used as a trigger opening a wide range of possibilities. The syntactic aspect should be realized, interpreted and respective action should be taken not in a prescribed way, but freely. Any compliance with the rules should be deliberate and free. Only thus will it lead to the development of the personality. I am not claiming that no rules are necessary, but that everybody should know why and what type of rules he is following and that the rules may be changed if necessary. We can find a similar concept in the communication theory, where there are two concepts of information communication: there is the transmission model of communication and the constructionist model of communication (Lanham, 2003). In the transmission model, information is sent from a sender to a receiver. This theory views communication including information recognition as a means for sending and receiving information where the set rules must be respected and kept.

The proponents of the other model claim that meaning is reflexively constructed, maintained and negotiated in the process of communication. Communication is in their opinion a social practice that transforms the communicators, their identities, reality framings, etc. This is close to Bateson's concept of information as a difference that makes a difference.

Realizing the limits and obstacles refers to a sort of compliance with the world. However, it is just the first step of the liberation and emancipation process which can't lead to a full emancipation, but to the formation of a new conditional meaning.

Information technologies allow easy distribution and communication of information. One of the expected advantages of this situation was the constraint of bureaucracy necessary for the governing of the state. However, the expectations have not been fulfilled and we are facing more bureaucratic practices in the support of security, fair treatment, etc., that lead to increased emphasis on conformation.

The easy availability of information was intended to free man from his limited perspective, but man hasn't used the chance yet and spends his time searching unimportant amusing content. Information should originally have informed us about the world, but we rather use it to leave the world and spend our lives in virtual environments. Information should have explained the world, but now causes its intransparency as the world is very complicated. That is different from the non-identity mentioned above. Intransparency means that man is deprived of his contact with the world as he can't influence anything and becomes passive.

References

Austin, J. (1962). *How to Do Things with Words: The William James Lectures Delivered at Harvard University in 1955* (eds. J. O. Urmson and Marina Sbisà), Clarendon Press, Oxford.

Biletzki, A. and Matar, A. (2016). Ludwig Wittgenstein, In *The Stanford Encyclopedia of Philosophy* (Fall 2016 Edition), Edward N. Zalta (ed.), Available online: https://plato.stanford.edu/archives/fall2016/entries/wittgenstein/.

D'Agostino, M. and Floridi, L. (2009). The enduring scandal of deduction, *Synthese* 167(2), 271–315.

Dreyfus, H. and Taylor, C. (2015). *Retrieving Realism*, Harvard University Press.

Duch, W. (1993). Syntactic and semantic information in finite systems, *UMK-KMK-Technical Report-1-1993*, Available online: https://fizyka.umk.pl/publications/kmk/93-P-info.pdf.

Dummett, M. (1959). Wittgenstein's philosophy of mathematics, *Philos. Rev.* 68, 324–348.

Gärdenfors, P. (2000). *Conceptual Spaces: The Geometry of Thought*, MIT Press, London.

Gillies, D. (1998). The Duhem thesis and the Quine thesis, in *Philosophy of Science: The Central Issues*, Curd, M. and Cover, J. A. (eds.), Norton, New York, USA, pp. 302–319.

Hjarvard, S. (2013). *The Mediatization of Culture and Society*, Routledge, New York.

Kripke, S. (1982). *Wittgenstein on Rules and Private Language*, Blackwell Publishing, Basil.

Kolmogorov, A. N. (1968). Logical basis for information theory and probability theory. *IEEE Trans. Inf. Theory* IT-14(5), 662–664. doi:10.1109/TIT.1968.1054210.

Kuhn, T. S. (1962). *The Structure of Scientific Revolutions*, First Edition, University of Chicago Press, Chicago, IL, USA.

Lanham, R. A. (2003). *Analyzing Prose*, 2nd edn., Bloomsbury Academic, London, UK.

McDowell, J. (1993). Meaning and intentionality in Wittgenstein's later philosophy, *Midwest Studies in Philosophy. Volume XVII: The Wittgenstein Legacy*, Notre Dame University Press, Notre Dame, IN, pp. 40–52.

Peregrin, J. (1988). Děláme světy jazykem? *Organon F 5, Příloha*, 40–48.

Peregrin, J. (2006). Extensional vs. intensional logic, in *Philosophy of Logic (Handbook of the Philosophy of Science 5)*, Jacquette, D. (ed.), Elsevier, Amsterdam, pp. 831–860.

Postman, N. (1985). *Amusing Ourselves to Death: Public Discourse in the Age of Show Business*, Penguin, USA.

Shannon, C. E. (1948). A mathematical theory of communication, *Bell Syst. Tech. J.* 27, 379–423 & 623–656.

Staten, H. (1984). *Wittgenstein and Derrida*, University of Nebraska Press.

Strate, L. (2014). *Amazing Ourselves to Death: Neil Postman's Brave New World Revisited*, Peter Lang, New York.

Weaver, W. (1949). The mathematics of communication, *Sci. Am.* 181(1), 11–15.

Weizsäcker, E. (2014). *Ernst Ulrich von Weizsäcker, Pioneer on Environmental, Climate and Energy Policies*, Springer.

Wittgenstein, L. (2001). *Philosophical Investigations*, Blackwell Publishing, Hoboken, NJ, USA.

Lightning Source UK Ltd.
Milton Keynes UK
UKHW020822250320
360861UK00001B/14

9 789813 277489